Electronics — Circuits and Systems

Electronics — Circuits and Systems

Third Edition

Owen Bishop

ELSEVIER

AMSTERDAM • BOSTON • HEIDELBERG • LONDON • NEW YORK • OXFORD
PARIS • SAN DIEGO • SAN FRANCISCO • SINGAPORE • SYDNEY • TOKYO

Newnes is an imprint of Elsevier

Newnes

Newnes is an imprint of Elsevier
Linacre House, Jordan Hill, Oxford OX2 8DP
30 Corporate Drive, Burlington, MA 01801

First published 1999
Reprinted 2000
Second edition 2003
Third edition 2007

British Library Cataloguing in Publication Data
A catalogue record for this book is available from the British Library

Library of Congress Cataloging-in-Publication Data
A catalog record of this book is available from the Library of Congress

ISBN: 978 0 7506 8498 9

For information on all Newnes publications
visit our web site at www.newnespress.com

Printed and bound in UK

07 08 09 10 10 9 8 7 6 5 4 3 2 1

Contents

Preface

This book is written for a wide range of pre-degree courses in electronics. The contents have been carefully matched to current UK syllabuses at Level 3 / A-level, but the topics covered, depth of coverage, and student activities have been designed so that the resulting book will be a student-focused text suitable for the majority of courses at pre-degree level around the world. The only prior knowledge assumed is basic maths and the equivalent of GCSE Double Award Science.

The UK courses covered by this text are:

BTEC National Engineering Pathways syllabus (2007), Units 5 (Electrical and Electronic Principles), 35 (Principles and Applications of Electronic Devices and Circuits), 62 (Microprocessor Systems and Applications), and the introductory stages of Units 51 (Industrial Process Controllers), 60 (Principles and Applications of Analogue Electronics), 68 (Principles and Applications of Microcontrollers), and 90 (Telecommunications Principles).

A-level (AS and A2) specifications from AQA, OCR and WJEC.

The book is essentially practical in its approach, encouraging students to assemble and test real circuits in the laboratory. In response to the requirements of certain syllabuses, the book shows how circuit behaviour may be studied with a computer, using circuit simulator software.

The book is suitable for class use, and also for self-instruction. The main text is backed up by boxed-off discussions and summaries, which the student may read or ignore, as appropriate. There are frequent 'Self Test' questions at the side of the text with answers given in Supplement B. Numeric answers to other questions and answers to the multiple choice questions are on the companion website.

The text has undergone a major revision to produce this third edition. Additions to the content include five new chapters. These cover electrical and magnetic fields, diodes, oscillators, integrated circuits, and industrial process control systems. Several other chapters have been expanded, to reflect the increasing importance of digital electronics and microcontroller systems. All chapters have been updated where necessary, to keep pace with the many recent developments in electronics.

The 'On the Web' panels in many chapters are a new feature of this edition. They are intended for students to make use of the wealth of relevant information available from that source. Also, this edition coincides with the launching of a companion website. This has a Power Point presentation of illustrations from the book for use by students and lecturers. It has the answers to numeric questions and to all the multiple choice questions. There are many more of these in this edition. There are pages of worked examples and questions for those who need extra support in maths.

The companion website also includes some novel features: a set of calculators for electronic formulae, animated diagrams to show electronic circuits in action, and series of interactive worksheets, with answers.

Owen Bishop

Practical circuits and systems

Circuit ideas

As well as being a textbook, this is a sourcebook of circuit ideas for laboratory work and as the basis of practical electronic projects.

All circuits in this book have been tested on the workbench or on computer, using a circuit simulator. Almost all circuit diagrams are complete with component values, so the student will have no difficulty in building circuits that will work.

Testing circuits

The circuit diagrams in this book provide full information about the types of components used and their values. Try to assemble as many as you can of these circuits and get them working. Check that they behave in the same ways as described in the text. Try altering some of the values slightly, predict what should happen, and then test the circuit to check that it does.

There are two ways of building a test circuit:

- Use a breadboarding system to build the circuit temporarily from individual components or circuit modules.

- Use a computer to run a circuit simulator. 'Build' the circuit on the simulator, save it as a file, and then run tests on it.

The simulator technique is usually quicker and cheaper than breadboarding. It is easier to modify the circuit, and quicker to run the tests and to plot results. There is no danger of accidentally burning out components.

Conventions used in this book

Units are printed in roman type: V, A, s, S, μF.

Values are printed in italic (sloping) type:

Fixed values	V_{CC}, R_1
Varying values	v_{GS}, g_m, i_D
Small *changes* in values	v_{gs}, i_d

Resistors are numbered, R1, R2, and so on. The *resistance* of a resistor R1 is represented by the symbol R_1. The same applies to capacitors (C1, C2) and inductors (L1, L2).

Significant figures

When working the numerical problems in this book, give the answers to three significant figures unless otherwise indicated.

Units in calculations

Usually the units being used in a calculation are obvious but, where they are not so obvious, they are stated in square brackets. Sometimes we show one unit divided by or multiplied by another.

Example

On p. 73, we state:

$R_1 = 14.3/2.63$ [V/μA] = 5.44 MΩ

A voltage measured in volts is being divided by a current measured in *micro*amperes.

Mathematically, this equation should be written:

$R_1 = 14.3/(2.63 \times 10^{-6}) = 5.44 \times 10^6$

Set out in this form, the equation is difficult to understand and to remember. To avoid this problem we quote the *units* instead of powers of 10. When the result is being worked out on a calculator, it is easy to key in the values (14.3, 2.63) and follow each by keypresses for 'EXP -6' or other exponents where required. The result, in Engineering or Scientific format, tells us its units. In this example the display shows 5.437262357^{06}. We round this to 3 significant figures, '5.44', and the '06' index informs us that the result is in megohms.

Companion website

The URL of the site is:

http://books.elsevier.com/companions/9780750684989.

Part 1 Circuits

An electrical circuit is a pathway for the flow of electric current.

In a **direct current** (or **DC**) circuit, the current flows in one direction. It flows from a point in the circuit that is at a high potential to a point in the circuit that is at a low potential. The circuit may be simple loop (like the one shown here) or it may be a network of two or more branches.

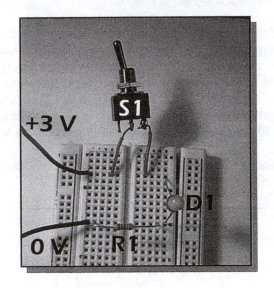

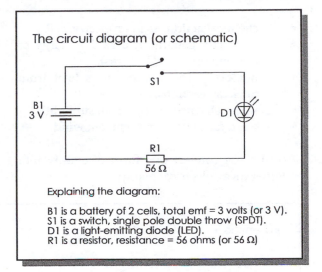

The circuit diagram (or schematic)

Explaining the diagram:

B1 is a battery of 2 cells, total emf = 3 volts (or 3 V).
S1 is a switch, single pole double throw (SPDT).
D1 is a light-emitting diode (LED).
R1 is a resistor, resistance = 56 ohms (or 56 Ω)

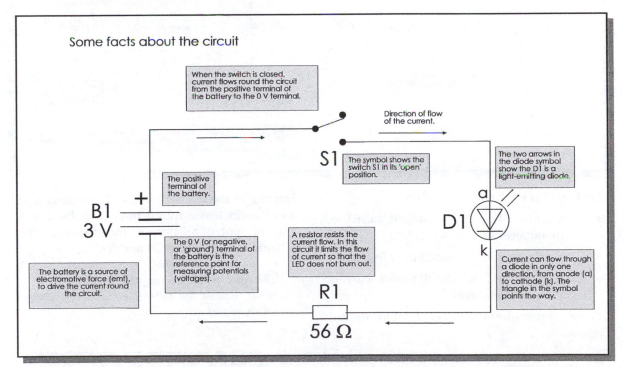

Some facts about the circuit

When the switch is closed, current flows round the circuit from the positive terminal of the battery to the 0 V terminal.

Direction of flow of the current.

The positive terminal of the battery.

The symbol shows the switch S1 in its 'open' position.

The two arrows in the diode symbol show the D1 is a light-emitting diode.

The 0 V (or negative, or 'ground') terminal of the battery is the reference point for measuring potentials (voltages).

A resistor resists the current flow. In this circuit it limits the flow of current so that the LED does not burn out.

The battery is a source of electromotive force (emf), to drive the current round the circuit.

Current can flow through a diode in only one direction, from anode (a) to cathode (k). The triangle in the symbol points the way.

A **conductor** is a material in which electric current can flow. All metals are conductors. List the conductors in this circuit. An **insulator** (or **non-conductor**) is a material in which current can not flow. List the insulators in this circuit (some answers on p. 370).

The diode D1 is made from a special kind of material called a **semiconductor** (see p. 3).

An electric current is a flow of electrically charged particles.

A current flows in a circuit if there is a potential difference (pd) between two points in the circuit. A pd may be produced in many different ways:

- chemical action: as in an electric cell.
- electricity generator: uses the energy of burning fuel (coal, oil, gas).
- nuclear power station: uses heat from nuclear reactions.
- solar cells: use energy from sunlight.
- wind farm: uses energy from wind.

All of these convert one form of energy into another form, electrical energy.

In the circuit on p. 1, the battery is the source of electrical energy. There is a pd between its terminals (see diagram below left), making current flow in the circuit. The electrical energy from the battery is converted by the LED and the resistor to other forms of energy (light and heat, see diagram below right).

Taking the 0 V (or 'negative') terminal of the battery as the reference point, the other terminal is at 3 V. The wire and switch are very good conductors so the potential at the anode of the LED (D1) is also 3 V. There is a **forward voltage drop** of about 2 V across a conducting LED so the potential at its cathode is about 1 V.

There is a drop of 1 V across the resistor, bringing the potential down to 0 V.

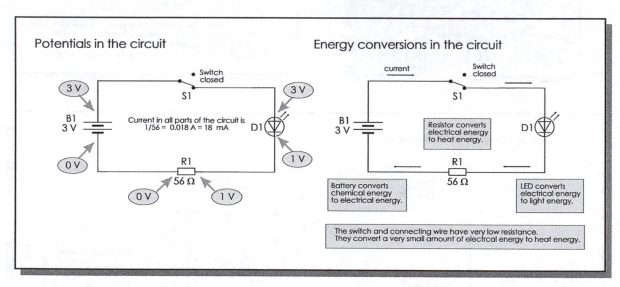

Each part of the circuit above has:

- a **current** flowing through it, measured in amperes (or **amps** for short).
- a **pd** across it, measured in **volts**.
- a **resistance** to the flow of current, measured in **ohms**.

These three quantities are related by this equation:

$$\text{current} = \frac{\text{pd}}{\text{resistance}}$$

The equation expresses Ohm's Law (see p. 363).

Starting at the 0 V terminal of the battery and moving clockwise round the circuit there is a rise of potential across the battery. The potential drops across D1 and R1.

The total potential (or voltage) drop in a circuit equals the potential rise across the source of emf.

Self test

There are some questions about volts, amps and ohms on p. 7 and on the companion website.

1 Diodes

A diode is made from semiconducting materials. It has two terminals, called the **anode** and the **cathode**. Current can flow easily from anode to cathode.

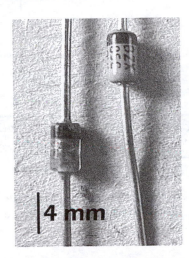

Typical low-power diodes. The diode on the left is a signal diode and the one on the right is a Zener diode. On both diodes the black band at one end indicates the cathode terminal.

Current and voltage

When there is a voltage (a pd) across a resistor, the two quantities are related as in the equation pd/current = resistance (see opposite). Ohm's Law applies.

This is not true for diodes. The circuit at top right has a milliammeter (mA) to measure the current through the diode, and a voltmeter (V) to measure the voltage across it. The power source is a power supply unit (PSU) that produces a DC voltage, variable from 0 V up to, say, 10 V.

The voltage is set to different values and the current is measured for each voltage. A graph of the results (the V-I graph) would be a straight line for a resistor, but is curved for a diode.

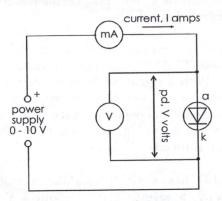

A circuit for measuring the voltage-current relationship of a forward biased diode.

The diode in the circuit above is connected so that current flows freely through it, from anode (a) to cathode (k). It is said to be **forward biased**.

The **V-I graph** obtained with this circuit is shown below. Its main feature is that no current flows though the diode until the forward bias of more than about 0.7 V. As the voltage is increased above 0.7 V, the current increases, slowly at first then more rapidly.

> **V and I**
>
> By convention, V is the symbol for the size of a voltage. I is the symbol for the size of a current. (p. 363)

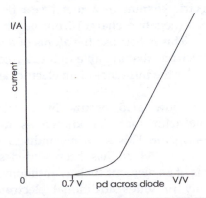

The V-I curve of a typical forward biased diode, made from silicon. A few types of diode are made from germanium. In these, conduction begins at a forward bias of about 0.3 V.

Reverse bias

If a diode is connected the other way round, so that its cathode is more positive than its anode, it is **reverse biased**.

Only a few nanoamps (1 nA = 10^{-9} A = 0.0000000001 A) flow through a reverse biased diode. With a signal diode or rectifier diode, the reverse voltage may be up to several hundred volts, depending on the type. But, if the voltage is too big, the diode breaks down and is destroyed.

An LED has a small breakdown voltage. Typically, a reverse bias of about 5 V will destroy it. Check that the LEDs are connected the right way round before switching on the power supply to a newly-built circuit.

LEDs may also break down when forward biased if the current through them is too big. The way to prevent this is described on pp. 14-15.

When reverse biased, a Zener diode breaks down at a fixed voltage, its Zener voltage, V_Z (see top right). This may be only a few volts, depending on the rating of diode.

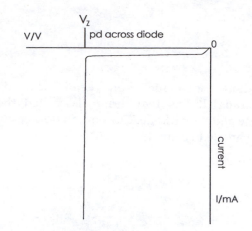

The V-I graph of a reverse biased Zener diode has a sharp 'knee' at the Zener voltage, V_Z.

Zener breakdown does not destroy the diode. This property is used in voltage regulator circuits (pp. 149-150).

Photodiodes are nearly always connected with reverse bias. The small leakage current through the photodiode is proportional to the amount of light falling on it.

Conduction in semiconductors

Extension Box 1

An electric current in a metal is a flow of electrons (negative charge) from negative to positive. This is because the atoms of a metal have electrons that are able to break free and flow under the influence of an electric field.

Electron flow also occurs in a type of semiconductor material known as **n-type semiconductor**. This is a semiconductor such as silicon, to which has been added small amounts of an element such as antimony. The antimony provides additional electrons to increase the conductivity of the material. It is called n-type because electrons carry negative charge.

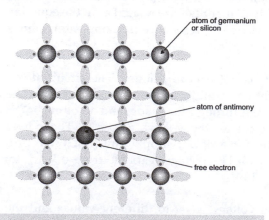

Doping pure silicon with antimony provides a free electron, which increases conductivity.

Conduction by holes

Another type of semiconductor is known as **p-type semiconductor**. In this, the added material is an element such as indium. The indium atoms have a 'vacancy' in their outer electron orbit where an electron can be taken in. Such a vacancy is known as a **hole**.

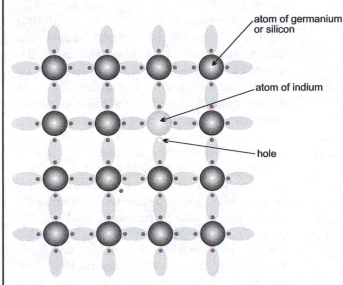

What happens when an electric field is present is shown below:

The first stage is illustrated in (a):

At (1) an electron escapes from a silicon atom and drifts in the electric field until it finds a hole at the indium atom. At (2), the hole this created is filled by an electron escaping from an atom further along the material.

The next stage is illustrated in (b).

At (3), the hole created at (2) is filled by an electron from the negative supply, having travelled along a metal wire. At (4) an electron escapes and passes into the metal wire of the positive supply. The hole created is again filled at (5) by an electron, so shifting the hole toward the negative end.

The holes apparently move from positive to negative, in the opposite direction to the electrons. So we can think of holes as positive charge carriers. For this reason, silicon doped with indium or boron is known as **p-type silicon**.

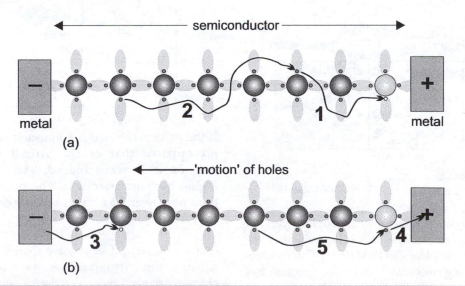

Conduction in p-type silicon. Holes move from positive to negative and so are equivalent to positive charge carriers.

Conduction in diodes   Extension Box 3

Consider a bar of silicon, doped so that half is p-type and half is n-type. The bar is not connected to a circuit, so no external electric field is applied to it. Electrons and holes are free to wander at random.

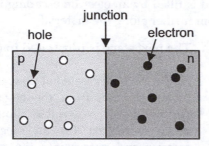

Some of the electrons from the n-type material wander across the junction, attracted by the holes in the p-type material. The holes in the region of the junction become filled. In the p-type region (to the left of the junction) the filling of holes means that the atoms have, on average, more electrons than they should. The atoms have become negative *ions*.

Similarly, in the n-type region near the junction the atoms have, on average, lost electrons leaving holes and so have become positive *ions*.

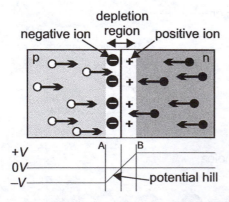

Although these ions are charged, they can not act as charge carriers. This is because they are *fixed* in their places in the lattice. The result is that the region on side A of the junction has overall negative charge and the region on side B has overall positive charge. There is a pd across the junction.

The effect of this pd is the same as if a cell was connected across the junction. This is not a real cell but we refer to its *effects* as a **virtual cell**. At a silicon p-n junction, the pd is about 0.7 V. At a germanium junction, it is about 0.2 V.

The region that contains ions has no mobile charge carriers and it is called the **depletion region**. We think of the pd across the depletion region as a **potential hill**, the potential rising as we pass from the p-type side to the n-type side. Eventually, as more carriers cross the junction, the hill becomes so steep that no more carriers can cross it.

If the semiconductor is connected into a circuit and a pd is applied to it, the junction can be **biased** in either direction. With **reverse bias**, the external source reinforces the action of the virtual cell, making the potential hill higher. This makes the depletion region wider than before, and there is even less chance of carriers finding their way across it. No current flows.

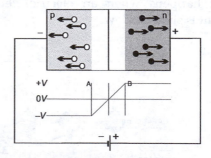

If the external source is connected so that its pd opposes that of the virtual cell, the junction is **forward biased.** The depletion region becomes narrower. The potential hill becomes lower. Electrons are more easily able to cross the junction.

If the external pd is greater than 0.7 V (for silicon), the virtual cell is overcome and charge carriers flow across the junction. There is a **forward voltage drop** of 0.7 V across the junction.

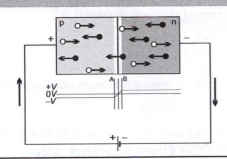

Conduction in diodes

The p-n junction has the unusual property that it allows conduction in one direction (p to n) but not in the other. We find pn junctions in diodes, MOSFETs (p. 69) and in BJTs (p. 83). They show one-way conduction and a forward voltage drop of approximately 0.7 V when forward biased.

Questions on circuits

1 Draw a schematic diagram of a door alert circuit built from a battery of 4 cells, a push-button, and a buzzer (see the chart of symbols on pp. 367-368). On the diagram mark the point in the circuit that is at the highest potential. Show the direction of flow of current.

2 Draw a schematic diagram of a door alert circuit that has two push-buttons, A and B. The buzzer is to sound when either A or B is pressed. What happens if both buttons are pressed at the same time?

3 A DC electric motor is powered by a 4.5 V battery and controlled by a SPST (single pole single throw) switch. There is a light emitting diode which comes on when the power supply is switched on. The diode needs a resistor to limit the current through it. Draw the schematic diagram of the circuit.

4 List the energy conversions that occur when the circuit of Question 3 is switched on.

5 Name four good conductors of electric current. Give an example of the use of each and state which one of the four is the best conductor.

6 Name four electrical insulators, giving an example of the use of each.

7 The pd across a torch lamp is 2.4 V and the current through it is 700 mA. What is the resistance of the lamp?

8 What is the current through a 220 Ω resistor if the pd across it is 5 V?

On the Web

A lot of information about electronics can be found on the World Wide Web. Where you see a panel like this one in the book, access the Web on a computer and visit some of the listed sites. Use a browser such as *Internet Explorer*, *Safari*, *Netscape* or *Mozilla Firefox*.

For some topics, go to a search engine such as *Google* or *Yahoo!* and type in a keyword, such as 'diode'.

In *Yahoo!*, 'diode' gave us over 7 million references. These included tutorials and explanations such as those from *Wikipedia* and *How Stuff Works* sites, as well as manuacturers' data sheets and suppliers' catalogues.

Yahoo! has an *Answers* site (http://answers.yahoo.com), which covers many topics, including electronics (under Engineering). There you can post questions or try to answer those asked by others.

Search the Web for information on diodes and related topics. It may help you with some of the questions on this page and overleaf.

Questions on diodes

1 Name the terminals of a diode. In which direction does the current flow when the diode is forward biased?
2 Describe, with the help of a graph, how the current through a forward biased diode varies as the pd across the diode is varied from 0 V to 5 V.
3 Using a data sheet or information obtained on the Web, list and comment on the characteristics of a named signal diode.
4 Repeat 3 for a high-power rectifier diode.
5 List the features of six different types of LED, and state an application for each.
6 Draw the symbols for (a) a signal diode, (b) a Zener diode, (c) a photodiode, and (d) a light-emitting diode. Label the anode and cathode of each symbol.
7 Search this book, or another electronics book, or the WWW, for a simple electronic circuit contaning a diode. Draw the circuit and describe how the properties of the diode are made use of in this circuit.

Multiple choice questions

1 An example of a conductor is:

    A   nylon.
    B   aluminium.
    C   rubber.
    D   glass.

2 An example of an insulator is:

    A   copper.
    B   steel.
    C   PVC (polyvinyl chloride).
    D   carbon.

3 The pd across a resistor is 15 V and its resistance is 120 Ω. The current through the resistor is:

    A   125 mA.
    B   8 A.
    C   1.25 A.
    D   105 mA.

4 A silicon diode is forward biased so that its anode is 0.5 V positive of its cathode. The current through the diode is:

    A   0.5 A.
    B   0 A.
    C   not known.
    D   enough to burn out the diode.

Companion site

A box with this icon tells you that the Companion Site for *Electronics — Circuits and Systems* has more about this topic.

The URL of the site is:

http://books.elsevier.com/companions/9780750684989.

The site has questions on electrical circuits and Ohm's Law. It includes Calculator windows to help you work out the answers.

2 Transistor switches

Transistors are used in one of two ways:

- as switches
- as amplifiers.

Transistor amplifiers are described later in the book. Transistor switches are described in this chapter, using three different types of transistor.

As we shall explain, the purpose of most transistor switches is to control an electrical device such as a lamp, a siren or a motor.

The switched device usually requires a current of several milliamperes, possibly several amperes. The current needed for operating the transistor switch is a lot smaller, often only a few microamperes. This makes it possible to control high-current devices from sensors, logic gates and other circuits with low-current output. The main limitation is that only devices working on direct current can be switched, but not devices powered by alternating current. These points are illustrated by the examples below.

Switching a lamp

The switching component in this circuit is a **metal oxide silicon field effect transistor**. This name is usually shortened to MOSFET. The words 'metal oxide silicon' refer to the fact that the transistor consists of a metal conductor, an insulating layer of oxide (actually silicon oxide), and a semiconducting layer of silicon.

The words 'field effect' refer to the way in which this type of transistor works. It works by placing an electric charge on the 'metal' part of the transistor, known as the **gate**. The charge on the gate results in an electric field, and the effect of this field is to control the amount of current flowing through the semiconductor layer. Current flows from the **drain** terminal to the **source** terminal.

This medium-power MOSFET is rated to switch currents up to 13 A. Note the metal tag for attaching a heat sink.

Drain current (I_D) flows from the drain to the source, if:

- the drain terminal is positive of the source terminal and,
- the gate is charged to a voltage that is sufficiently positive of the source.

In this way, the MOSFET acts as a **voltage-controlled switch**. The voltage at which the transistor switches on is called the **threshold voltage**.

A MOSFET of the type just described is known as an **n-channel enhancement MOSFET**. There are other types (pp. 16 and 69), but this type is by far the most commonly used.

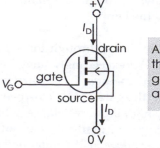

A MOSFET has three terminals: gate, drain and source

Sensor

The light-sensitive component in this circuit is a **photodiode.** Like any other diode, it conducts in only one direction.

The can of this photodiode has a lens top. The diode is visible through this as a square chip of silicon.

Usually a photodiode is connected with **reverse bias.** This would mean that no forward current flows through it. However, a small **leakage current** passes through it. The leakage current is only a few nanoamps in darkness, but rises to several microamps when light falls on the photodiode.

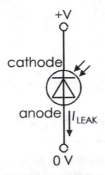

Leakage current passes through the photo-diode only in the light.

Switching circuit

This circuit uses a MOSFET transistor for switching a lamp. We might use this to control a low-voltage porch lamp which is switched on automatically at dusk.

The sensor is a photodiode, which could be either of the visible light type or the type especially sensitive to infra-red radiation. It is connected so that it is reverse-biased. Only leakage current passes through it. The current varies according to the amount of light falling on the diode. In darkness, the current is only about 8 nA but it rises to 3 mA or more in average room lighting.

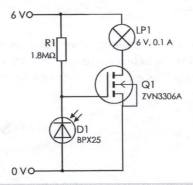

The MOSFET switches on the filament lamp when the light intensity falls below a given level.

The current change is converted into a voltage change by resistor R1. In light conditions, the current is relatively large, so the voltage drop across the resistor is several volts. The voltage at the point between R1 and the diode is low, below the threshold of the transistor. The transistor is off and the lamp is not lit.

At dusk, light level falls, the leakage current falls, and the voltage drop across R1 falls too. This makes the voltage at the gate of the transistor rise well above its threshold, which is in the region of 2.4 V. The transistor turns fully on. We say that it is **saturated.** When the transistor is 'on', its effective resistance is only 5 Ω. Current flows through the lamp and transistor, lighting the lamp.

The current flowing down through the sensor network (R1 and D1) is exceedingly small — no more than a few microamps. Little current is available to charge the gate of Q1. However, a MOSFET has very high input resistance, which means that it requires virtually no current to turn it on. It is the *voltage* at the gate, not the current, that is important.

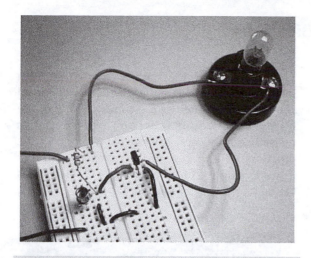

A breadboarded version of the light-sensitive MOSFET switch. The photodiode and resistor are on the left. The low-power MOSFET is the small black package on the right.

A circuit such as this does not normally need a heat sink (pp. 130-131) on the transistor, even if the current being switched is several amps. The circuit is always in either of two states, 'off' or 'on'. When the transistor is off, there is no current and no heating effect. When the transistor is on, it is *fully* on. Its resistance may be only an ohm or two, or even less. Little energy is dissipated within the transistor and it never gets hot.

System design

Electronic systems often consist of three stages. The first stage is the **input** stage. This is the stage at which some feature of the outside world (such as a temperature, the light level, or a sound) is detected by the system and converted into some kind of electrical signal.

The last stage is the **output** stage in which an electrical signal controls the action of a device such as a lamp, a motor, or a loudspeaker.

Connecting the input and output stages is the **signal processing** stage. This may comprise several sub-stages for amplifying, level-detecting, data processing, or filtering.

As a system, this circuit is one of the simplest. It has three sections:

Because the gate requires so little current, a MOSFET is the ideal type of transistor for use in this circuit. If we were to use a BJT (see p. 12), we might find that the sensor network was unable to provide enough current to turn it on.

The lamp specified for this circuit has a current rating of 100 mA. The transistor has a rating of 270 mA, so it is not in danger of burning out. There is no need to use a power MOSFET such as the one shown in the photo on p. 9.

Variations

A useful improvement is to reduce R1 (to 820 kΩ) and wire a variable resistor (about 1 MΩ) in series with it. Connect the gate of Q1 to the wiper of the variable resistor. This allows the switching level of the circuit to be adjusted.

The circuit may be made to operate in the reverse way by exchanging the resistor and photodiode. It may be necessary to alter the value of the resistor.

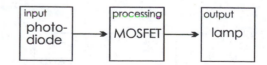

The **input** section consists of the photodiode.

The **processing** section consists of R1 and Q1. R1 generates a voltage, which is fed to the gate of Q1, which is either turned 'off' or 'on'.

The **output** section consists of the lamp, which is switched by Q1.

Three-stage systems are very common. There are more examples in this chapter and elsewhere in this book.

Light-sensitive alarm

Compared with the circuit on p. 10, this circuit uses a different type of light sensor, and a different type of transistor. It has a similar three-stage system diagram:

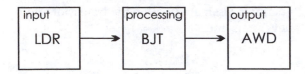

- **Input** is accepted by a **light-dependent resistor** (or LDR). This is made from a semiconductor material such as cadmium sulphide. The resistance of this substance varies according the the amount of light falling on it. The resistance of an LDR ranges between 100 Ω in bright light to about 1 MΩ in darkness.
- **Processing** uses a voltage divider (R1/R2, *see* p. 17) that sends a varying voltage to the **bipolar junction transistor**, Q1. The voltage at the point between the resistors is low (about 10 mV) in very bright light, but rises to about 5 V in darkness. With LDRs other than the ORP12, voltages will be different and it may be necessary to change the value of R1 or replace it with a variable resistor.
- **Output** is the sound from an **audible warning device**, such as a siren.

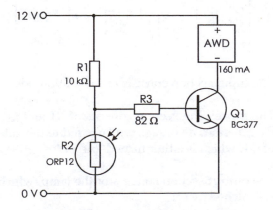

The audible warning device sounds when the light intensity falls below a given level.

The function of the circuit is to detect an intruder passing between the sensor and a local source of light, such as a street light. When the intruder's shadow falls on the sensor, its resistance increases, raising the voltage at the junction of the resistors. Increased current flows through R3 to the base of Q1. This turns Q1 on and the siren sounds.

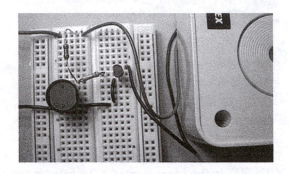

The quadruple siren to the right of the breadboard emits an unpleasantly loud scream when the LDR on the left is shaded.

This circuit uses a BJT; almost any npn type will do, provided it is able to carry the current required by the audible warning device, which in this example is a piezo-electric siren. A typical ultra-loud (105 dB) warbling siren suitable for use in a security system takes about 160 mA when run on 12 V. The BC337 transistor is rated at 500 mA.

BJT switches

Like a MOSFET, a BJT has three terminals. Because the BJT works differently, they have different names: base, collector and emitter.

A BJT is connected so that its collector terminal is several volts positive of the emitter terminal. When current (I_B) flows into the base terminal, a much bigger current (I_C) flows into the collector terminal. The combined currents I_B and I_C flow out of the emitter terminal.

The combined currents make up the emitter current I_E, and:

$$I_E = I_B + I_C$$

The action of a BJT depends on the fact that I_C is alway very much larger than I_B. For a typical BJT it is about 100 times larger. Thus, small variations in a small current result in large variations in a large current.

When used as a switch, a BJT is a **current-controlled switch**. This contrasts with a MOSFET, which makes a voltage-controlled switch. The amount of current needed to switch on the transistor in this circuit depends on the amount of **collector current** needed to drive the siren. This depends on the **current gain** of the transistor.

The siren used in the demonstration circuit requires 160 mA. The BC377 has a current gain of between 100 and 600. Taking the minimum gain of 100, the base current needs to be:

$$I_B = I_C/\text{gain} = 160/100 = 1.6 \text{ mA}$$

Inside an npn BJT there is a pn junction between the base and emitter terminals. This junction has the properties of a diode. In particular, when it is forward biased, there is a voltage drop of about 0.7 V between the base and emitter terminals. In practice, the voltage drop may be 0.5 V to 0.9 V, depending on the size of I_B. Therefore, when Q1 is in the 'on' state, its base is, on average, 0.7 V positive of its emitter.

In a BJT, a small current controls a much larger current.

Now we can calculate what voltage from the R1/R2 potential divider is needed to switch on Q1 and sound the siren. We need a minimum of 1.6 mA flowing through R3. Given that R3 is 82 Ω, the voltage across R3 must be:

$$\text{current} \times \text{resistance} = 1.6 \times 10^{-3} \times 82 = 1.31 \text{ V}$$

If the voltage at the R1/R2 connection is less than 1.31 V, no current flows through R3 and Q1 is off. I_C is zero and the siren does not sound. If the voltage is greater than 1.31 V, Q1 turns on and becomes **saturated**. I_C rises to 160 mA and the siren sounds. There is a state between Q1 being off and being on and saturated, but we will leave discussion of this until later.

This circuit is able to use a BJT because the sensor network is able to deliver a current of the size required to turn on Q1. It is also possible to use a MOSFET. In this case, R3 is not required as almost no current flows to the gate of a MOSFET.

Heater switch

The heater switch is a circuit for maintaining a steady temperature in a room, a greenhouse, or an incubator. The sensor is a **thermistor**, a semiconductor device. Its resistance decreases with increasing temperature, which is why it is described as a **negative temperature coefficient** (ntc) device (see p. 367). Thermistors are not ideal for temperature *measuring* circuits, because the relationship between temperature and resistance is far from linear. This is no disadvantage if only one temperature is to be set, as in this example.

Thermistors have the advantage that they can be made very small, so they can be used to measure temperatures in small, inaccessible places. They also quickly come to the same temperature as their surroundings or to that of an object with which they are in contact. This makes them good for circuits in which rapid reading of temperature is important.

In many applications, these advantages offset the disadvantage of non-linearity.

The switched device in this example is a **relay**. This consists of a coil wound on an iron core. When current passes through the coil the core becomes magnetised and attracts a pivoted armature. The armature is pulled into contact with the end of the core and, in doing so, presses two spring contacts together. Closing the contacts completes a second circuit which switches on the heater. When the current through the coil is switched off, the core is no longer magnetised, the armature springs back to its original position, and the contacts separate, turning off the heater.

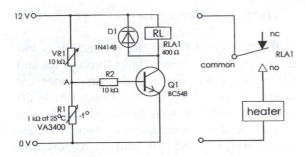

A relay is necessary to switch a heater which runs from the 230 V AC mains.

Most relays have contacts capable of switching high voltages and currents, so they may be used for switching mains-powered devices running on alternating current. Current ratings of various types of relay range from 1 A up to 40 A. Various arrangements of switch contacts are available, including simple **change-over contacts** (see drawing above) and relays with three or more independent sets of contacts. Changeover contacts are normally-open (no) or normally closed (nc). They can be wired so that a device can be turned off at the same time that another is turned on. Relays are made in an assortment of sizes, including some only a little larger than a transistor, suitable for mounting on printed circuit boards.

Self test

What is the minimum contact rating for a relay required to switch a 1000 W, 230 V floodlamp?

Thermistor

Thermistors are available in several resistance ratings, the one chosen for this circuit having a resistance of 1 kΩ at 25°C. VR1 is adjusted so that the voltage at point A is close to 0.6 V when the temperature is close to that at which the heater is to be turned on. The transistor is just on the point of switching on. As temperature falls further, the resistance of the thermistor R1 increases (see p. 367), raising the voltage at A, and increasing i_B, the base current. This switches on the transistor and current passes through the relay coil. The relay contacts close, switching on the heater.

Note that the heater circuit is electrically *separate* from the control circuit. This is why a relay is often used for switching mains-voltage alternating current.

After a while, as the heater warms the thermistor, R1 decreases, the voltage at point A falls and the transistor is turned off. The relay coil is de-energised and the heater is turned off.

This kind of action is called **negative feedback**. It keeps the temperature of the room more or less constant. The action of the circuit, including the extent to which the temperature rises and falls, depends on the relative positions of the heater and thermistor and the distance between them.

Protective diode

The diode D1 is an important component in this circuit.

At (a) below a transistor is controlling the current through an inductor. When the transistor switches off (b), current ceases to flow through the coil and the magnetic field in the coil collapses. This causes an **emf** to be induced in the coil, acting to *oppose* the collapse of the field. The size of the emf depends on how rapidly the field collapses. A rapid switch-off, as at (b), collapses the field instantly, resulting in many hundreds of volts being developed across the transistor. This emf produces a large current, which may permanently damage or destroy the transistor.

> **Emf**
>
> This stands for electromotive force. Emf is a potential difference produced by conversion of another form of energy into electrical energy. Ways of producing emf include chemical (dry cell), mechanical (dynamo, rubbing a plastic rod with a woollen cloth), and magnetic (collapsing field, car ignition system).

An LED needs a current of a few tens of milliamps. Often 20 mA or 25 mA makes the LED shine brightly. Like any other diode, an LED has a forward voltage drop across it when it is conducting. For most LEDs the drop is about 2 V.

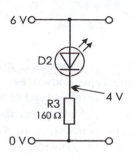

An LED normally has a resistor in series with it.

The LED will burn out if the voltage across it is much more than 2 V. If the LED is being powered by a supply greater than 2 V, we need a resistor (R3) in series with it. This drops the excess voltage.

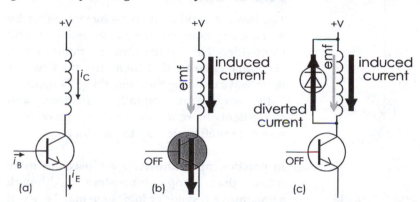

(a)   (b)   (c)

If a diode is wired in parallel with the inductor, as at (c), it protects the transistor by conducting the induced current away to the power line. It is a **protective diode**.

Indicator lamp

An indicator lamp is an optional but useful addition to this heater circuit. It could also be added to the other switching circuits we have studied. The idea is to have a lamp that comes on when the circuit is switched on. We could use a filament lamp (such as a 6 V torch bulb) but a light-emitting diode is generally preferred. This is because, for a given brightness, LEDs do not take as much current as filament lamps. Also, filament lamps

In the diagram above, the supply is 6 V, so the excess voltage is 4 V. If we decide to have a current of 25 mA through the LED, the same current passing through R3 must produce a drop of 4 V across it. So its resistance must be:

$$R_3 = 4/0.025 = 160 \ \Omega$$

Select the nearest standard value. In this case, a 160 Ω resistor can be used, though a 150 Ω resistor would be close enough.

In general, if the supply voltage is V_S and the current is to be i_{LED}, the series resistor R_S is:

$$R_S = (V_S - 2)/i_{LED}$$

Certain types of LED, such as flashing LEDs, have built-in resistance and do not need a series resistor.

> **Self test**
>
> 1 Calculate the value of the series resistor when:
>
> a the supply voltage is 9 V and the current required is 20 mA.
>
> b the supply voltage is 24 V and the current required is 15 mA.
>
> 2 If the supply is 12 V and the series resistor is 1.5 kΩ, what current flows through the LED?

Overheating alert

Previous circuits in this topic have used BJTs of the type known as **npn** transistors. Here we use the other type of BJT, a **pnp transistor**. The action of a pnp transistor is similar to that of an npn transistor. In particular, a small base current controls a much larger collector current. However the polarities are reversed:

- The transistor is connected with its emitter *positive* of its collector.
- Base current flows *out of* the base terminal instead of flowing into it.
- When the transistor is 'on', the base terminal is 0.7 V *negative* of the emitter.

Here is a switching circuit using a pnp BJT:

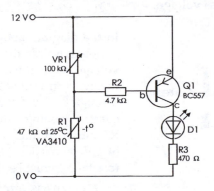

Compare this circuit with the circuit on p. 14. Note the symbol for a pnp transistor.

The circuit is set by adjusting VR1 until the LED is just extinguished. The base (b) potential is then a little above 11.3 V (that is, the base-emitter voltage drop is a little less than 0.7 V). If the thermistor (R2) is warmed, its resistance decreases. This reduces the voltage at the point between R1 and R2. Current flows out of the base, turning the transistor on. When the transistor is on, the collector current flows in through the emitter (e), out of the collector (c) and on through the LED and series resistor R3. Thus, the LED comes on whenever the temperature of R1 rises. By adjusting VR1, the circuit can be set to switch on the LED at any required temperature in the region of 25°C.

Inverse action

The switching action is the opposite to that of the npn transistor circuit on p. 14. There the relay comes on when temperature of the thermistor falls. In this circuit the LED comes on as temperature rises.

We could have achieved the equivalent action with an npn transistor in the relay circuit by exchanging VR1 and the thermistor. The values of the resistors would have needed altering but the action would be to turn on the relay with *falling* temperature.

Pnp or npn?

The inverse action can be achieved either by exchanging components or by using a pnp transistor. This illustrates a point about electronic circuits, that there are often two or more ways of doing the same thing. Sometimes both ways are equally effective and convenient. In other instances we may have reason to prefer one way to the other.

In practice, npn transistors are used far more often than pnp transistors. Although sometimes a resistor or logic gate may be saved in a circuit by using pnp, it is usually just as convenient to swap components to invert the action.

The main reason for using pnp is when we want a given signal to produce an action and its inverse *at the same time*. Then we use a pair of transistors, npn and pnp, with equal but opposite characteristics. Examples are seen in the power amplifiers described on pp. 128-129.

MOSFET types

MOSFETs too are made that work with polarities opposite to those of the n-channel MOSFETs described in this chapter. These are known as **p-channel MOSFETs** (pp. 69-70).

Schmitt trigger

In the two previous circuits, the resistance of the thermistor changes relatively slowly, so that the circuit spends several minutes in a state intermediate between 'off' and 'fully on'. The transistor is partly, not fully, switched on. In the heater switch circuit, the coil of the relay is not fully energised and the armature tends to vibrate rapidly, switching the heater on and off several times per second. This leads to sparking at the contacts, which may become fused together so that the relay is permanently 'on'. In an LED switching circuit, the LED gradually increases in brightness as temperature increases. It gives no clear indication of whether or not the temperature is too high. The circuit below avoids these problems, giving a sharp trigger switching action.

This is necessary because we have *two* inverting transistors in this circuit. The resistance of R3 must be greater than that of the relay coil, so that more current flows through R6 when Q2 is on, and less flows when Q1 is on.

Snap action

As temperature falls and Q1 begins to turn off, Q2 begins to turn on. The increasing current through R6 raises the voltage at the Q1 emitter. As the Q1 base voltage continues to fall its emitter voltage starts to rise, rapidly reducing the base current and turning Q1 off very rapidly. If Q1 turns off faster, then Q2 turns on faster. The circuit spends much less time in its intermediate state.

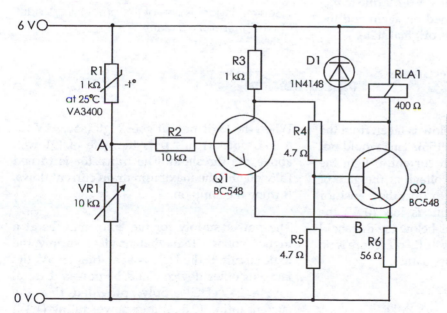

The Schmitt trigger circuit has a 'snap' action, which is suitable for switching heaters, indicator lamps and many other devices.

This **Schmitt trigger** consists of two npn transistors. Q1 is switched by the sensor network. Q2 is switched by Q1 and in turn switches the relay.

The thermistor and variable resistor are arranged so that the voltage at point A *falls* as temperature falls. This is the opposite action to that in the circuit on p. 14.

Hysteresis

In this circuit the temperature at which the relay is turned on or off depends on whether the temperature is rising or falling.

The action of the circuit is such that the 'turn-on' temperature is lower than the 'turn-off' temperature. We call this action **hysteresis**.

The circuit works differently in the two states:

- **The heater is off**, the temperature of R1 is relatively high but falling, and its resistance is low, but rising. The voltage at A is relatively high but falling. Q1 is on and Q2 is off. Current through R6 and the voltage across it are fairly small. For example, measurements on this circuit found the voltage at B to be 0.4 V. This is due to the current flowing through R3 and Q1 and finally through R6. As the voltage at A falls to 1.1 V (0.4 V + 0.7 V due to the base-emitter voltage of Q1), the circuit starts to change state, turning the heater on.

- **The heater is on**, and this increases the current through R6 because the resistance of the relay coil is lower than that of R3. The increased current makes the voltage at B rise to, say, 0.9 V. Now the voltage at A has to rise to 1.6 V (0.9 V + 0.7 V due to v_{BE}) before Q1 can be turned on again and the circuit can return to its original state.

In terms of heating, the heater comes on when the room has cooled to a set temperature, say 15°C. We call this the **lower threshold**. The heater stays on, perhaps for 10–15 minutes, until the room has warmed to a higher temperature, say 17°C, then goes off. We call this the **higher threshold**. The difference betweeen the two thresholds, in this case two degrees, is the hysteresis of the circuit. This is a much better action than having the heater continually switching on and off several times a minute at each little rise or fall in temperature.

Any of the switching circuits in this chapter could be modified to have a Schmitt trigger between the sensor and the switched device.

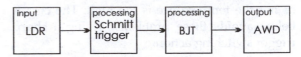

A Schmitt trigger is one of the processing stages of this light-sensitive alarm.

Logical control

The input of the circuit below is taken from the output of a logic gate (p. 156). This would not be able to supply enough current to light the lamp, but can raise the voltage at the gate of the MOSFET sufficiently to turn the transistor on. When the logic output is low (0 V) the voltage at the gate of Q1 is below the threshold (2.5 V) and the transistor is fully off. There is no drain current and the lamp is unlit.

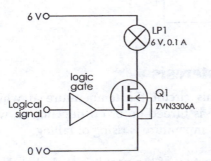

This is how a lamp may be controlled by a logic circuit, such as a microcontroller.

When the logic output goes high (say, +5V for a TTL output) this puts the gate of Q1 well above the threshold. The transistor is turned fully on and the maximum drain current flows, turning the lamp on.

The power supply for the lamp may be at a higher voltage than that used to supply the logic circuit. If the logic is operating on 5V, the lamp or other device could be powered from, say, a 24V DC supply, provided that the current rating (0.5 A) and power rating (1 W) of the transistor are not exceeded.

This circuit could also be used to switch other devices, including a **solenoid**. A solenoid is a coil wound on a former. There is an iron **plunger** or **armature** which slides easily in and out of the coil. Usually, the plunger is held by a light spring so it rests only partly inside the coil when there is no current flowing (see photo, on the next page).

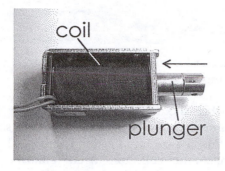

When a current is passes through the coil of this solenoid, the plunger moves strongly in the direction shown by the arrow.

When current flows through the coil, the magnetic field strongly attracts the plunger, which is drawn forcefully into the coil.

There is usually some kind of mechanism linked to the plunger. For example, the plunger may be attached to a sliding bolt, so that energising the coil causes a door to be bolted. In this way the door may be bolted by computer control.

As with the relay circuit on p. 14, a solenoid switching circuit requires a diode to protect the switching transistor.

Current gain — Extension Box 4

The gain of a BJT is expressed as the ratio of the collector current to the base current. This is known as the **large signal current gain**, h_{FE}:

$$h_{FE} = I_C/I_B$$

Because h_{FE} is a current divided by a current, it has no unit. It is just a *number*.

Example

When the base current to a given BJT is 270 µA, the collector current is 40.5 mA. What is the current gain of the BJT?

$$h_{FE} = I_C/I_B = 40.5/270 \ [mA/µA]$$
$$= 150$$

The current gain is 150.

Self test

If the gain of a BJT is 240, and I_B is 56 µA, what is I_C?

Memo

There is also the **small signal current gain**, h_{fe} (small letters for the suffix 'fe'). A *small change* i_b in the base current causes a relatively *small change* i_c in the collector current. Then, $h_{fe} = i_c/i_b$.

In most cases, h_{fe} has a value equal or close to that of h_{FE} (see p. 73).

Transconductance — Extension Box 5

A MOSFET is a voltage controlled device. Its 'gain' is expressed as the *change* in the drain current for a given *change* in gate-source voltage. This is known as the **transconductance**, g_m:

$$g_m = i_d/v_{gs}$$

Because g_m is a current divided by a voltage it has the units of conductance (the inverse of resistance). The unit of conductance is the **siemens**, symbol **S**.

Example

When the gate-source voltage of a given MOSFET increases by 2 V, the drain current increases by 1.7 A. What is the transconductance of the MOSFET?

$$g_m = i_d/v_{gs} = 1.7/2 = 0.85 \ S$$

The transconductance is 0.85 S, or 85 mS.

Note that the calculations in Box 4 and this box assume that the collector-emitter pd and the drain-source pd are constant, and are sufficient to drive the current through the transistor.

DATA — Bipolar junction transistors

The transistors are listed by **type number**. Many of the circuit diagrams in this book specify type number BC548. This is an easily available, cheap, general-purpose BJT, made by several different manufacturers. As can be seen from the table, there are several other types of BJT that would work in these circuits just as well as the BC548.

The type number is not particularly useful when you are selecting a transistor. The easiest way is to refer to the final column of the table and select by application. The majority of transistors are described as 'general-purpose' (GP) possibly with some additional features mentioned. One of these types will suit most projects.

All transistors in this table are npn except the BC557, which is pnp.

Probably the most important item is the **maximum collector current**. The data sheet lists this in a column headed I_C **(max)**. This is the maximum collector current (almost equal to the emitter current) that the BJT can pass when it is saturated. It is equal to the current passing through the load. The next column lists the **maximum power rating,** P_{tot}.

The **current gain** specified in the table is the small signal current gain, h_{fe}, mentioned on p. 11. The values of h_{FE} is almost the same as h_{fe}, so these values can be used in transistor switch calculations.

Type	Case	I_c (max)	P_{tot}	h_{fe} at I_c (mA)		F_t at I_c (MHz at mA)		Applications
BC107	TO-18	100 mA	300 mW	110-450	2	300	10	GP small signal amplifier
BC108	TO-18	100 mA	300 mW	110-800	2	300	10	GP small signal amplifier
BC109	TO-18	100 mA	300 mW	200-800	2	300	10	Low noise, small sig. amp.
BC109C	TO-18	100 mA	300 mW	420-800	2	300	10	Low noise, high gain, amp
BC337	TO-92	500 mA	625 mW	100-600	2	100	10	Output stages
BC548	TO-92A	100 mA	500 mW	110-800	2	300	10	Low noise, small sig. amp.
BC557	TO-92	100 mA	625 mW	110-800	2	280	10	**pnp** output transistor
BC639	TO-92	1 A	1 W	40-250	150	200	10	Audio output
BD139	TO-126	1.5 A	8 W	40-250	150	250	50	Power GP
BD263	TO-126	4 A	36 W	750	1.5 A	7	1.5 A	Med. power Darlington
BUX80	TO-3	10 A	100 W	30	1.2 A	8		High current switch
MJE3055	TO-220	10 A	75 W	20-70	4 A	2	500	Power output
2N2222	TO-18	800 mA	500 mW	100-300	150	300	20	High speed switch
2N3641	TO-105	500 mA	350 mW	40-120		250	50	GP, amplifier, switch
2N3771	TO-3	30 A	150 W	15-60	1.5 A	0.2	1 A	Power output
2N3866	TO-39	400 mA	1 W	10-200	50	500	50	VHF amplifier
2N3904	TO-92	200 mA	310 mW	100-300	10	300	10	Low power GP

DATA — BJTs (continued)

It is measured with the collector current at the stated value. Usually, a range is specified because transistors of the same type may vary widely in gain. The BC548, for instance, has a gain of 110 to 800. Sometimes transistors are graded by the manufacturer and sold in gain groups. A transistor sold as a BC548A, for instance, has a gain in the range 110 to 220. BC545B ranges fron 200 to 450, and BC548C ranges from 420 to 800.

The fourth important feature of a transistor is its **maximum frequency, F_t**. A transistor being used as a switch responds quickly enough for most applications.

When working at radio frequencies, it may not be able to respond quickly enough to changes in the signal current. The gain falls off at high frequencies. F_t is defined as the frequency for which the gain falls to 1. For many transistors, F_t is rated in hundreds of megahertz and, in the table, is quoted in megahertz at a given collector current.

Normally, we do not have to worry about F_t. It is mainly in radio circuits and computer circuits that such high frequencies are used. But it can be seen from the table that high-power transistors have much lower F_t than other types.

DATA — MOSFETs

The table below lists a selection of low-cost, low-voltage MOSFETs. Maximum drain-source voltages range between 60 V and 100 V, except for the IRF630 which is 200 V.

The most useful figures for choosing a transistor are maximum drain current and drain-source source resistance when 'on'.

Transconductance is of less important when selecting, but you will need it when calculating outputs.

In general, transistors with low I_D and low g_m are considerably cheaper. In this table, the cheapest n-channel MOSFET is the ZVN3306A and the dearest is the IRFD024.

The threshold voltage of a given type may vary considerably between different transistors of the same type. Usually it is between 2 V and 4 V. This is why we have not quoted threshold voltages in this table.

Type	Case	n/p	I_D (max)	P_{tot}	R_{DS} (max)	g_m (S)
BUZ73L	TO-220	n	7 A	40 W	0.40 Ω	5.00
IRF630	TO-220	n	9 A	75 W	0.40 Ω	3.00
IRF9540	TO-220	p	19 A	150 W	0.20 Ω	5.00
IRFD110	TO-250	n	1 A	1.3 W	0.54 Ω	1.30
IRFD024	TO-250	n	2.5 A	1.3 W	0.1	0.90
ZVN2106A	Eline	n	450 mA	1.5 W	2 Ω	0.40
ZVN3306A	Eline	n	270 mA	625 mW	5 Ω	0.15
ZVP2106A	Eline	p	280 mA	700 mW	5 Ω	0.15
2N7000	TO-92	n	200 mA	400 mW	5 Ω	0.10

BJTs or MOSFETs?

When choosing between BJTs and MOSFETs, keep these points in mind:

BJTs need appreciable current to drive them; MOSFETs need none.

For switching: MOSFETs have very low 'on' resistance and very high 'off' resistance.

For switching: MOSFETs are faster (but BJTs are usually fast enough).

As amplifiers: BJTs have linear input/output relationship. MOSFETs are not perfectly linear, which may cause distortion of large signals.

MOSFETs need no resistor in their gate circuit, which simplifies construction.

MOSFETs are liable to damage by static charge. They need careful handling.

Activities — Switching circuits

1 On a breadboard or on stripboard, build the switching circuits illustrated in this chapter and test their action. Do not connect any of the relay circuits to the mains. Instead, use a power pack running at, say, 15 V, to power the switched device.

2 There are a number of simple ways in which these circuits can be altered to change their action:

- Modify the lamp switching circuit (p. 10) so that the lamp is turned on when light level increases.

- Add a variable resistor to the circuit on p. 12 to make it possible to set the light level at which the switch operates.

- Try increasing and decreasing the resistance of R6 in the circuit on p. 17 to see what effect this has on the hysteresis.

- Look at the effect of changing the value of R3.

- Build a flip-flop based on a pair of CMOS NAND gates (p. 187) and use this as on pp. 10-11 to drive the MOSFET and filament lamp.

3 Use the circuits as a guide when designing and building switching circuits with the following functions:

- A photodiode and a BJT to switch a lamp on when light falls below a set level.

- A light dependent resistor and a MOSFET to switch an LED on by detecting the headlamps of an approaching car at night.

- A circuit to sound a siren as a garden frost warning.

- A circuit to switch on an electric fan (a small battery-driven model) when the temperature falls below a given level. This circuit is better if it has hyteresis.

- Any other switching function that you think is useful.

4 When you have designed a circuit, either build it on a breadboard, or 'build' and test it on a circuit simulator, using a computer.

Questions on transistor switches

1 Describe the action of a MOSFET switch used with a photodiode as sensor for switching on a filament lamp.

2 Design a fire-alarm circuit for switching a warning lamp on when the temperature rises above a given level.

3 Design a "Dad's home!" circuit for switching on a small piezo-electric sounder when the sensor detects approaching car headlamps at night. Calculate suitable component values.

4 Explain why it is essential to include a protective diode in a circuit that switches an inductive device.

5 Describe the operation of a simple change-over relay, illustrating your answer with diagrams.

6 Draw a diagram of a Schmitt trigger circuit based on BJTs and explain how it works.

7 Design a Schmitt trigger circuit based on MOSFETs.

8 Design a circuit for operating a magnetic door release automatically when a person approaches the doorway. The release operates on a 12 V supply and requires 1.5 A to release it.

9 Compare the features of BJTs and MOSFETs when used as switches.

10 What is a light dependent resistor? What is the effect of shining light on it? How can an LDR be used to operate a BJT switch to turn on a lamp at night, when it is dark?

11 Why is a thermistor unsuitable for a use in temperature measuring circuit?

12 On pp. 20-21 are data tables for BJTs and MOSFETs. Use these tables to select transistor types suitable for the switching circuits listed on p. 22. Access manufacturers' catalogues on CD-ROM or the Internet if you need further information. For each circuit, give your reasons for selecting the transistor. You can assume that all the transistors listed are suitable for the operating voltages of the listed circuits:

a Select alternative transistors for the switching circuits described in this chapter. Note the modifications, if any, to component values that may be necessary to enable the circuit to work with the chosen transistor.

b A light-dependent resistor is the sensor for a circuit in which a MOSFET switches a current of 20 mA through an LED. The power supply is 6 V. Select a suitable MOSFET and design the circuit.

c The circuit is a morning 'wake-up' alarm. A photodiode is the sensor and the circuit sounds an electronic buzzer when the light intensity rises above a given level. The buzzer takes 15 mA at 6 V. Select a BJT or MOSFET and design the circuit.

d A frost alarm uses a thermistor as sensor and a BJT to turn on a filament lamp when the temperature falls below a given level. The 3.6 W lamp runs on 12 V. The switching temperature is adjustable. Select the transistor and design the circuit.

e A 6 V motor takes 300 mA and is to be switched on by a high output level (5 V) from a logic circuit. Select the transistor and design the circuit.

f The water level in a tank is monitored by a pair of probe wires placed closely side by side. When the probes are covered the resistance between them is 1 MΩ. The input stage of the system is shown below. Find a suitable value for R1, and design the remainder of the circuit.

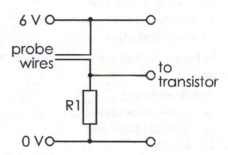

The input stage of the circuit for question 12f.

g A greenouse window is opened by a solenoid that runs on a 12 V supply. The coil resistance is 14 Ω. Design a switching circuit to energise the solenoid when the temperature rises above a given level.

As further work on these questions, build each circuit you have designed, to check that it works.

Extension questions

13 Why is the transconductance of a MOSFET expressed in siemens, but the gain of a BJT has no unit?

14 A MOSFET has a g_m of 500 mS. If the gate-source voltage increases by 50 mV, what is the change in I_D?

15 A BJT is to switch on a small DC motor that takes 400 mA at 12 V. The current gain of the transistor is 250. What minimum base current is needed?

16 Given a MOSFET with a g_m of 1.6 S, what change of gate voltage is needed to decrease I_D by 400 mA?

Multiple choice questions

1 When used as a switch, a BJT is connected in:

    A the common-emitter connection.
    B the common-collector connection.
    C the common-base connection.
    D none of the above.

2 Which of these features of a MOSFET is not of importance in a transistor switch?

    A high input resistance.
    B low 'on' resistance.
    C medium transconductance.
    D low input capacitance.

3 A light-dependent resistor is usually made from:

    A silicon oxide.
    B p-type semiconductor.
    C cadmium sulphide.
    D carbon.

4 A thermistor is not ideal for a temperature measuring circuit because:

    A its resistance is too high
    B its response is not linear.
    C it responds too slowly to changes of temperature.
    D it has a negative temperature coefficient.

5 The maximum current that a MOSFET can switch when operating on a 9 V supply and with an 'on' resistance of $2\,\Omega$:

    A depends on the device being switched
    B depends on its maximum current rating.
    C is 4.5 A.
    D is 222 mA.

6 The voltage level at which a Schmitt trigger switches on is different from that at which it switches off. We call this:

    A negative feedback
    B current gain
    C hysteresis
    D positive feedback.

Transistor switches

The Companion site (see p. viii) has several animated diagrams that illustrate the action of transistor switches.

Visit the site and see how these circuits work.

3 Potential dividers

A potential divider is a resistor network that produces a fixed or variable potential (voltage). This potential is lower than the potential of the supply. The potential divider is probably the most often used circuit module. There are many instances of potential dividers being used in this book. These include the circuits on pp. 10, 12, 14, 16, and 17 of Chapter 2.

Essentially, a potential divider consists of two resistors connected in series.

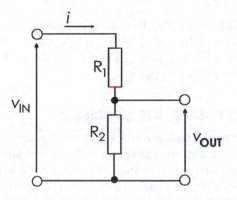

Built on a breadboard it might look like this:

Current and pd

Assuming that no current flows out of the divider at the junction of R1 and R2, the same current, i, flows through both resistors.

Ohm's law applies to each resistor. For R2:

$$i = v_{OUT}/R_2$$

Ohm's Law also applies to the two resistors when connected in series:

$$i = v_{IN}/(R_1 + R_2)$$

Combining these two equations:

$$v_{OUT}/R_2 = v_{IN}/(R_1 + R_2)$$
$$v_{OUT} = v_{IN} \times R_2/(R_1 + R_2)$$

We use this equation to calculate v_{OUT}, given v_{IN} and the values of the two resistors.

Example
Calculate v_{OUT}, given that $v_{IN} = 9$ V, $R_1 = 1.6$ kΩ, and $R_2 = 2$ kΩ.

$$v_{OUT} = 9 \times 2000 / (1600 + 2000)$$
$$v_{OUT} = 5 \text{ V}$$

The output of the potential divider is 5 V.

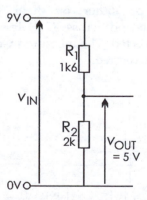

When the source voltage is 9 V, the output voltage is 5 V.

The effect of the load

The calculation opposite assumes that no current flows out of the output terminal. This is *almost* true when we use a digital meter to measure v_{OUT}. We find that the measured value of v_{OUT} is almost correct, allowing for possible tolerance errors in the resistor values.

However, current is drawn from the output in most practical circuits. This is what happens:

Current i passes through R_1. At the output of the divider, a current i_{LOAD} passes through the load while the remainder i_2 passes on through R_2. The diagram shows that R_2 and the load (R_{LOAD}) are resistances in parallel.

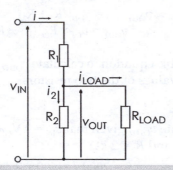

When a load is connected to the divider, it is in parallel with R2.

If R2 and the load are resistances in parallel, their combined resistance is less than the resistance of either of them separately. Therefore the pd across R2 is less when the load is connected. The measured value of v_{OUT} is less than expected.

Example

A load with resistance 3 kΩ is connected to the potential divider described in the previous example. What is the new value of v_{OUT}?

The combined resistance of R_2 and the load is:

$$R = \frac{2000 \times 3000}{2000 + 3000} = 1200 \ \Omega$$

In effect, the divider now consists of R_1 in series with 1.2 kΩ, so:

$$v_{OUT} = 9 \times \frac{1200}{1600 + 1200} = 3.86 \ V$$

The output voltage is reduced to 3.86 V.

In the example, the output voltage of the divider falls by 1.14 V when the load is connected. The amount of the fall depends on the resistance of the load.

Increasing the precision

The usual technique for preventing excessive drop in output voltage is to have a relatively large current flowing through the divider. Then the current drawn by the load has relatively less effect. As a rule-of-thumb, make the divider current about ten times the current taken by the load.

Example

In the previous example, the current drawn by the 3 kΩ load is:

$$i_{LOAD} = 5/3000 = 1.67 \ mA$$

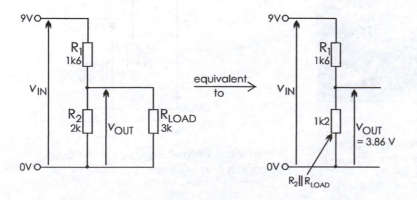

Connecting a load to the divider lowers the resistance across which v_{OUT} appears.

The current flowing through the divider must be at least ten times that amount, which is 16.7 mA. For a current of this size, R_1 and R_2 in series must total no more than R = $v_{IN}/0.0167$ = 539 Ω. We have to specify two resistors that have the same ratio as in the example, but total no more than 539 Ω.

Unfortunately, using resistor values totalling only 539Ω means that the divider takes a current of over 25 mA. In a battery powered device we might not be able to afford to waste so much current.

Measurement errors

Unlike a digital testmeter, an analogue moving-coil testmeter needs appreciable current to drive it. Typically, the coil resistance is 200 kΩ when switched to the 10 V range. A cheap analogue meter may have a coil resistance of only 20 kΩ. If a potential divider or other network of high-value resistors is tested with an analogue meter, serious errors may occur. This is because the coil of the meter acts as a load on the network. Current is drawn by the meter and this results in a voltage drop at the point where voltage is being measured.

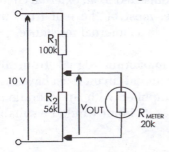

> An analogue meter may give low voltage readings on a high-resistance network.

Example

In the diagram, the actual value of v_{OUT} is:

$$10 \times 56\,000/156\,000 = 3.59 \text{ V}$$

But the meter has a coil resistance of 20 kΩ, so R2 has 20 kΩ in parallel with it. Their combined resistance is:

$$R = \frac{56\,000 \times 20\,000}{56\,000 + 20\,000} = 14.7 \text{ k}\Omega$$

The value of v_{OUT}, as read with the meter connected to the divider, is:

$$10 \times 14700/11470 = 1.28 \text{ V}$$

The meter reading is 2.31 V too low.

The example demonstrates that a moving-coil meter is far from accurate when used to measure voltages in a network of high resistances. Using a digital meter gives no such problems, as it has an input resistance of many megohms and takes virtually no current from the network.

Variable potential dividers

The diagrams show three ways of building a potential divider with variable output voltage.

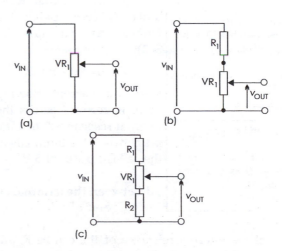

When calculating v_{OUT}, we think of the variable resistor as two resistors joined at the point of contact of the wiper.

Internal resistance

When analysing circuits we have so far ignored the fact that the source itself has resistance. Often this is so small that it can be ignored, but there are times when it can not.

To understand the effect of the internal resistance of a cell (and also the meaning of the term emf), think of the cell as two components, connected in series.

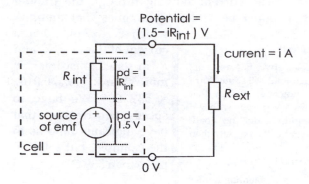

When a current flows there is a fall of potential (or voltage drop) across the internal resistance.

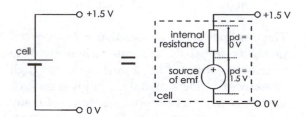

A cell can be thought of as a source of emf (the chemical energy of the cell) in series with a resistance (the internal resistance of the cell).

If there is no circuit connected to the cell, as in the diagram above, no current flows through the cell. So the current through the internal resistance is zero. Using the Ohm's Law equation (p. 363):

$$V = IR = 0 \times R = 0$$

With no current flowing there is zero pd across the internal resistance. The total pd between the terminals of the cell is $1.5 + 0 = 1.5$ V.

pd between the terminals = emf of cell

---

pd and volts

The unit of pd (symbol, V) and emf (symbol, E) is the volt (symbol , V).

Spot the difference? (p. 363)

Current and amps

The unit of current (symbol, I) is the amp (symbol, A).

Let the internal reistance of the cell be R_{int}. If the cell is connected to an external circuit, with resistance R_{ext}, a current flows through this and also through the internal resistance. These two resistances are in series and the circuit (top right) is a potential divider.

The fall in potential across the internal resistance (sometimes called the 'lost volts') means that the:

pd between the terminals < emf of cell

Taking this circuit as a potential divider, it divides the emf and its 'output' is at the positive terminal of the cell or other source. The equation for a potential divider (p.25) is:

$$v_{OUT} = v_{IN} \times R_2/(R_1 + R_2)$$

In the case of the circuit above this becomes:

$$terminal\ pd = emf \times R_{ext}/(R_{int} + R_{ext})$$

If R_{int} is small compared with R_{ext}, as is the case with batteries and other power supplies, the terminal potential is only slightly less than the emf. The voltage drop can be ignored.

If the source has high internal resistance, such as the crystal in a piezo-electric microphone (p. 68), and it is connected to an external circuit of low resistance, most of the emf from the crystal will be lost in its internal resistance.

To obtain the maximum signal from the microphone the external circuit must have high resistance. An amplifier with an operational amplifier (p. 91) as its input would be suitable for this.

Activities

1 Set up several potential divider circuits, with different input voltages and using various pairs of resistors. Calculate the expected output voltage and check the result with a digital multimeter.

   Use the resistor colour code (see Supplement A, p. 366) to select suitable resistors from the E24 series. If possible, use resistors with 1% tolerance.

2 Extend Activity 1 by experimenting with loads of various resistances.

3 Measure the internal resistance of an alkaline cell. Connect a digital testmeter across the terminals of the cell to measure its emf of the cell with no load attached (the current flowing to a digital meter is virtually zero). The measure the pd of the cell when a low-value resistor is connected across it. A 10 Ω, 0.5 W resistor can be used. **Warning!** it will become very hot! Take the reading quickly and then disconnect. Calculate the internal resistance using the equation opposite.

Questions on potential dividers

1 A load of 300 Ω is connected in parallel with the lower resistor (R2) of a potential divider. The resistance of R2 is 1.2 kΩ. What is the combined resistance of R2 and the load?

2 A potential divider consists of two resistors in series. R1 is 4.5 kΩ and R2 is 3.3 kΩ. The potential applied to the divider is 12 V. What is the combined resistance of R1 and R2? How much current flows through the divider with no load connected? What is the potential across R2?

3 A potential divider is built from two resistors of 36Ω and 24Ω. The supply potential is 8 V. What two output potentials are obtainable from this divider?

4 State Ohm's Law and use it to prove that the combined resistance of several resistors in series is equal to their sum.

5 Use Ohm's law to obtain a formula for calculating the combined resistance of three resistances connected in parallel.

6 Three resistors, 47 Ω, 120 Ω, and 91 Ω, are connected in series with each other and with an unknown fourth resistor. When a pd of 17 V is applied across the 4-resistor chain, a current of 50 mA flows through it. What is the pd across each resistor? What is the resistance of the fourth resistor?

7 Suggest practical uses for each of the three types of variable divider.

8 A potential divider consists of a 390 Ω resistor and a 470 Ω resistor. The load is 2.5 kΩ, connected across the 470 Ω resistor. What is the output potential of the divider when the input is 10 V?

9 Design a potential divider, using resistors of the E24 series, to produce a 4V output from a 10 V input. It must be able to supply up to 1.8 mA to the load without a serious drop in voltage.

10 Design a potential divider, using E24 series resistors, in which the output potential is approximately 45% of the input.

11 A variable potential divider produces an output ranging from 6.75 V to 9 V when the input potential is 9 V. What values and types of resistor are being used?

12 For each of the variable dividers shown on p. 27, describe in words the range of output voltages obtainable.

13 Design a divider to provide a variable output of from 0 V to 5 V, given a constant input of 16 V. The resistance of the variable resistor is 100 kΩ.

14 A battery has an emf of 6 V. The pd across its terminals falls to 5.88 V when the battery is connected to a 120 Ω resistor. What is its internal resistance?

15 A cell has an emf of 1.2 V and an internal resistance of 0.5 Ω. What is the pd across its terminals when it is connected to a 22 Ω resistor?

4 Capacitors

A capacitor consists of two metal plates parallel with each other and very close together. There is an insulating layer between them, called the **dielectric**. In many types of capacitor this is a thin sheet of plastic, but some capacitors have a dielectric of air or other insulating material.

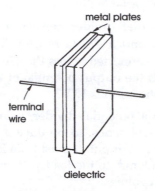

metal plates

terminal wire

dielectric

A capacitor consists of two metal plates with an insulating dielectric sandwiched between them.

Because of the dielectric, current can not flow from one plate to the other. When the capacitor is connected to a DC source, electrons accumulate on the plate connected to the negative supply terminal.

The negative charge repels electrons from the atoms of the other plate. These electrons flow away to the positive terminal of the DC source. This leaves the plate positively charged.

If the capacitor is disconnected from the supply, the charges remain. The capacitor stores the electric charge indefinitely.

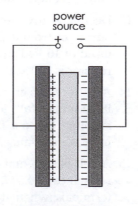

power source

+ −

A capacitor stores charge received from a DC supply.

Capacitance

The ability of a capacitor to store charge is expressed as its **capacitance**, symbol **C**. The unit of capacitance is the **farad**, symbol **F**.

If a capacitor stores **Q** coulombs of charge when the pd between its plates is **V** volts, its capacitance in farads is:

$$C = Q/V$$

Example

A capacitor is charged by connecting it to a 3 V DC supply. It then holds 1.5 coulombs of charge. What is its capacitance?

$$C = Q/V = 1.5/3 = 0.5 \text{ F}$$

In practice, the farad is too large a unit to be commonly used in electronics. Capacitors rated in farads are found only as special types for backing up the power supply to computers.

> **Self test**
>
> 1 A capacitor holding a charge of 0.1 C has 24 V across it. What is its capacitance?
>
> 2 A 0.2 F capacitor is charged with 4 C. What is the voltage across it?

Most of the capacitors we use are rated in:

- microfarads (10^{-6} F), symbol μF,
- nanofarads (10^{-9} F), symbol nF, or
- picofarads (10^{-12} F), symbol pF.

Capacitors in parallel and series

The capacitance of a capacitor is proportional to the area of its plates. If we connect capacitors in parallel, this is equivalent to adding their areas together.

Therefore, for several capacitances $C_1, C_2, \ldots C_n$ in parallel, the combined capacitance C is given by:

$$C = C_1 + C_2 + \ldots + C_n$$

The combined parallel capacitance is equal to the sum of the individual capacitances.

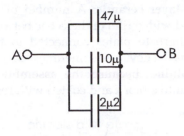

The combined capacitance equals the sum of the capacitances.

Example

What is the combined capacitance of the three capacitors in the diagram above?

Summing capacitances gives:

$$C = 47 + 10 + 2.2 = 59.2 \ \mu F$$

This recalls the formula for the combined resistance of resistors *in series*.

Conversely, the formula for capacitances *in series* has the same form as that for resistances *in parallel*. For capacitances C_1, C_2, ..., C_n in parallel the combined capacitance C is given by:

$$\frac{1}{C} = \frac{1}{C_1} + \frac{1}{C_2} + ... + \frac{1}{C_n}$$

Selecting capacitors

There are several factors to consider when selecting a capacitor:

- **Capacitance:** this is obviously the most important factor. Capacitors are made in a standard range of capacitances, based on the E24 range (p. 366).

- **Tolerance:** some applications, such as timing circuits, may require a high-precision capacitor. If we use low-precision types, we must make sure that the circuit will still work if the actual value of the capacitor is very different from its nominal value.

- **Working voltage:** the capacitor must not break down if the circuit places high voltages across it.

- **Temperature coefficient:** This tells us how much the capacitance varies with temperature. We must allow for this in a circuit that is intended to operate in extreme conditions. Tempco (as it is often called) may be expressed as the percentage variation in value over the working range of temperature, or as the variation in parts per million per degree Celsius.

- **Polarisation:** Some types of capacitor have specified positive and negative terminals. They must not be used in circuits in which the voltage across them may reverse in polarity.

- **Leakage current:** In some types, leakage may occur through the dielectric. The capacitor may not hold its charge for long enough.

There are many types capacitor including the following:

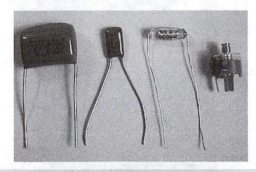

Types of capacitor: polyester (100 nF and 1 nF), polystyrene (1 nF) and a trimmer (0 to 65 pF).

Polyester: Metal foil plates with polyester film between them, or a metallised film is deposited on the insulator. This makes a long narrow 'sandwich' that is rolled to make it more compact and is then coated with plastic to insulate it. Polyester capacitors are available in the range 1 nF to 15 μF, and with working voltages from 50 V to 1500 V. Tolerance is 5%, 10% or 20%. They have a high temperature coefficient (or tempco). Compared with most other types, polyester capacitors have high capacitance per unit volume. This feature, together with their relatively low price makes polyester capacitors one of the most commonly used and a first choice in most applications.

Polystyrene: These have a similar structure to polyester capacitors but the dielectric is polystyrene. They are made only in low values, usually 10 pF to 47 nF. Typically, their tolerance is 5% or 10% but high precision types are available with a tolerances of 1% and 2%. Working voltages are 30 V to 630 V. They have a high negative tempco of -125 parts per million per degree Celsius. The precision types are suitable for tuning and filter circuits.

Polycarbonate: These cover the range 100 pF to 10 µF, and have working voltages up to 400 V DC. Their tolerance is 5% or 10%. Tempcos are about 100 ppm/°C, and are fairly constant over the operating range of 55°C to 100°C. This makes them preferred to polyester capacitors for filtering and timing circuits.

Multi-layer ceramic: A number of plates is stacked with a film of dielectric between them and alternate plates connected to the same terminal. They are sometimes described as 'monolithic' because the assembly is then fused into a block and coated with resin.

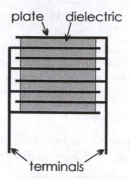

plate   dielectric

terminals

Multilayer ceramic capacitors have high capacities for their size (see photo, opposite).

5 mm

Types of capacitor: polycarbonate (680 nF). The capacitor is also marked with its tolerance (J = ±5%) and working voltage (100 V).

Polypropylene: These cover the range 100 pF to 10 µF but their main feature is their high working voltage. Types with working voltages up to 3000 V are produced. In addition they are good at handling pulses. These features make polypropylene capacitors useful in circuits in which operating voltages are typically high. These include power supply circuits, power amplifiers, particularly valve amplifiers and TV circuits.

The capacitors are made in a range from 1 pF to 1 µF and with tolerances from 2% to 20% (though as wide as -20% and +80% for one type). The tempco varies with the material used for the dielectric. The dielectric of those with the lowest capacitances (below 1 nF) is NPO or COG, with zero or very low tempco. Working voltage is around 50 V. Medium-capacitance types (about 1 nF to 100 nF) have X7R dielectric and a higher tempco of ±15% over the range -55°C to 125°C. High-capacitance types (100 nF to 1µF) have higher tempcos.

The range of characteristics is complicated and you should look in manufacturers' data tables, for more detailed information.

The NPO/CGO types are highly suitable for tuning circuits and filters. The other types are more suitable for coupling and decoupling (pp. 34–35) mainly because of their small size.

Ceramic disc: These are single-layer ceramic capacitors with high capacitance for their size. They are made in the range 1 pF to 220 nF, with working voltage usually 50 V, and a wide range of tolerances. They are widely used for many purposes, including decoupling.

Types of capacitor: metallised polyester film, (4.7 nF), multi-layer ceramic (10 nF), ceramic disc (470 p).

Aluminium electrolytic: These are made by rolling two strips of aluminium foil with a strip of absorbent paper between them. This is soaked in an electrolyte solution, and the assembly is sealed in a can. A voltage is applied between the two plates, causing a very thin layer of aluminium oxide to be deposited on one plate. The oxide is an insulator and acts as the dielectric. On one side it is in contact with the aluminium foil on which it was formed. On the other side it is in contact with the electrolyte.

Because of the thinness of the oxide layer, the capacitance of an aluminium electrolytic capacitor is extremely high. Most types of electrolytic capacitor have the surface of the aluminium etched to increase its surface area and thus further increase capacitance.

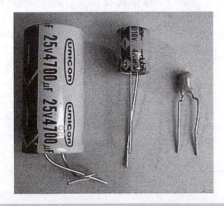

Types of polarised capacitor: electrolytic (4700 µF and 470 µF), and tantalum (1 µF). Note the leads twisted together on the large-value capacitor.

Electrolytic capacitors are made to cover the range 1 µF to 10 000 µF, and special types are available outside this range.

Electrolytics have a wide tolerance range, usually ±20% or more. They also show appreciable change of capacitance with age and usage. If a capacitor is not used for some months its capacitance gradually decreases. It is restored if the circuit is switched on and a voltage is placed across it for a while. Tempcos are large, typically being 20% or more over the operating temperature range.

Working voltages range from 10 V to 100 V. Capacitors of a given value are made with three or more different voltages. As explained below, choosing a suitable working voltage can be important.

Most other types of capacitor have very low leakage; their resistance is rated in gigaohms. Leakage through electrolytics is much more significant. Typically, the leakage current in microamps is $0.01VC$ to $0.03VC$, where V is the working voltage and C is the capacitance in microfarads. A 220 µF capacitor with a working voltage 25 V, for example, has 55 µA to 165 µF leakage current.

If the capacitor is operated at about 40% of its working voltage, the leakage of current is reduced to about 10%. On the other hand, a capacitor with higher working voltage is more expensive and is larger. It takes up more space on the circuit board.

Electrolytics are made with two different arrangments of the terminal wires. In the **radial** arrangement, illustrated by the electrolytic capacitors on the left, the two wires are at the same end of the capacitor. In the **axial** arrangement, the wires are at opposite ends.

Radial capacitors stand vertically on the circuit board and so take up less board space than axial capacitors. However, it may not be possible to use radial types if there is insufficient height for them between adjacent boards or between the board and the wall of the enclosure.

There are two safety precautions to be observed when using electrolytics:

- They must be connected into the circuit with the correct polarity. Markings on the can indicate the *negative* terminal wire. If the capacitor is connected the wrong way round and power is applied, gas is formed inside the can and it will explode.

- The amount of charge that can be stored in a large electrolytic may give an unpleasant, possibly dangerous, shock. Store such capacitors with their terminal wires twisted together, as in the photo. This warning also applies to capacitors in circuits. They may hold a possibly lethal charge long after the equipment has been switched off. Always give them time for the charge to leak away.

Electrolytics are useful for storing charge, including smoothing the current from power supply units. They are also used where a large pulse is to be generated such as in the starting circuits of fluorescent lamps, and photographic flashes. Their wide tolerance and poor stability make them unsuitable for timing and filtering circuits.

Tantalum bead: These are polarised capacitors. They range from 100 nF to 150 µF in capacitance, with working voltages from 3 V up to 35 V. They have wide tolerance (20%) but lower leakage (maximum $0.01VC$) than electrolytics. Their main advantage is that they are smaller than electolytics of the same capacity.

Variable capacitors: Small trimmer capacitors are used for making fine adjustments to tuning circuits. The type shown in the photo on p. 31 has two sets of metal plates. One set is fixed and the other set is rotated by a screwdriver. Plastic film between the plates acts as dielectric. As the rotating set is turned, the area of overlap between this set and the fixed set is varied. This varies the capacitance.

Larger variable capacitors with air as the dielectric are sometimes used in tuning radio receivers.

Coupling

Capacitors are often used for carrying signals from one part of a circuit to another. In the circuit below, which is a MOSFET amplifier circuit, a signal is carried from the microphone MIC1 to the gate of the MOSFET, Q1.

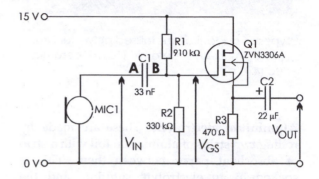

The MOSFET amplifier has coupling capacitors on its input and output sides.

MIC1 is a crystal microphone. When there is no sound, there is no pd across the crystal and both terminals of MIC1 are at 0 V. Plate A of capacitor C1 is at 0 V. When MIC1 receives a sound, an alternating pd of a few millivolts amplitude is generated. The voltage of A rises (goes positive) and falls (goes negative) by a few millivolts.

When A goes positive, it attracts electrons to plate B, leaving the gate of Q1 more positive. When A goes negative, it repels electrons from plate B, making the gate of Q1 more negative. In this way, the signal from MIC1 is transferred to the gate.

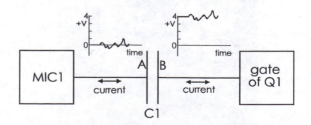

After passing unchanged through the capacitor, the signal now 'rides' on a DC level of 4 V.

The point to note is that the gate is NOT at 0 V when there is no signal. R1 and R2 act as a potential divider, which holds the gate at 4 V. The transistor is on and a constant drain current I_D passes through it.

When a signal arrives, the voltage at the gate alternates a few millivolts above and below 4 V. Q1 is turned a little more on and a little less on. I_D increases and decreases a little compared with its steady no-signal level. The signal reappears as variations in the voltage across R3.

This circuit would not work without the coupling capacitor C1. If we were to connect MIC1 directly to the gate, the voltage at the gate would be pulled down close to 0 V, well below the threshold voltage. The signal of a few millivolts amplitude would not bring V_{GS} above the threshold and Q1 would stay permanently off. The signal would be lost.

The signal from this circuit would normally be passed on to another stage of amplification. It has a second capacitor C2 to couple it to the next stage.

Coupling is a convenient way of joining two parts of a circuit together. The DC voltages on either side of the capacitor are not necessarily equal but the signal can pass freely across.

We sometimes use small transformers as coupling devices but they are less often used than capacitors. The main reason is that, except for those used in radio-frequency circuits, transformers are heavy and bulky compared with capacitors. Also they are liable to pick up and generate electromagnetic interference.

Decoupling

Coupling capacitors transfer wanted signals. Decoupling capacitors are used to prevent unwanted signals from passing from one part of a circuit to another. As an example, consider the audio amplifier system, shown in the next column as a block diagram.

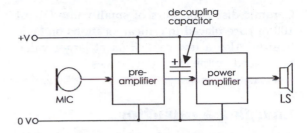

The capacitor damps out signals passing back from the power amplifier to the pre-amplifier.

Sounds picked up by the microphone go to the pre-amplifier, which provides voltage gain. The amplified signal goes to the power amplifier which provides current gain and feeds a powerful signal to the loudspeaker. The signal going to the loudspeaker is a fluctuating current of several amps in magnitude. The power amplifier draws this current from the power lines. The varying current drain on the power supply causes its output voltage to vary slightly. Variations in the voltage supplied to the pre-amplifier make its output vary. In other words, there is a signal added to that coming from the microphone, which passes into the system and is itself amplified.

The effect of the feedback through the power line is to cause distortion. It may also lead to a type of oscillation known as 'motor-boating' because the sound resembles that of a motor-boat. The way to avoid this problem is to wire a large-value capacitor across the power supply lines. This is wired between the two amplifiers. The capacitor absorbs fluctuations in the supply voltage and so prevents these from getting back to the sensitive part of the circuit, the pre-amplifier.

A similar effect is found in logical circuits. When devices such as LEDs, lamps and motors, are switched on or off, there is a sudden change in current requirements. The same occurs if several logic gates change state at the same time. Pulses are generated on the supply lines and these may cause some of the logical elements to behave erratically. For example, memory units may change state in unpredictable ways.

Ceramic disc capacitors of small value (about 100 nF) are placed in critical locations on logic boards. Also a few capacitors of larger value are placed on the main power lnes.

Charging a capacitor

When switch S1 is closed, a capacitor C is charged from a *constant* voltage supply V_S, through a resistance R.

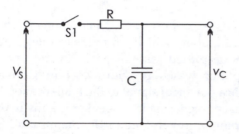

The voltage v_C across the capacitor is measured at regular intervals after S1 is closed.

The **time constant** of the circuit is defined as RC.

Before going on to use this quantity we will look more closely at the way it is derived.

Consider the relationships between the electrical quantities of charge (Q), current (I), potential difference (V), resistance (R), and the quantity time (T).

From Ohm's Law we have $R = V/I$.
From the definition on p. 30 we have $C = Q/V$.

Substituting these two equalities in the definition of the time constant gives:

$RC = (V/I) \times (Q/V) = Q/I$

But, from the definition of the coulomb, we have $Q = IT$.

Therefore, $RC = Q/I = IT/I = T$

The quantity obtained by multiplying a resistance by a capacitance is a *time* — the **time constant**.

The graph below is plotted for the circuit on the left, given that $V_s = 10$ V, $R = 100$ Ω, and $C = 1$ µF. The time constant is $RC = 100$ µs.

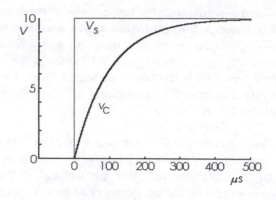

S1 is turned on at 0 seconds and v_C (grey curve) begins rising to 10 V, as the capacitor charges.

The shape of the curve is described as **exponential**. This is because the equations for the curves contain the **exponential constant**, e.

There are three points on this curve that are of special interest. The first of these is shown below:

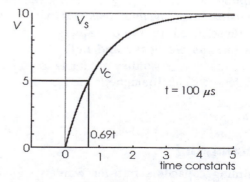

The plot is the same as before, but the x-axis is marked in time constants instead of microseconds.

The graph shows that v_C, the voltage across the capacitor reaches half the supply level after 0.69 time constants.

A little further up the curve we find that, after 1 time constant, v_C has risen to $0.63V_S$.

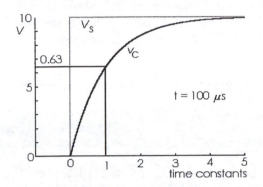

The exponential curve has the same shape for all values of R and C.

Finally, note that the graph is plotted for 5 time constants after the voltage step. It is seen that the curve levels out *almost* completely during this time. In theory, the capacitor never reaches full charge but, for all practical purposes, we can say that it takes 5 time constants to fully charge the capacitor.

For proofs of these relationships, see Extension Box 6, p. 38.

Discharging a capacitor

If a capacitance C is already charged to V_S and is discharged through a resistance R, the voltage across the capacitor falls as in the graph at top right.

The voltage across the capacitor falls to $0.5V_S$ after 0.69 time constants. This is the same time as for a charging capacitor to rise to the half-charge point. In 1 time constant v_s falls to 3.7 V (that is, the voltage across it *falls* by $0.63V_C$). After 5 time constants, the capacitor is almost completely discharged.

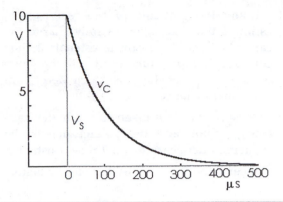

The discharge curve also is exponential.

Time

When a current is passed through a resistor, a voltage develops across it instantly, its size according to Ohm's Law. When a current is passed into a capacitor, the voltage across it changes gradually. It takes **time**.

In this way, using capacitors requires us to take time into account when we analyse the behaviour of the circuit. In the next chapter we study ways in which we use capacitors in time-dependent circuits.

Activity — Capacitors

Set up the capacitor charging circuit on p. 36, but omitting S1. Make $R = 100$ kΩ and $C = 10$nF. For v_{IN} use a signal generator, set to produce square waves at 50 Hz, and an amplitude of 1 V. Observe v_{OUT}, using an oscilloscope. The waveform should be similar to this:

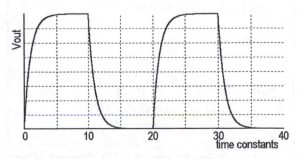

Charging and discharging cycles.

1   Using the graticule of the oscilloscope, estimate the time taken to fully charge the capacitor. Use this result to estimate 1 time constant. Compare this result with the value of the time constant as calculated from component values.

2   Measure the time taken to reach the half-charge point, and the voltage across the capacitor after charging for 1 time constant.

3   Repeat 1 and 2 for the discharging capacitor.

Discharging — Extension Box 7

The exponential equation for a discharging capacitor is:

$$v_C = V_S e^{-t/RC}$$

The three results stated above can be derived from this, using methods similar to those used for a charging capacitor.

Charging — Extension Box 6

The equation for charging a capacitor is:

$$v_C = V_S(1 - e^{-t/RC})$$

1 At the half-charged point:

$$v_c = 0.5V_S$$

$$e^{-t/RC} = 0.5$$

$$\tfrac{-t}{RC} = \ln 0.5 = -0.69$$

$$t = 0.69RC$$

The capacitor is half charged after 0.69 time constants.

2 After 1 time constant:

$$t = RC$$
$$v_C = V_S(1 - e^{-1})$$
But $e^{-1} = 1/e = 0.37$
$$v_C = V_S(1 - 0.37)$$
$$v_C = 0.63V_S$$

The capacitor is charged to 0.63 of the supply after 1 time constant.

3 After 5 time constants:

$$t = 5RC$$
$$v_C = V_S(1 - e^{-5})$$
But $e^{-5} = 0.0067$
$$v_C = V_S(1 - 0.0067)$$
$$v_C = 0.993V_S$$
$$v_c \approx V_S$$

The capacitor is almost fully charged after 5 time constants.

Dielectrics — Extension Box 8

The dielectric keeps the plates of the capacitor a fixed distance apart and so gives its capacitance a fixed value. The dielectric also prevents the plates from touching and short-circuiting. This applies to all types of capacitor except those with air as the dielectric.

More important, the dielectric greatly increases the capacitance. The drawing below shows two capacitors of identical size and both are charged to the same pd. The capacitor on the left has air as its dielectric, while the other has a different dielectric. The charge on the other capacitor is much greater than that on the air-filled capacitor.

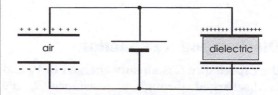

Both capacitors are charged to the same pd but the dielectric increases the charge that can be stored.

In other words, the dielectric increases the capacitance.

Dielectric (continued)

The effect of the dielectric is expressed as a ratio between the capacitance with air as the dielectric and with another substance as the dielectric:

$$\varepsilon_r = \frac{\text{capacitance with dielectric}}{\text{capacitance with air}}$$

The ratio ε_r is called the **relative permittivity** of the dielectric, or the **dielectric constant**. The value of ε_r is expressed in farad per metre. Typical values for ε_r are 2 F/m for polythene and 5 F/m for mica, and some substances have much higher values.

Another important property of a dielectric is its **dielectric strength**. This is the largest pd that can exist in the dielectric without resulting in ts breakdown. This is expressed in millions of volts per metre. Air has a low dielectric strength of 3×10^0 V/m. Polythene rates at 50×10^0 V/m and mica at 150×10^0 V/m.

Questions on capacitors

1 A capacitor holding a charge of 0.25 C has 15 V across it. What is its capacitance?

2 A 4700 µF capacitor is charged to 15 V. How much charge does it hold?

3 Express 2.2 µF in nanofarads.

4 Express 1800 pF in nanofarads.

5 What is the combined capacitance of 47 nF and 2.2 nF when connected (a) in series, and (b) in parallel?

6 What is the combined capacitance of 10 nF, 680 pF and 22 nF connected in series?

7 What type of capacitor would you choose for (a) decoupling, (b) a precision timer circuit, (c) a circuit for generating the carrier frequency of a radio transmitter (d) a high-voltage power supply?

8 Describe the structure of an aluminium electrolytic capacitor. What safety precautions are needed when working with such capacitors?

9 Describe the different types of multi-layer ceramic capacitor and their applications.

10 List three different materials used as the dielectric of capacitors.

11 Explain how a capacitor is used for (a) coupling and (b) decoupling.

12 Select a circuit that has a capacitor in it and explain what the capacitor does.

13 A 22 nF capacitor is charged through a 47 kΩ resistor. Calculate the time constant.

14 In Question 13, what are we assuming about the voltage source? What are we *not* assuming about the voltage source?

15 In Question 13, how long does it take for the discharged capacitor to almost fully charge? How long does it take to charge to half the supply voltage? (Proofs not required)

16 Given that the capacitor in Question 13 is charged from a constant 9 V DC source, what is the voltage across the capacitor after 1.034 ms?

17 A 2200 µF capacitor is charged to 18V. It is then discharged through a 100 Ω resistor. How long does it take to discharge almost completely? What current flows through the resistor as discharging begins? What should be the wattage of the resistor?

Extension questions

18 A 2.2 µF capacitor is charged to 5 V. A 4.7 µF capacitor is charged to 9 V, as shown in the diagram. The switch between the capacitors is then closed. What is the voltage across the two capacitors?

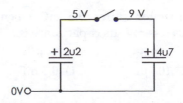

19 Define the time constant of an RC network. In what unit is it measured?

20 Approximately how many time constants does it take for a capacitor to charge from zero volts to the supply voltage when charged through a resistor from a constant voltage source? Prove your statement by reference to the equation:

$$v_C = V_s(1 - e^{-t/RC})$$

21 Beginning with the equation quoted in Extension Box 4, show how long it takes for a capacitor C charged to 12 V to discharge through a resistor R to 4.44 V.

22 Calculate how long it takes for a capacitor to discharge through a resistor to 0.7 of its original charge.

Multiple choice questions

1 An electrolytic capacitor is used when it is necessary to have:

    A high working voltage.
    B low leakage.
    C high capacitance.
    D stable capacitance.

2 The best type of capacitor for a tuning circuit is:

    A NPO/CGO multi-layer ceramic.
    B tantalum bead.
    C polyester.
    D polypropylene.

3 For use in a high-voltage power supply circuit, the most suitable type of capacitor is:

    A polyester.
    B ceramic disc.
    C polypropylene.
    D polycarbonate.

4 A capacitor stores 0.5 mC when the pd across it is 5 V. Its capacitance is:

    A 100 nF.   B 25 F.
    C 100 μF.   D 0.5 mF.

2 In the circuit below, the capacitor has zero charge. Then the switch is closed and current flows from the constant voltage source. The pd across the capacitor:

    A gradually falls.
    B rises quickly, then more slowly.
    C instantly changes to Vs.
    D rises slowly, then faster.

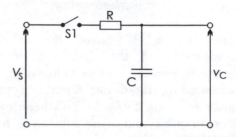

6 In the circuit above, R = 100 kΩ and C = 47 μF. The time taken for the pd across the capacitor to rise from 0 V to V_s is:

    A the time constant.
    B depends on V_s.
    C is 4.7 s.
    A is approximately 23.5 s.

7 When a charged capacitor is discharged through a resistor the current through the resistor:

    A is constant.
    B is high at first, gradually reducing.
    C decreases at a constant rate.
    A stops after 5 s.

8 When a capacitor is used to transfer a signal between two parts of a circuit that are operating at different voltages, the capacitor is said to be:

    A coupling the circuits.
    B filtering the signal.
    C decoupling the circuits.
    D amplifying the signal.

There are more calculations on capacitance and impedance on the Companion website.

5 Using capacitors

Chapter 4 demonstrates that capacitors introduce time into the action of a circuit. The delay circuit below uses a capacitor in this way.

Delay

The circuit below has a potential divider R2/R3 which has an output voltage 4.8 V. The operational amplifier (see Chapter 11) compares this voltage with the voltage across C1.

The circuit is triggered by pressing S1 to discharge the capacitor fully. This brings the non-inverting input (+) of the op amp down to 0 V. The output of the op amp swings down to 0 V and the LED goes out.

When S1 is released, current flows through R1 and charges C1. As soon as the voltage across it reaches 4.8 V, the output of the op amp swings almost to 6 V and the LED comes on. With the values of R1 and C1 shown in the diagram, the delay is 75 s. Extension Box 9, on p. 50 shows how this result is calculated.

Diagram labelling

From now on, resistor and capacitor values are written in a shortened form on circuit diagrams (see p. 329).

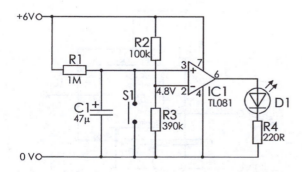

The time for which the LED is off depends on the values of R1 and C1.

Monostable multivibrator

This kind of circuit, usually known by the shortened name of **monostable,** is a useful pulse generator or pulse extender.

The monostable circuit (overleaf) consists of two transistor switches. The output from each switch is the input to the other. A system diagram shows that the connection from Q2 to Q1 is direct (through R3), but the connection from Q1 to Q2 passes through a delay stage. The delay stage is provided by C1 and R2 — another example of charging a capacitor through a resistor.

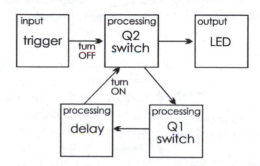

Without the delay, Q1 would turn Q2 on again immediately. You would not see even a flash of light from the LED. With the delay unit present, there is a delay of a few seconds while C1 charges through R2. Q2 is turned on when C1 has charged to the right level.

This circuit is stable when Q1 is off and Q2 is on. It remains indefinitely in that state. It is unstable in the reverse state, with Q1 on and Q2 off. After the delay, it goes back to the other state.

The behaviour of the circuit is more complicated than in the previous example, the base of Q2 being pulled down to a negative voltage when the circuit is triggered. The length of the pulse is approximately $0.7R_2C_1$. With the values in the diagram, the pulse length is $0.7 \times 6.8 \times 470$ [kΩ × μF] = 2.2 s.

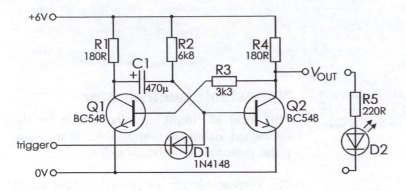

Cross-connected BJT switches form a monostable pulse generator.

The charging time t depends on the values of R1 and C1 according to the equation:

$$t = 1.1R_1 C_1$$

R_1C_1 is the time constant, and the factor 1.1 is obtained by a calculation similar to those set out on p. 29. Note that t does not depend on supply voltage. This is one of the main assets of the 555, making it suitable for circuits powered by battery or other unregulated supplies.

555 timer IC

One of the most popular monostables is based on the 555 integrated circuit. It operates on a wide range of supply voltages, from 4.5 V to 16 V.

The original 555 uses bipolar circuitry, but there are several newer

> **Bipolar devices:** conduction is by electrons and holes, for example, in npn and pnp transistors.

devices, notably the 7555, based on CMOS. The advantages of the CMOS versions is that they use far less current, produce longer timing periods, cause less disturbance on the power lines when their output changes state, and operate on a wider range of supply voltages (3 V to 18 V).

The 7555 output is able to sink or source only

> **Sink or source current:** to take in or to provide current.

100 mA, in comparison with the 200 mA of the 555, but this is rarely a drawback. The 7555 has 2% precision compared with 1% for the 555.

The 555 and 7555 are also available as dual versions, the 556 and 7556, with two timers contained in a 14–pin package.

The diagram on the right shows the 555 connected as a monostable. The length of the output pulse depends on the time taken to charge the capacitor to two-thirds of the supply voltage.

The control pin is sometimes used to adjust the exact length of the period. If this facility is not required, the control pin is either connected to the 0 V line through a low-value capacitor (for example 10 nF).

With the 7555, the control pin is left unconnected.

> **Self Test**
>
> 1 A 555 monostable circuit has R1= 2.2 kΩ and C1=1.5 µF. What is the length of the output pulse?
>
> 2 Design a 555 monostable circuit using a 220 nF capacitor to give an output pulse of 25 ms.

The 555 needs the decoupling capacitor C3 connected across the supply terminals, but decoupling is not necessary with the 7555.

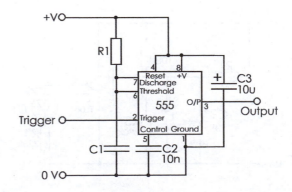

The basic 555 monostable circuit. C2 and C3 are not required with the 7555 CMOS IC.

The trigger (set) and reset pins are normally held high. A low-going pulse (which needs only fall to two-thirds of the supply voltage) triggers the monostable. The device is not retriggerable, but the output pulse can be ended at any time by a low-going pulse on the reset pin. In the figure, the reset pin is permanently wired to the positive supply (high) so that resetting is not possible.

The action of the monostable is as follows:

1) In the quiescent state, the timing capacitor (C1) is charged to one-third of the supply voltage. The output is low (0 V). Current flowing through R1 after C1 is charged is passed to ground by way of the discharge pin. The threshold pin monitors the voltage across C1.

2) When the circuit is triggered, its output goes high and the discharge pin no longer sinks current. Current flows to the capacitor, which begins to charge.

3) When the capacitor is charged to two-thirds of the supply voltage, as monitored by the threshold pin, the output goes low. The charge on the capacitor is rapidly conducted away through the discharge pin until the voltage across the capacitor has fallen again to one-third of the supply voltage.

The 555 timer has the advantage that the voltage levels are precisely monitored by the internal circuitry, so that timing has high precision. Most often it is limited by the precision of the timing resistor and capacitor. This is particularly the case when an electrolytic capacitor is used to give a long period. They have wide tolerance (20%) and their capacitance varies with age and usage.

Astable multivibrator

The 555 can be used as an astable simply by connecting the trigger pin to the threshold pin and adding another resistor. The circuit is triggered again as soon as the capacitor is discharged at the end of its cycle. The capacitor is charged and discharged indefinitely and the output alternates between high and low.

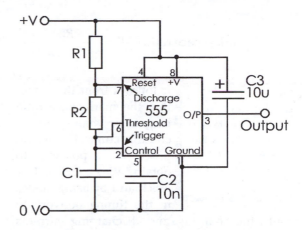

The 555 timer IC can also be used as an astable multivibrator..

The action of the 555 astable is divided into two phases.

1 Charging: Output is high while the capacitor is charged from one-third to two-thirds the supply voltage, through R1 and R2 in series. The time taken is:

$$t_H = 0.69(R_1 + R_2)C_1$$

$(R_1 + R_2)C_1$ is the time constant for charging. The factor 0.69 is obtained by a calculation similar to those set out on p.39. Note that t does not depend on supply voltage.

2 Discharging: Output is low while the capacitor is discharged through R2, current flowing into the discharge pin. The time taken is:

$$t_L = 0.69R_2C_1$$

The total time for the two phases is:

$$t = t_H + t_L = 0.69C_1(R_1 + 2R_2)$$

The frequency of the signal is:

$$f = \frac{1}{t} = \frac{1}{0.69C_1(R_1 + 2R_2)}$$

$$f = \frac{1.44}{C_1(R_1 + 2R_2)}$$

The **mark-space ratio** is the ratio between the period when output is high and the period when it is low. Using the equations for t_H and t_L we obtain the result shown overleaf.

$$\text{mark - space ratio} = \frac{t_H}{t_L} = \frac{R_1 + R_2}{R_2}$$

Self Test

A 555 astable has R_1=10 kΩ, R_2 = 6.8 kΩ and C_1 = 150 nF. What is the frequency and the mark-space ratio of the output signal?

It follows from this result that the mark-space ratio is greater than 1 in all circuits connected as on p. 43. It is possible to obtain ratios equal to 1 or less than 1 by using diodes in the timing network to route the charging and discharging currents through different resistors.

Alternating signals

So far in this chapter, we have seen what happens when a capacitor is charged from a *constant* voltage source. The circuits we have looked at have many useful timing applications. Now we will look at a capacitor being charged from an *alternating* voltage source.

The alternating voltage is a **sinusoid** — what is often called a **sine wave**. A sinusoid is a signal that, when plotted to show how the voltage or current varies in time, has the shape of a sine curve.

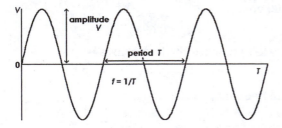

A sinusoid is defined by its amplitude V and period T. Its frequency f is equal to 1/period.

There is good reason to choose a sinusoid as our example of an alternating signal. It can be shown that all alternating waveforms, such as square waves, sawtooth waves and the complex waveforms made by musical instruments, can be produced by adding together two or more sinusoids.

As an example, we will consider what happens to an analogue signal consisting of two sinusoids added together:

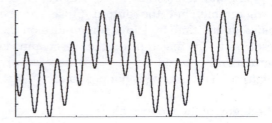

The drawing was done using two signal generators in a computer simulation. The signal was obtained by adding two sinusoids, one of 1 kHz, one of 8 kHz, and both with the same amplitude of 1 V. The mixture of the two frequencies is clearly shown.

The simulation was then extended to include this circuit:

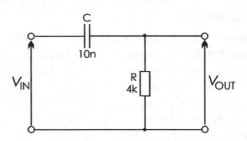

V_{IN} is the mixed signal consisting of 1 kHz and 8 kHz sinusoids. The signal received as V_{OUT} is plotted below, to the same scale as before:

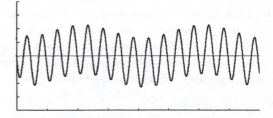

The high-frequency (8 kHz) signal is present, with the same amplitude as before. The low-frequency (1 kHz) signal has been much reduced in amplitude. Here we have a circuit that affects different frequencies differently. Because it reduces the amplitude of the low frequency signal but passes the high frequency signal almost unchanged, we call this a **highpass filter**.

Lowpass filter

If we exchange the capacitor and resistor, we obtain this circuit:

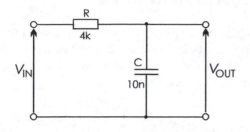

Exchanging the capacitor and resistor result in the inverse action.

If we feed in the same mixed signal as before, the output signal looks like this:

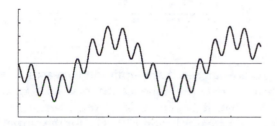

The 8 kHz signal is reduced in amplitude but the 1 kHz signal passes through unchanged. The circuit is a **lowpass filter**.

Resistance and reactance

R is the symbol for **resistance**. This is a property of any material that conducts current. Its unit is the ohm, and it is not affected by frequency.

X_C is the symbol for the **reactance** caused by capacitance. A capacitor has very high resistance because its dielectric is a non-conductor. But alternating signals are able to pass across from one plate of the capacitor to the other. Reactance is a measure of how easily they can pass. In this way, reactance is similar to resistance. The unit of reactance is the ohm, the same unit as that of resistance.

Unlike resistance, reactance depends on the frequency of the signal. The higher the frequency, the smaller the reactance and the easier the signal passes through the capacitor.

For a given capacitance, C, the reactance is:

$$X_C = 1/2\pi fC.$$

Impedance

Z is the symbol for **impedance**. This quantity includes both resistance and reactance. For a resistance, $Z = R$. For a capacitance, $Z = X_C$. The unit of impedance is the ohm.

Example

What is the impedance of a 330 nF capacitor at 2.5 kHz?

$$Z = X_C = 1/(2\pi \times 2.5 \times 330) \quad [1/(kHz \times nF)]$$
$$Z = 193 \ \Omega$$

The inpedance of the capacitor is 193 Ω at 2.5 kHz.

Note that we must specify the frequency at which reactance or impedance are to be calculated. It is meaningless to say simply that 'The impedance of the capacitor is 193 Ω.'

Self test

a What is the reactance of a 22 μF capacitor at 150 Hz?

b What is the impedance of a 390 pF capacitor at 500 kHz?

c What is the resistance of a 2.2 kΩ resistor at 40 kHz?

d At what frequency does a 220 nF capacitor have an impedance of 2.2 kΩ?

Impedance and frequency

A highpass filter can be re-drawn as a potential divider, made up of two impedances:

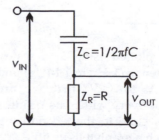

Z_C depends on the frequency of v_{IN}, but Z_R does not.

Consider what happens if the input is a pure sinusoid with *single* frequency f, and with amplitude v_{IN}. The output is a sinusoid with:

- the same frequency, but
- an amplitude v_{OUT} that is less than or almost equal to v_{IN}, and varies with f.

> **Self test**
>
> **a** Find the reactance of a 150 nF capacitor at 10 Hz, 100 Hz and 1 kHz.
>
> **b** At what frequency is the reactance of a 33 pF capacitor equal to 500 kΩ?
>
> **c** A highpass filter has R = 470 Ω, and C = 2.2 μF. What are the impedances of the resistor and capacitor at 10 kHz? At 500 kHz?

At the highest frequencies, Z_C is low (see equation p. 45). As a result, most of v_{IN} is dropped across the *resistor*. This makes v_{OUT} almost as large as v_{IN}. The signal is passed at almost its full amplitude.

At low frequencies, Z_C is high. Most of v_{IN} is dropped across the *capacitor*. Only a small voltage is dropped across the resistor, so the amplitude of v_{OUT} is small.

Frequency response

The way in which the amplitude of the output of a filter varies with the frequency of the signal is called the **frequency response**. The graph below plots the output of a highpass filter as the frequency of the input signal is varied over the range 1 Hz to 1 MHz.

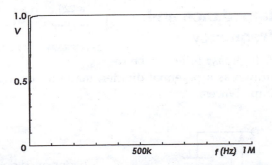

The graph above is plotted using *linear* scales. The ranges of frequency (1 Hz to 1 MHz) and of amplitudes (0 V to 1 V) that we are interested in, are both very wide. The result is that the curve runs close to the left and top sides of the plot. Such a graph tells us little.

The graph below provides the same information in a more easily visualised form.

Frequency is plotted on a *logarithmic* scale. Equal steps along the x-axis give tenfold (ten *times*) increments of frequency.

Amplitude is plotted on a logarithmic scale, using **decibels** (see box, opposite). Every step of −20 on the decibel scale is a reduction of amplitude of one tenth.

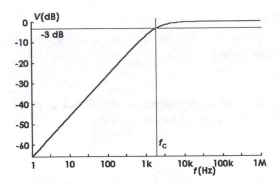

Now we can see that amplitude increases at a steady rate with frequency up to about 1 kHz. After that, it begins to level off. There is no further increases beyond 10 kHz, for the output amplitude is then virtually equal to that of the input signal. The signal is passing through the filter with almost no loss. The frequency response is exactly what would be expected for a highpass filter.

Cut-off point

The frequency response plot above has a line drawn across it at the **−3 dB level**. This is the level at which the power of the output signal is exactly half that of of the input signal (*see* box). This level is also known as the **half-power level**. The point at which the half-power level cuts the response curve marks the **cut-off point** or **cut-off frequency**, f_C.

The curve plotted above is that of a highpass filter in which C = 100 nF and R = 820 Ω. Reading from the plot, f_C is approximately 2 kHz.

The exact value of f_C is calculated from this formula:

$$f_C = 1/(2\pi RC)$$

Example

For the highpass filter referred to above, $R = 820$ Ω and $C = 100$ nF. Calculate the cut-off frequency.

$$f_C = 1/(2\pi \times 820 \times 10^{-9}) \quad [1/(W \times nF)]$$
$$= 1940 \text{ Hz}$$

The cut-off point is at 1.94 kHz.

This confirms the estimate of 2 kHz made from the frequency distribution diagram.

Lowpass response

Below is the frequency distribution of a lowpass filter (circuit p. 35) which has $R = 4$ kΩ and $C = 10$ nF. The frequency was swept from 10 Hz up to 1 MHz.

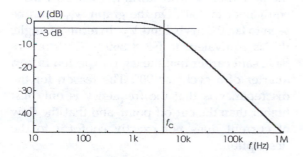

This frequency response curve is typical of a lowpass filter

Low-frequency signals pass through the filter without loss, up to about 3 kHz. Above this frequency, amplitude falls off to reach about -48 dB at 1 MHz.

Reading from the graph, the cut-off frequency is about 4 kHz. This is confirmed by using the formula:

$$f_C = 1/(2\pi \times 4 \times 10) \quad [1/(kW \times nF)]$$
$$= 4 \text{ kHz}$$

Decibels

Decibels express the *ratio* between two *identical* quantities. In electronics, the decibel is most often used to express the ratio, n, between two powers, two voltages or two currents. If, for example, a signal of power P_1 is amplified to power P_2, the amplification in decibels is calculated from:

$$n = 10 \log_{10}(P_2/P_1)$$

Example

A 50 mW signal is amplified to 300 mW:

$$n = 10 \log_{10}(300/50)$$
$$= 10 \log_{10} 6 = 7.78$$

The power amplification is 7.78 dB.

Power is proportional to voltage squared or to current squared, so we multiply by 20 when calculating n from voltages or currents.

Example

A signal of amplitude 60 mV is amplified to 500 mV:

$$n = 20 \log_{10}(500/60)$$
$$= 20 \log_{10} 8.333 = 18.4$$

The voltage amplification is 18.4 dB.

If there is power loss, as in a filtered signal, the decibel value is negative. The loss at the half-power level is:

$$n = 10 \log_{10} 0.5 = -3.01 \text{ dB}$$

As explained on p. 36, the **–3 dB point** is often used to define the cut-off frequency of amplifiers and filters.

The amplitude of a signal may also be stated with reference to a standard signal. If the strength of a voltage signal is quoted in dBm, it means that the standard is a signal of 1 mW across a 600 Ω load. P = V2/R so its amplitude is 780 mV. If the amplitude of a given signal is 60 mV, its dBm strength is:

$$n = 20 \log_{10}(60/780) =$$
$$20 \log_{10} 0.077 = -22.3 \text{ dBm}$$

Phase

Consider a lowpass filter fed with a 1 kHz signal, amplitude 1 V:

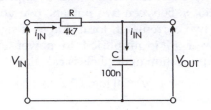

We measure V_{IN} and V_{OUT}, using an oscilloscope and display them on the same screen:

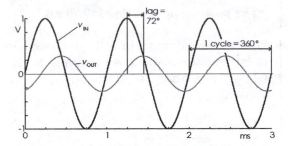

There are two important points to notice:

- The amplitude of V_{OUT} is less than that of V_{IN}.

- V_{OUT} does not reach its peaks and troughs at exactly the same time as V_{IN}.

The reduction in amplitude is to be expected because calculation shows that the cut-off frequency is 400 Hz, and the signal frequency is a little higher than this.

The fact that V_{OUT} reaches its peaks and troughs after V_{IN} is something that we have not discussed before. Measurement on the graph reveals that V_{OUT} is about 0.2 ms later than V_{IN} in reaching the same stage in its cycle. This is due to the delay that capacitance always introduces in the operation of circuits. Capacitors take *time* to charge and discharge.

The capacitor is charging during the first (positive) half of the cycle of V_{IN}. But it does not reach its maximum charge until about the *end* of that cycle.

Conversely, the capacitor is being discharged during the second (negative) half of the cycle of V_{IN}. But it does not become completely discharged until the end of that half-cycle. As a result of these delays, the *peaks and troughs* of V_{OUT} coincide with the *ends* of the half-cycles of V_{IN}. Putting it another way, V_{OUT} lags about a quarter of a cycle behind V_{IN}.

Cycles and phase

It is conventional to measure differences of phase in terms of angle. When the sine of an angle is plotted, the curve is repeated every 360°. One complete cycle is equivalent to 360°, as illustrated in the diagram. We measure phase differences in terms of degrees of angle.

Angles are also measured in radians, where one complete cycle is 2π rad. So phase differences can also be expressed in radians. However, in this discussion we will use only degrees.

In the diagram, where the frequency is 1 kHz, the period of the waveform is 1 ms. So 1 ms is equivalent to 360°. On the graph, V_{OUT} can be seen to lag 0.2 ms behind V_{IN}. In terms of angle, this is equivalent to $0.2 \times 360 = 72°$. Yet we have said earlier that the lag is expected to be a quarter of a cycle, or 90°. The reason for the discrepancy is that the frequency is only *just* higher than the cut-off point, and that the flow of current does not exactly conform to the description.

The way that the phase lag varies with frequency can be shown by measuring the frequency response of the filter and plotting the phase lag over the range of frequencies:

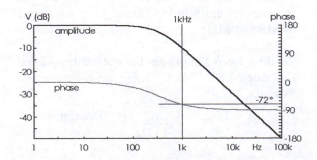

The plot of amplitude is typical for a lowpass filter. The -3dB point is 400 Hz. At 1 kHz, amplitude has begun to fall off.

The phase plot shows that there is little lag at low frequencies, where the period of the sinusoid is long compared with the time taken to charge and discharge the capacitor. Then the lag increases (the curve goes more negative) until it becomes -72° at 1 kHz. Above that, the lag increases and eventually settles to -90° at frequencies higher than about 10 kHz. This agrees with the prediction made above.

Current

From the discussion above, it is expected that the current flowing into and out of the capacitor will vary in a similar way to the output voltage. Here is a plot of V_{IN} and I_{OUT} against time:

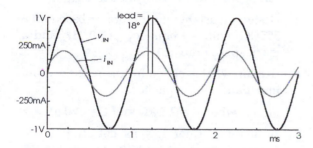

The plot shows that the curve of i_{OUT} is a sinusoid with frequency 1 kHz and amplitude 20 mA. The peaks and troughs of this curve are ahead of those of V_{IN}, with a phase difference of 18°. We say that current **leads** V_{IN} by 18°.

Current increases and begins to charge the capacitor as V_{IN} swings positive. However, the current curve begins to climb less rapidly as charge accumulates on the capacitor (compare with charging a capacitor, p. 36).

On the negative swing of V_{IN}, current reverses, to charge the capacitor in the opposite direction. Again, because charge accumulates on the capacitor, current reaches its peak earlier than V_{IN}.

Noting that 72° + 18° = 90°, we would expect to find that current leads V_{OUT} by 90°. That this is so is shown in this plot:

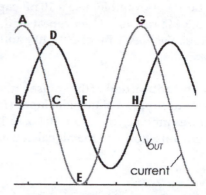

Current leads V_{OUT} by 90° in a lowpass R-C filter.

Current is at its positive maximum (A) when it begins to charge the capacitor (B). Current falls as the capacitor gradually accumulates charge. Eventually, current is zero (C) when the capacitor is fully charged (D). Then the current reverses and the capacitor begins to discharge. Current reaches its negative maximum (E) as the discharged capacitor begins to charge in the opposite direction (F). The next half cycle is a repeat of the above, except that polarities are reversed. The current reaches its positive maximum again (G) as the capacitor begins to charge in the original direction.

Highpass phase response

Interchanging the resistor and capacitor reverses some of the effects described above. At low frequencies, V_{OUT} has a phase **lead** of 90°. This drops as frequency increases. At 1 kHz, with the values for R and C that we used before, the lead is 18°. It falls to zero at high frequencies.

V_{OUT} is the result of current flowing through the resistor to ground. By Ohm's Law, current is proportional to voltage at every instant. This means that the changes in current are exactly in phase with V_{OUT}.

Activities — Filters

Investigate the action of some highpass filters set up on a breadboard or on a computer simulation. For example, use a 10 nF capacitor and a 3.9 kΩ resistor. Then repeat with other capacitors, such as 1 nF and 10 µF, and other resistors.

For a breadboarded filter, use a signal generator to supply a sinusoidal input signal at frequencies ranging from, say, 10 Hz to 1 MHz. Make the amplitude of the signal constant at, say, 1 V.

A breadboarded highpass RC filter, ready for testing.

As you vary the frequency of the input, use an oscilloscope to observe the output signal. Measure its amplitude at a series of frequencies, such as 10 Hz, 100 Hz, 1 kHz, and so on. In what way does amplitude vary with frequency?

Build or simulate low-pass filters and investigate their action in the same way as above.

Questions on using capacitors

1   Design a circuit to switch an LED off for 0.5 s when a button is pressed (use an op amp or the 555 timer IC).

2   Design a circuit to switch an LED on for 20 s when a button is pressed (use an op amp or the 555 timer IC).

3   What is the action of a monostable multivibrator? Describe the circuit of a monostable.

Delays   **Extension Box 9**

Referring to the circuit on p. 41 and the calculations on p. 39, $V_S = 6$ and $v_C = 4.8$, which gives:

$$\frac{v_C}{V_S} = (1 - e^{-t/RC}) = \frac{4.8}{6} = 0.8$$

$$-e^{-t/RC} = -0.2$$

$$-t/RC = \ln 0.2 = -1.6$$

$$t = 1.6RC$$

The delay is 1.6 time constants.

Given that R = 1 MΩ and C = 47 µF, the time constant is 47, and the delay is 1.6 × 47 = 75 s.

4   Describe a monostable based on the 7555 IC. It produces a pulse about 5 s long and can be reset by pressing a push-button.

5   Design an astable multivibrator based on a 555 IC, with a frequency of 10 Hz, and a mark-space ratio of 1.4.

6   Explain what is meant by the reactance and impedance of a capacitor.

7   Find f_C when $R = 2.2$ kΩ and $C = 470$ µF.

8   A highpass filter has its cut-off point at 33 Hz. If the resistor is 2.2 kΩ, what is the value of the capacitor?

9   A highpass filter has $R = 470$ Ω and $C = 4.7$ µF. What is its cut-off frequency? What is the impedance of the capacitor at this frequency?

10   Describe in outline the way in which a lowpass resistance/capacitor filter works.

11   Sketch the frequency response curve of a highpass filter in which $R = 330$ Ω and $C = 2.2$ nF, and mark the value of f_C.

12   Design the circuit of an RC bandpass filter, that is, one that filters out low and high frequencies but passes frequencies in the middle range. The pass-band should run from 10 kHz to 100 kHz. Check your design with a simulated circuit or on a breadboard.

6 Fields

When one object interacts with another object with which it is not in physical contact, it is because of a **field**. Think of a field as something that produces a force at a distance. If one or both objects are free to move, they will do so as a result of the force of the field.

Gravitation is an everyday example of a field. The force of gravity acts at a distance; for example, the Moon's gravity raises the ocean tides on Earth.

Two other kinds of field are important in everyday life — **electric fields** and **magnetic fields**.

Electric fields

An electrically charged object is surrounded by an electric field. The charge may be positive or negative. The field affects other charged objects. If both have the same charge, they are repelled from each other and may move apart. If the two objects have opposite charges, they are attracted to each other.

> **Polarity**
>
> The terms *positive* and *negative* mean only that the two kinds of field are opposite. They do not mean that the positive field has something that is lacking from the negative field.

Magnetic fields

A charged object produces an electric field whether it is moving or stationary but, when it *moves*, it produces an additional field, a **magnetic field**.

The magnetic field around a bar magnet, for example, is the result of the spinning and orbiting of the charged electrons in its atoms. Each atom is equivalent to a tiny magnet, each one having a north and south pole. The atoms may be aligned in random directions, so their fields cancel out for the bar as a whole.

But, if a sufficiently high proportion of them are aligned in the same direction, their fields combine and the material is said to be **magnetised**.

Magnetic flux, or **lines of force**, run from the north pole of the magnet to the south pole of the same or another magnet. They show the direction in which a compass needle aligns itself in the magnetic field.

> **Magnetic poles**
>
> An object can have a single type of electric charge, positive *or* negative, but magnetic poles do not exist separately. Magnetised objects always have a **north pole** *and* a **south pole**.

In the same way that two electric fields may interact to produce attractive or repulsive forces, two unlike poles attract each other, while two like poles repel each other.

> **Flux**
>
> Means 'flow'. The flux in a wire of solder helps the solder to flow when melted.

Fields and motion

When a charged object moves in a magnetic field it is subject to a force. The direction of the force depends on Fleming's left-hand rule.

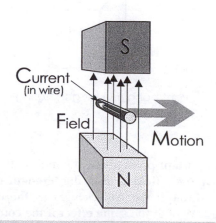

Illustrating the directions of fields and force in Fleming's left-hand rule (see overleaf).

For example, if a current of electrons is moving along in a wire and the wire is in a magnetic field, each electron experiences a force. The resulting force on all the electrons appears as a force acting on the wire.

The left-hand rule

Hold your thuMb, First finger and seCond finger at right-angles. Point your First finger in the direction of the electric Field and your seCond finger in the direction of the Current. Your thuMb points in the direction of the Motion.

Practical applications of this effect include electric motors and loudspeakers.

When a force moves a charged object in a magnetic field, it generates an electric field (an emf). The direction of the current obeys Fleming's right-hand rule.

For example, when a coil of wire is moved in a magnetic field, the emf generated causes an electric current to flow in the coil. This is called **electromagnetic induction**.

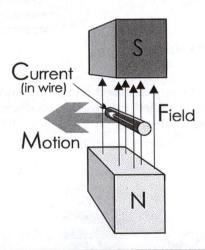

Fleming's right-hand rule for generating current.

The important point about the motion through the magnetic field is that the strength of the field through the coil *changes* as the coil is moved. We get the same effect if the coil stays in the same place and the field becomes stronger or weaker.

Describing electric fields

Electric field strength, E. An electric field acts on a charge Q with force F:

$$E = F/Q \text{ newton per coulomb}$$

This is numerically equal to the pd V between two points in the field d metres apart:

$$E = V/d \text{ volts per metre}$$

Electric field flux, φ. An electric field can be described by drawing lines that show the direction moved by charged objects within the field. These **lines of electric flux** are imaginary.

The electric field flux is proportional to the number of field lines passing at right-angles through a given area.

Describing magnetic fields

Magnetic flux density, B. A magnetic field produced by a unit charge moving at v m/s, in an electric field of E V/m has flux density:

$$B = v \times \frac{1}{c^2} E$$

c is the velocity of light in m/s, and equals 3×10^8 m/s. **B** is expressed in tesla (T).

Magnetic field strength, H. This is also expressed in tesla. B and H are related by the equation:

$$B = \mu H$$

In this equation, μ is the **magnetic permeability**, in henries per metre. The henry is described on p. 55. In a vacuum, μ = 1, so B and H have the same value.

Force on a wire. When a current flows through a wire in a magnetic field, the force on the wire is:

$$F = BIL$$

F is the force in newtons, B is the flux density in teslas, I is the current in amps, and L is the length of the wire, in metres.

Magnetomotive force. Electromotive force (emf, p. 1) produces an electric flux or current. Magnetomotive force (mmf) produces magnetic flux. The unit of emf is the ampere-turn. When a direct and constant current of 1 A fkows in a coil of 1 turn, the magnetic flux through the coil is 1 ampere-turn.

Note that the word *force* in the terms *electromotive force* and *electrocmagnetic force* does not mean that emf and mmf are mechanical forces. An electric cell, for example, produces emf as a result of chemical action, with no physical forces involved.

Magnetic reluctance: The lines of forces of a magnet are closed loops. Each loop is a magnetic circuit, equivalent to an electric circuit.

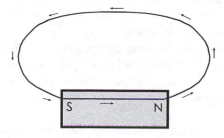

A single line of force produced by a bar magnet. The arrows indicate the direction of the field (the direction of a compass needle).

The reluctance of the magnetic circuit is equivalent to the resistance of an electric circuit. In an electric circuit or part of an electric circuit the resistance is R = emf/current. In a magnetic circuit the reluctance is:

$$S = mmf / magnetic\ flux$$

If there are objects with low reluctance in the field, the lines of force tend to crowd together and pass through the objects, avoiding the high-reluctance space between them. This is equivalent to the way in which electric current takes the route of least resistance.

When lines of force are concentrated through a low-reluctance material, the material develops north and south poles. It behaves as a magnet for as long as it is part of a magnetic circuit.

This explains the action of a solenoid. The plunger is made of soft iron, which has low reluctance, and it is not permanently magnetised. When the coil is energised, its magnetic field becomes concentrated through the plunger. The plunger becomes strongly magnetised. The plunger is strongly attracted toward a region of high magnetic flux, so as to crowd even more lines of force into it. The result is that the plunger is *pulled* into the coil. A reed switch operates on a similar principle.

On the Web

There are several unexplained terms and devices mentioned in this chapter. Examples: Hall effect sensors, series and shunt motors, and reed switches.

For more information, search the Web using the terms or the device names as keywords.

Use the Web to find out the properties and uses of ferromagnetic materials such as ticonal, alcomax, and magnadur.

Ferromagnetism

This is the commonest type of magnetism. Bar magnets, compass needles, and the coating on magnetic hard disks all have ferromagnetism. It is a property of iron, cobalt, and compounds of these and other metals such as nickel, chromium and magnesium.

These materials have ferromagnetic properties because the spins of the electrons cancel out much less than in other materials. Groups of them align readily with an external magnetic field, forming magnetic domains. They align permanently so that the material becomes magnetised, and remains so, even when the external field is removed.

A piece of ferromagnetic material is magnetised by placing it in a magnetic field. The stronger the field, the more domains align themselves with it and the more strongly the material become magnetised. This is the basis of the hard disk drive of a computer. The relationship between the strength of the external field (H) and the magnetic flux density in the material (B) is illustrated in the graph below.

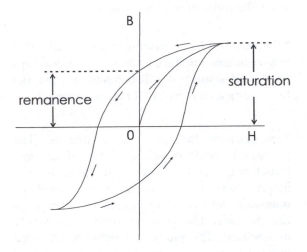

The effect of a magnetic field (*H*) of the magnetisation (*B*) of ferromagnetic material.

The diagram represents the effect on B of a sequence of changes in H. Follow the arrows. We begin at the origin (0) with unmagnetised material (B = 0) and no external field (H = 0).

As the external field is increased, B increases too, but the graph is not a straight line. It eventually levels out and further increase o\of H does not increase B. The material is **saturated**.

If H is gradually reduced, B falls too. But when H is reduced to zero, B is still greater than zero. The material has become magnetised.

Some of the magnetic field has remained (that is, some of the domains are permanently aligned). This remaining field is the **remanence**

If the field is then reversed and increased in the reverse direction, B falls and eventually becomes saturated in the reverse direction. From that state, returning the field through zero and into the original direction will restore the original saturation. The effect is that changes in B *lag behind* changes in H. This lagging behind is called **hysteresis**.

Reversing the direction of magnetisation requires energy, the amount of energy being proportional to the area within the hysteresis curve (or loop). Magnetic materials differ in the amount of energy needed to reverse the magnetisation, as shown in the drawing below.

Hard magnetic materials need a strong field to magnetise them but have high remanence, so they make strong magnets. Once magnetised, they are not readily demagnetised by stray magnetic fields and so are suitable for permanent magnets. Examples of such materials are ferrites, alloys such as Alcomax and neodymium-iron-boron (for the strongest msgnets) and ceramic materials such as Magnadur.

Soft magnetic materials can be strongly magnetised by an external field but it takes little energy to demagnetise them or reverse their state. Examples are soft iron (used in electromagnets), cobalt alloys and chromium dioxide (used in magnetic hard disks).

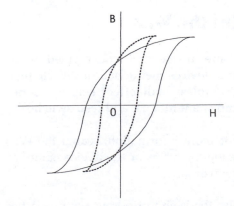

Typical B-H hysteresis loops of hard (continuous lines) and soft (dotted lines) magnetic materials.

7 Inductors

When a current is passed through a coil of wire, a magnetic field is generated. This is the principle on which solenoids, electric motors and loudspeakers work.

The converse happens too. When there is a change in the magnetic field through a coil, an emf is generated in the coil. This causes current to flow through the coil. This is known as **electromagnetic induction.**

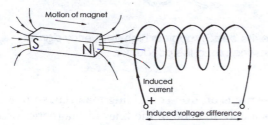

There are three important points to note:

- **Emf is induced** in the coil only when the magnetic field is *changing* (Faraday's First Law). Current flows only if we move the magnet toward or away from the coil, or move the coil toward or away from the magnet. If both are still, there is no emf and no current.

- The **size** of the emf depends on the *rate of change* of the magnetic field (Faraday's Second Law). The faster we move the magnet or coil, the greater the emf and current.

- The **direction** of the induced current is such as to *oppose* the motion of the magnet or coil (Lenz's Law). In the drawing above, the north pole of the magnet is moving toward the coil. The direction of the current produces a north pole at the end of the coil that is nearer the magnet. Like poles repel, so the effect is to try to prevent the magnet from coming nearer. In practice, it does not prevent the motion but means that a little extra force is needed to move the magnet toward the coil. The extra energy reappears as the energy of the current made to flow in the coil.

Self induction

When the current through a coil changes, the magnetic field through the coil changes. This changing field acts just like a magnet being moved around near the coil — it induces another current in the coil. The direction of the current opposes the change in current through the coil. This effect, in which a coil induces current *in itself*, is called **self induction**.

The size of the induced current depends on the number of turns in the coil and other factors. The symbol for inductance and self-inductance is **L**, and its unit is the **henry**, symbol **H**. Inductance can be defined in various ways, the most usual of which is:

$$E = -L \bullet \frac{dI}{dt}$$

E is numerically equal to the emf produced in a coil of self-inductance L henries when the current changes at the rate of 1 A per second. The negative sign indicates that the induced emf opposes the change of current.

Self Test

The current in a coil begins to fall at the rate of 6 A/s. If L = 840 mH, what emf is produced in the coil?

The same three points mentioned earlier apply also to self-induced currents:

- If there is *no change* in the current flowing through the coil, there is no induced current. DC flows through the coil without being opposed. It is affected only by the resistance of the wire of the coil.

- The size of the induced current is proportional to the *rate of change* of the magnetic field. If there is a very rapid change of current through an inductive component, such as when the current is switched off, self induction produces a very large current which may damage other components in the circuit (see p. 15).

- The *direction* of the current is such as to oppose the changing current. If the current through the coil is increasing, the induced current acts to prevent the increase. Conversely, if the current is decreasing, the induced current acts to prevent it from decreasing. The result of this is that, if an alternating current is passed through the coil, its amplitude is reduced. This effect, which resembles resistance is called the **reactance** of the inductor.

The inductance, and hence the reactance of a coil are much increased if the coil is wound on a **core** of magnetic material. This acts to concentrate the line of magnetic force within the core and so increase the strength of the magnetic field.

Reactance

The reactance of an inductor has the symbol X_L and is expressed in ohms. If a signal of frequency f is passed through an inductor of inductance L, its reactance is:

$$X_L = 2\pi f L$$

The equation shows that X_L increases with increasing f. At zero frequency (DC), X_L is zero. Direct current flows freely through an inductor coil. X_L is high at high frequency so radio-frequency signals are much reduced in amplitude when passed through an inductor coil.

Inductive reactance can also be expressed as inductive impedance, Z_L. Impedance is discussed on p. 35.

Types of inductor

Many kinds of component have self-inductance. Solenoid coils, loudspeakers and other devices that include a coil are examples. Even the leads of a capacitor, resistor or transistor, though not coiled, have significant self-inductance at radio frequencies. Inductance in these cases may be ignorable, or may cause problems.

Inductance is a useful property of certain other components. These include:

Chokes: They are used to block high-frequency signals from passing through from one part of a circuit to another. Low-frequency signals or DC voltage levels are able to pass through. Large chokes look like transformers, but have only one coil. Small chokes consist of beads or collars made of ferrite, threaded on to the wire that is carrying the high-frequency signals.

conductor

Ferrite beads suppress high-frequency signals passing along the conductor.

Ferrite is an iron-containing material, so it acts as a core to contain the lines of magnetic force around the wire. Sometimes a choke is made by winding the wire around a ferrite ring.

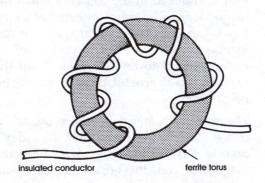
insulated conductor   ferrite torus

High-frequency signals in a conductor may be damped by winding the wire around a ferrite torus.

Tuning coils: Used in radio transmittters and receivers to tune a circuit to a particular radio frequency.

The coil is wound on a plastic former. It may have a core of ferrite or iron dust ceramic that can be screwed in or out of the coil to tune it.

When connected to a capacitor, it forms a tuned network that resonates at a particular frequency, as explained below.

The inductor may be a few turns of wire wound on a plastic **former,** as on the right. A tuned inductor may have more than one winding, and is equivalent to a transformer that operates best at one particular frequency.

This VHF tuning coil with 3.5 turns of wire has a self inductance of 0.114 μH.

One or more coils are wound on a plastic former. The iron-dust core is threaded, as is the inside of the former. The core is moved into or out of the coil by turning it with a special non-magnetic tool so as not to introduce spurious magnetic effects. This allows the self inductance of the coil to be adjusted precisely. Inductors of this type are used in the same way as capacitors, for coupling two parts of a circuit (p. 34). However, they are suited only to radio-frequency signals.

Energy transfers

Resistance is fundamentally different from capacitance and inductance, even though they are all forms of impedance. When a current passes through a resistance, some of its energy is converted irreversibly to other forms. In most cases it is converted to heat, though lamps and LEDs produce light and motors produce motion. The loss appears as a loss of potential across the resistance.

When current passes into a capacitor, the charge is stored on its plates. The stored energy appears as a difference of potential between the plates. The charge is released later when the capacitor is discharged. There is no loss of energy during charging and discharging. In theory at least, no heat or other form of energy is released.

Similarly, when the current through an inductor increases, electrical energy is stored in the form of a magnetic field. When the current decreases, the field wholy or partly collapses, releasing energy as the induced emf. Again, there is no loss of energy in this process. No heat or other form of energy is produced.

L-C network

The impedance of a capacitor decreases with increasing frequency. The impedance of an inductor is the opposite of this. Combining a capacitor and an inductor in the same network results in an interesting action. Here is the circuit:

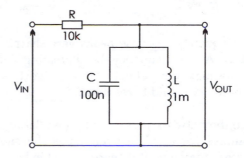

V_{IN} is a sinusoid, frequency f. At low frequencies the inductor has low impedance. The signal passes easily through it, so V_{OUT} is small. At high frequencies the capacitor has low impedance. The signal passes easily through it, so V_{OUT} is again small. There is one frequency, between high and low, at which the inductor and capacitor have equal impedances. At this frequency the signal passes less easily through them and V_{OUT} is a maximum.

Calling this frequency f, the impedance of C is:
$$X_C = 1/2\pi f C$$
At the same frequency, the impedance of L is:
$$X_L = 2\pi f L$$
When $X_C = X_L$:

$$\frac{1}{2\pi f C} = 2\pi f L$$

$$f^2 = \frac{1}{4\pi^2 LC}$$

$$f = \frac{1}{2\pi\sqrt{LC}}$$

With the values of C and L shown in the diagram:

$$f = 1/(2\pi \times \sqrt{(1 \times 100)}) \quad [1/(\text{mH} \times \text{nF})]$$
$$f = 15.9 \text{ kHz}$$

To confirm these calculations, the frequency response of the circuit is:

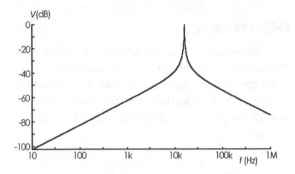

The frequency response peaks very sharply at 16 kHz. At this frequency the alternating signal has the maximum effect on the network, and the network is said to **resonate.**

Once the network is resonating (oscillating), it continues for a while even when the signal source is removed. But energy is lost in heating the conductors and also in the dielectric of the capacitor and the armature of the inductor. Eventually the oscillations die out. The network can be kept oscillating by supplying it with relatively small pulses of energy in phase with its oscillations. This is similar to the way in which the pendulum of a clock is kept swinging by small amounts of energy transferred to it from the spring or weights through the escapement mechanism.

When the network is oscillating, it cycles through four states. Starting with the top left diagram in the next column, the capacitor is fully charged and no current is flowing. Then the capacitor discharges though the inductor, generating a magnetic field in its coil. This builds up until the capacitor is fully discharged (top right).

At this stage, the current ceases and the field collapses, but self-induction results in the generation of an emf in the inductor, tending to keep the current flowing.

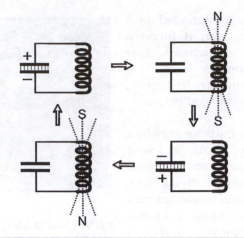

A capacitor and inductor connected in parallel form a resonant network. The energy in the network is stored alternately as charge on the capacitor or as a magnetic field in the inductor. The network alternates between the two states at a rate depending on the capacitance and inductance.

This current charges the capacitor in the opposite direction (bottom right), and the field in the inductor gradually decays.

The capacitor now discharges, generating a magnetic field but in the opposite direction (bottom left). As this field collapses, an emf is generated which re-charges the capacitor in the original direction.

As explained earlier, storage and release of energy in capacitors and inductors takes place with very little loss of energy. Little energy from outside is needed to keep the network oscillating. The only requirement is that the frequency of the outside source of energy must be very close to that of the resonant frequency of the network.

The resonant property of a parallel capacitor/inductor network is made use of in oscillators and in the tuning circuits of radio receivers.

RL filters

An RL filter has an inductor as the reactive component instead of a capacitor.

RL filters are very rarely used, mainly because of the weight and bulkiness of inductors. There is also the problem that large inductors may need magnetic shielding to prevent them from picking up or causing electromagnetic interference within the circuit, or with adjacent circuits. However, a combination of resistance and inductance in certain circuits may give rise to unexpected filtering effects of which the circuit designer should be aware.

Here is an RL lowpass filter:

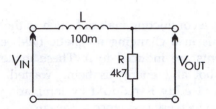

As might be expected from the fact that inductors respond to frequency in the reverse way to capacitors, this lowpass filter has a similar layout to an RC highpass filter. Conversely, the RL highpass filter looks similar to an RC high-pass filter.

The cut-off frequency of both types of RL filter is given by:

$$f_C = \frac{R}{2\pi L}$$

Phase response is also the inverse of that of the corresponding RC filters. In the lowpass RL filter, V_{IN} lags V_{IN} by 90° at high frequencies. The current signal is in phase with V_{OUT}.

Other features of RL filters are given in the summary table below.

Type	RC filters				RL filters			
Response	Lowpass		Highpass		Lowpass		Highpass	
Frequency	Low	High	Low	High	Low	High	Low	High
Circuit								
Cut-off frequency	$1/2\pi RC$				$R/2\pi L$			
V_{OUT} phase	0	lag	lead	0	0	lag	lead	0
I_{OUT} phase	lead	0	lead	0	0	lag	0	lag

The lowest two rows indicate phase with respect to V_{IN}. Maximum lags and leads are 90°. A zero entry indicates that the signals are in phase.

Transformers

A transformer consists of two coils wound on a core. The core is made of layers of iron.

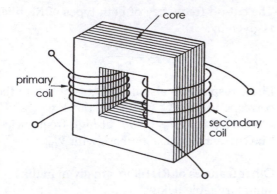

core

primary coil

secondary coil

The coils normally have many more turns than are shown in the drawing. When a current is passed through the primary coil it produces a magnetic field. The core provides a path for the lines of magnetic force so they almost all pass through the secondary coil.

Induction occurs only when there is a *change* of magnetic field. So a transformer does not work with DC. When AC flows through the primary coil there is an alternating magnetic field. This induces an alternating current in the secondary coil. Using a low remanence material, such as soft iron miminises the energy lost in hysteresis.

When a conducting material, such as the iron core, is in a changing magnetic field, **eddy currents** are induced in it. These make the core hot and energy is being wasted. The effect of eddy is reduced by laminating the core to increase resistance to the currents.

The effect is greater at high frequencies. This is why ferrite, which has high electrical resistance, is used as the core material of radio-frequency transformers.

Transformer rules

Frequency: The frequency of the induced AC equals that of the inducing AC.

Amplitude: If V_P is the amplitude of the voltage in the primary coil, and V_S is the amplitude of the voltage in the secondary coil, then:

$$\frac{V_S}{V_P} = \frac{\text{secondary turns}}{\text{primary turns}}$$

Example

A transformer has 50 turns in the primary coil and 200 turns in the secondary coil. The amplitude of the primary AC is 9 V. What is the amplitude of the secondary AC?

Rearranging the equation above gives:

$$V_S = V_P \times \frac{\text{secondary turns}}{\text{primary turns}}$$

$$V_s = 9 \times \tfrac{200}{50} = 36 \text{ V}$$

The amplitude of the secondary current is 36 V.

These calculations assume that the transformer is 100% efficient.

Electric generator

In an electric generator, a current is induced in the coil as it is rotated in a magnetic field. In the diagram the field flows from north to south. On the lower side of the coil, motion is to the left. Applying the right hand rule, current flows from C to D. The result of induction is that current flows from terminal A towards terminal B, and B is therefore the positive terminal of the generator.

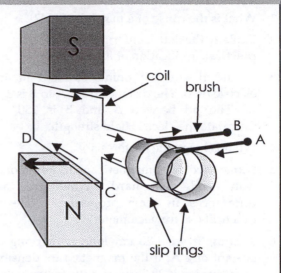

Half a revolution later the situation is reversed. Current now flows from B to A, and A is positive. As the coil rotates, A is positive of B and B is positive of A, alternately. The generator produces **alternating current**.

The size of the current depends on the *rate of change* of the magnetic flux passing through the coil. This is greatest when the coil is vertical (as in the diagram) and least when it is horizontal.

It can be shown that the rate of change is proportional to the sine of the angle between the plane of the coil and the vertical. The AC has the form of a sine wave, with a frequency of 1 cycle per rotation of the coil.

Electric motor

The left-hand rule is the basis of operation of a DC electric motor. This has a permanent magnet to provide the field. Other types of motor have coils to provide this.

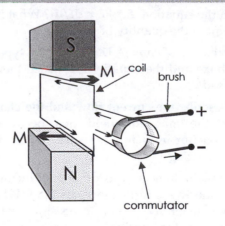

The motor in the diagram has a pair of commutator rings that reverse the direction of current flowing in the coil every half-revolution. This keeps the current flowing in a fixed direction relative to the magnetic field, so the force on the wire is always in the same direction.

Magnetic screening   Box 14

Sensitive parts of a circuit can be shielded from stray magnetic fields by enclosing them in a magnetic shield. .

This is made of an alloy such as mumetal (nickel-iron-copper-molybdenum) which has very high magnetic permeability. The lines of force are concentrated in the shield and do not penetrate it.

Questions on fields

1 What is the cause of a magnetic field?

2 Explain the left-hand rule and describe a practical application of it.

3 A and B are two points situated in an electric field. The distance from A to B is 2.6 m. The pd betwen A and B is 150 V. Calculate the electric field strength.

4 What is magnetic hysteresis?

5 Name two ferromagnetic materials, one 'soft' and one 'hard'. What are the differences in their properties? State a practical use for each material.

6 A straight wire 20 cm long is carrying a current of 2 A. If the magnetic flux density in the region is 25 T, what is the force on the wire? In which direction does the force act?

Questions on inductors

1 How can we demonstrate that an emf is induced in a coil only when the magnetic field through the coil is changing?

2 What is the significance of the negative sign in the equation: $E = -L \times dI/dt$? What is the unit of the quantity L?

3 What is a choke? Describe two types of choke and the situations in which they are used.

4 Describe the current flow and the changes in the magnetic field that occur in a capacitor-inductor network when it is resonating.

5 Explain how a resistor and inductor are connected to form a lowpass filter. Why do we generally prefer a resistor-capacitor network for filtering?

Multiple choice questions

1 The size of the current induced in a coil by a moving magnet does *not* depend on:

A the inductance of the coil
B the strength of the magnetic field
C whether the magnet is moving toward of away from the coil
D the rate of change of magnetic field strength.

2 If a 470 Ω resistor and 22 µH inductor are connected as in the diagram below, the network acts as a:

A lowpass filter
B highpass filter
C resonant network
D potential divider

3 The cut-off frequency of the network in Q. 2 is:

A 3.4 MHz
B 15.4 Hz
C 340 kHz
D 10.7 MHz.

4 A resonating LC network needs very little external energy to keep it oscillating because:

A the impedance of the network is a maximum
B the reactances of the capacitor and inductor are equal
C the energy is alternately stored as charge and as a magnetic field
D almost no energy is lost as it alternates between stored charge and magnetic field.

There are more questions on inductance on the Companion website.

8 MOSFET amplifiers

The common-source MOSFET amplifier with which this chapter begins is a typical transistor amplifier.

Common-source amplifier

The amplifier is built around an n-channel enhancement mode MOSFET. Given a fixed supply voltage V_{DD}, the size of the current i_D flowing through the transistor depends on v_{GS}, the voltage difference between the gate (g) and the source (s). Current flows in at the drain (d) and out of the source (s).

> **Source and drain**
>
> The names refer to the flow of electrons into the transistor at the source and out through the drain. The current i_D is conventional current, which is taken to flow in the opposite direction.

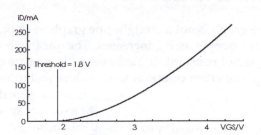

> As the gate-source voltage is increased above the threshold, the drain current increases more and more rapidly.

Transconductance

The amount by which i_D increases for a given small increase in v_{GS} is known as the transconductance of the transistor. We measure transconductance by using the circuit below.

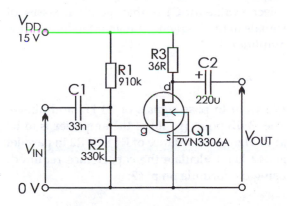

> The source terminal is common to both the input and output sides of this MOSFET amplifier, so it is known as a common-source amplifier.

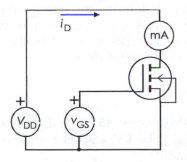

The drain voltage V_{DD} is held constant while the gate voltage is varied over a range of values. The meter registers the drain current i_D for each value of v_{GS}. The graph above shows typical results, obtained from a ZVN3306A.

Example

From the graph, find g_m when v_{GS} is 3.5 V.

By measurement on the graph (or by reading scale values on the simulator) as v_{GS} rises by 0.5 V, from 3.25 V to 3.75 V, i_D rises by 70 mA.

The graph shows how v_{GS} and i_D are related. Below a certain voltage, known as the **threshold voltage**, there is no current through the transistor. As voltage is increased above the threshold, a current begins to flow, the current increasing with increasing voltage.

Applying the formula for g_m:

$$g_m = 70/0.5 \text{ [mA/V]} = 140 \text{ mS}$$

The typical transconductance quoted for the ZVN3306A in data sheets is 150 mS.

The graph is not a straight-line graph. It slopes more steeply as v_{GS} increases. This means that g_m is not constant. It increases with increasing v_{GS}. The effect of this is to introduce distortion into the signal. The only way to minimise this is to keep the input signal amplitude small, so that the transistor is operating over only a small part of the curve. The smaller part of the curve is closer to a straight line and there is less distortion.

Self test

1 When v_{GS} is increased by 0.2 V, the increase of i_D through a given transistor is found to be 25 mA. Calculate g_m.

2 The graph opposite shows that when $v_{GS} = 4$ V, then $g_m = 0.164$ S. What increase in v_{GS} is needed to increase i_D by 15 mA?

Biasing

In the common-source amplifier, the gate of Q1 is held at a fixed (quiescent) voltage so that the transistor is passing a current, even when there is no signal. The gate is biased by R1 and R2, acting as a potential divider.

Example

Given that $V_{DD} = 15$ V, and the quiescent voltage is to be 4 V, calculate suitable values for R1 and R2.

Because MOSFETs require hardly any gate current, we can use high-value resistors in the potential divider (p. 25). If we decide on 330 kΩ for R2, then :

$$R_1 = \left(\frac{15}{4} \times 330k\right) - 330k = 907.5k$$

Self test

Find a pair of standard resistor values totalling about 1 MΩ for biasing a transistor to +3.8 V when the supply voltage is 18 V.

The nearest E24 value is 910 kΩ.

We have chosen biasing resistors of fairly high value because we want the amplifier to have a high input resistance. This is an advantage because it does not draw large currents from any signal source to which it may be connected.

Coupling

Capacitor C1 couples the amplifier to a previous circuit or device (such as a microphone). Sometimes it is possible to use a directly wired connection instead of a capacitor. But making a direct connection to another circuit will usually pull the voltage at the gate of Q1 too high or too low for the transistor to operate properly. With a capacitor, the quiescent voltages on either side of the capacitor can differ widely, yet signals can pass freely across the capacitor from the signal source to the amplifier.

C1 and the biasing resistors form a highpass filter, which might limit the ability of the amplifier to handle low frequencies. We must select a value for C1 so that the filter passes all signals in the frequency range intended for this amplifier.

Example

We want to pass signals of 20 Hz and above. The -3 dB point (p. 46) of the amplifier is to be at 20 Hz. The resistance of R1 and R2 in parallel is 242 kΩ. Calculate the capacitance required, using the formula on p. 47:

$$C = \frac{1}{2\pi f R} = \frac{1}{2\pi \times 20 \times 242 \times 10^3} = 32.9 \times 10^{-9}$$

The nearest available value is 33 nF.

Output voltage

The transistor converts an input voltage (v_{GS}) into an output current (i_D). This current is converted to an output voltage by passing it through the drain resistor R3.

By Ohm's Law:

$$v_{OUT} = i_D R_3$$

Ideally the no-signal (quiescent) output of the amplifier should be halfway between 0 V and V_{DD}. This allows the output to swing widely in either direction without clipping or distortion. The value of R3 is chosen to obtain this.

Example

Given that the gate is held at 4 V, the v_{GS}-i_D curve shows that i_D is then equal to 210 mA. We must arrange that this current flowing through R3 causes a voltage drop of $V_{DD}/2$. The voltage drop required is 7.5 V:

$$R_3 = \frac{7.5}{0.21} = 35.7\ \Omega$$

> **Self test**
>
> If the quiescent i_D is 100 mA and V_{DD} is 12 V, what is the best value for the drain resistor?

A 36 Ω resistor is the nearest E24 value.

Output resistance

Current flowing to the output must pass from the positive line and through R3. In other words, the output resistance is equal to the value of the drain resistor R3. In this amplifier, R3 has a very low resistance. This is an advantage because it means that the amplifier is able to supply a reasonably large current to any circuit or device (such as a speaker) to which it is connected. It has low output resistance. If the circuit to which it is supplying current takes a large current, there is a relatively low drop of voltage across R3. It can supply a reasonably large *current* without an unduly large fall in output *voltage*.

There is a highpass filter at the output, formed by R3 and C2. Although this is the 'other way up' compared with the highpass filter on p. 44, it still acts as a highpass filter. In this filter, the low frequency signals are absorbed into the positive supply line instead of the 0 V line, but the effect is just the same.

To comply with the previous specification, the cut-off frequency still must be at 20 Hz. This may entail changing capacitor values.

Example

Using the same equation as before:

$$C = \frac{1}{2\pi f R} = \frac{1}{2\pi \times 20 \times 36} = 221 \times 10^{-6}$$

A 220 µF aluminium electrolytic capacitor is required.

Testing the amplifier

When a sinusoid signal of amplitude 100 mV and frequency 1 kHz is fed to the amplifier, a plot of the input and output signals looks like this:

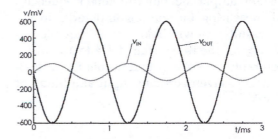

The output signal is a sinusoid of the same frequency, but with an amplitude of approximately 600 mV. The voltage gain of the amplifier is:

$$A_V = 600/-100 = -6$$

The negative sign indicates that the output is negative when the input is positive and positive when the input is negative. Another way of expressing this is to say that this is an **inverting amplifier**.

The expected voltage gain is calculated as follows, given that $g_m = 150$ mS and $R_3 = 36\ \Omega$.

If v_{IN} increases by an amount $v_{in} = 100$ mV, the corresponding increase in i_D is:

$$i_d = g_m \times v_{in} = 0.150 \times 0.1 = 0.015\ A$$

The current increases by 0.015 A, so the voltage across R3 increases by:

$$v_{out} = i_d \times R_3$$
$$= 0.015 \times 36$$
$$= 0.54\ V$$

> **Subscripts**
>
> We use capital-letter subscripts for absolute values, and small-letter subscripts for *changes* in values (see p. viii).

The calculation shows that, when v_{IN} increases, the voltage across R3 *increases* too. Since the positive end of R3 is connected directly to the +15 v line, the voltage at the other end of R3 must *fall* by 0.54 V. This explains why this is an inverting amplifier. Putting it another way, the output is 180° out of phase with the input.

The expected voltage gain is -0.54/0.1 = -5.4. This compares well with the gain obtained from the test measurements plotted on p. 65.

Frequency response

The frequency response of an amplifier is obtained by inputting sinusoidal signals of constant amplitude but different frequencies and measuring the output amplitude. In the test results shown below the frequency is varied over the range from 1 Hz to 100 MHz. The graph plots the output amplitude at each frequency, given constant input amplitude of 100 mV.

The amplifier maintains an amplitude of 600 mV for all frequencies between about 100 Hz and 10 MHz. Amplitude falls off below 100 Hz because of the action of the highpass filters at the input and output. The point where the -3 dB line crosses the curve occurs when the frequency is 20 Hz. This is as expected, because we calculated the values of C1 and C2 to have this effect.

The amplitude also falls off at very high frequencies, above about 10 MHz, because of capacitance effects within the transistor. The curve cuts the -3dB line at 9 MHz. The half-power response of the amplifier is thus found from 20 Hz to 9 MHz. We can say that the bandwidth of the amplifier is approximately 9 MHz.

Phase

The frequency response plot also provides information on the variation of phase with frequency.

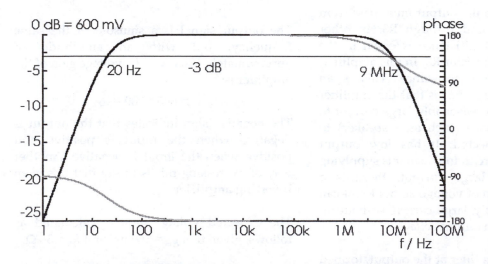

For most of the range, input and output signals are exactly 180° out of phase, because the amplifier is an i n v e r t i n g amplifier. Phase differences depart from 180° at the lowest and h i g h e s t f r e q u e n c i e s because of the effects of the capacitances of C1, C2 and the transistor.

This Bode Plot shows that the amplitude (black) of the common-source amplifier falls off at low and high frequencies. The plot also shows how phase (grey) varies with frequency.

The graph, known as a Bode Plot, is plotted on logarithmic scales (p. 46) so that a wide range of frequencies and amplitudes can be shown in a single diagram.

Activity — MOSFET c-s amplifier

Using the diagram on p. 63 as a guide, build a single-stage common-source amplifier.

Connect a signal generator to the input and an oscilloscope to the output. Find the voltage gain of the amplifier at 1 kHz, as explained under the heading 'Testing the amplifier'. Then try varying the frequency of the input signal in steps over the range 10 Hz to 100 MHz. Keep the input amplitude constant at 100 mV. Measure the output amplitude at each frequency. Plot a graph of output amplitude against frequency, using logarithmic scales.

Try altering R1 and R2 to see the effect of biasing. For example, make $R_2 = 100$ kΩ. Then make $R_2 = 1$ MΩ.

Try altering C1 or C2 to see the effect, if any, on the frequency response of the amplifier.

Activity —
MOSFET transconductance

Set up the circuit on p.63 for measuring transconductance. Use the MOSFET that you used in the Activity above. Use a power pack set at 15 V for V_{DD} and a variable power pack for v_{GS}. The graph on p. 63 shows a suitable range of values to try.

Plot a graph of your results and, from this, calculate two or three values of g_m at different values of v_{GS}.

Common-drain amplifier

The MOSFET amplifier at top right is known as a common-drain amplifier because the drain terminal is common to both input and output circuits. There is a resistor between the source and the output line and the output is taken from the source terminal. Biasing and coupling arrangements are the same as for the common-source amplifier, except that R3 is bigger, for a reason that will be explained later. This means that the value of C2 needs to be re-calculated if we still want to provide a highpass filter with cut-off point at 20 Hz.

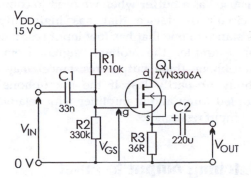

A common-drain MOSFET amplifier has unity voltage gain but high current gain.

Voltage gain

When the output voltage is v_{OUT}, the drain current is given by:

$$i_D = v_{OUT}/R_3$$

From the definition of transconductance we know that:

$$i_d = g_m v_{gs}$$

It can be seen in the circuit diagram that the voltage at the gate of Q1 equals v_{IN} and the voltage at the source equals v_{OUT}. Therefore the gate-source voltage, v_{GS}, is equal to their difference, and:

$$i_D = g_m(v_{IN} - v_{OUT})$$
$$v_{OUT}/R_3 = g_m(v_{IN} - v_{OUT})$$
$$\text{voltage gain} = \frac{v_{OUT}}{v_{IN}} = \frac{R_3 g_m}{1 + R_3 g_m}$$

It is clear from the equation that the voltage gain is less than 1. If R_3 is appreciably greater than $1/g_m$, the '1' in the denominator may be ignored and the equation approximates to:

$$\text{voltage gain} = 1$$

Because the amplifier has unity voltage gain, it is called a **voltage follower**. Its usefulness depends on the high input resistance (which may be several hundred kilohms) and its low output resistance (which may be only a few hundred ohms).

It is used as a **buffer** when we need to connect a circuit or device that has high output resistance to one that has low input resistance. For example, the voltage signal from a microphone (high output resistance) may be wholly or partly lost if the microphone is coupled to an audio amplifier with relatively low input resistance.

Matching output to input

At (a) below, v_S represents the piezo-electric or magnetic transducer in the microphone that converts sound energy into electrical energy. R_{OUT} represents the output resistance of the microphone unit. The microphone is connected to a audio amplifier that has input resistance R_L.

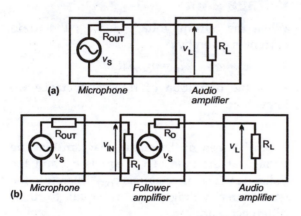

(a)   *Microphone*   *Audio amplifier*

(b)   *Microphone*   *Follower amplifier*   *Audio amplifier*

> In (a) the microphone is not able to supply enough current to drive the audio amplifier. In (b) we use a voltage follower amplifier to match the output resistance of the microphone to the input resistance of the audio amplifier.

If R_L is less than R_{OUT}, the voltage drop across R_L is less than that across R_{OUT}. In other words, most of the signal voltage is 'lost' in R_{OUT} and only a fraction of it appears across R_L, ready to be amplified. Only a reduced signal can eventually appear from the amplifier.

In (b) a follower amplifier is connected as a buffer between the microphone and the audio amplifier.

The follower amplifier has high input resistance R_I and low output resistance R_O. Because R_I is much greater than R_{OUT}, most of v_S appears across R_I and only a small part of it is 'lost' across R_{OUT}. We now have transferred most of the original signal to the follower amplifier. Here its voltage is not amplified, so we may still represent the signal by v_S.

Now the resistance through which the signal has to pass is R_O, which is very much *smaller* than R_L. This being so, only a small portion of v_S is 'lost' across R_O and most of it appears across R_L. Most of the signal from the microphone has been transferred through the follower amplifier to the audio amplifier, where it can be amplified further and perhaps fed to a speaker.

This technique for using a follower amplifier for transferring the maximum signal is known as **impedance matching**.

Current gain and power gain

Although the voltage gain of a MOSFET follower amplifier is only about 1, it has high current gain. The input current to the insulated gate is only a few picoamps, while the output current is measured in milliamps or even amps. So the current gain of MOSFET voltage amplifiers is very high. It is not possible to put an exact figure on the gain because the current entering the gate is a leakage current, and may vary widely from transistor to transistor.

Because current gain is high and power depends on current multiplied by voltage, the power gain of MOSFET amplifiers is very high.

Activity — MOSFET c-d amplifier

Using the circuit on p. 67 as a guide, build a single-stage MOSFET common-drain amplifier. Use the component values quoted in the diagram. Connect a signal generator to its input and an oscilloscope to its output. Observe the input and output signals. Find the gain of the amplifier at 1 kHz.

Next, try varying the frequency of the signal in steps over the range 10 Hz to 100 MHz. Keep the amplitude at 100 mV. Measure the output amplitude at each frequency.

Plot a graph of output against frequency. It is best to use a logarithmic scale for frequency, as on p. 66. There is no need to convert the output amplitudes to the decibel scale.

Try increasing the amplitude of the input signal and observe the amplitude of the output signal. Find out what is the largest input amplitude that this circuit can follow without distortion.

Try altering R1 and R2 to see the effect of biasing. For example, make $R_2 = 100$ kΩ. Then make $R_2 = 1$ MΩ.

MOSFET applications

MOSFETs are used in ON-OFF transistor switches (pp. 9-10) and in amplifiers (this chapter). They are also used as analogue switches (top right). The analogue signal passes (in either direction) when the control input (gate) is high.

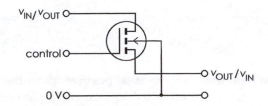

In the analogue switch, the substrate of the MOSFET is connected to 0 V.

A MOSFET can also act as a voltage-controlled resistor. The voltage output of an attentuator circuit varies with the control voltage. Attenuators are used to provide automatic gain control (AGC) in amplifiers. They are also used in amplitude modulation (AM) circuits.

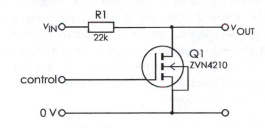

A MOSFET attenuator works like a potential divider in which one of the resistors (Q1) is variable.

Types of MOSFET

In **n-channel** MOSFETs (right) the drain current is conducted as electrons through n-type material. The transistor is made of p-type semiconductor with isolated electrodes of n-type connected to the source and drain terminals. Conduction from source to drain can not occur because conduction from p-type to n-type at the drain electrode can not occur. This is because the p-n junction there acts as a reverse biased diode.

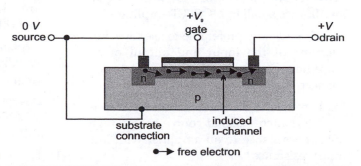

The effect of the field at the gate is to create a channel of n-type silicon connecting the two existing n-type regions.

Types of MOSFET (continued)

When the gate is at a positive potential, electrons are attracted toward the region of the gate. This creates a continuous zone of n-type electrons joining the two electrodes. Now the conduction path is completely n-type, as shown in the diagram, and flow of electrons from source to drain can occur.

The greater the potential of the gate, the more electrons are attracted and the lower the resistance to the flow of drain current.

In **p-channel** MOSFETs, the current is conducted as holes through p-type material. The transistor is made of n-type material, with p-type electrodes. A negative potential at the gate repels electrons from that region. This creates a continuous region of p-type material joining the electrodes and a 'current' of holes can flow.

In either type of MOSFET there is only one type of charge carrier (electrons or holes), which gives these transistors the description **unipolar**. Both n-type and p-type MOSFETs work by the creation of a channel when a suitable potential is applied to the gate. These MOSFETs are therefore described as **enhancement** MOSFETs.

It is possible to manufacture n-channel MOSFETs which have n-type substrate. A conductive channel is present in these when the gate and source are at the same potential. The channel is *reduced* in width as the gate is made more negative of the source and electrons are repelled from its vicinity. These transistors are known as **depletion** MOSFETs. They are rarely used and are not commonly listed in suppliers' catalogues.

Questions on MOSFET amplifiers

1 Explain what is meant by trans-conductance. How is the output current of a MOSFET converted to an output voltage in a common-source amplifier?

2 Why is it important to keep signal amplitudes small in MOSFET amplifiers?

3 Why is it usually preferable to use capacitor coupling at the input and output of an amplifier, rather than use wired connections?

4 Explain how a common-drain amplifier is used to match a signal source with a high output resistance to an amplifier with low input resistance?

5 Explain what the -3 dB point means.

6 What is the advantage of the high input resistance of MOSFET amplifiers? Give two practical examples of this.

7 Why do we usually try to make the quiescent output voltage of an amplifier equal to half the value of the supply voltage?

8 Outline the practical steps in testing the frequency response of an amplifier. Why are the results usually plotted on logarithmic scales?

9 Explain why the amplitude of an amplifier usually falls off at low and high frequencies.

10 Design an amplifier using an n-channel MOSFET with operating voltage 10 V, and able to amplify signals of 200 Hz and over.

11 The supply voltage of a common-drain MOSFET amplifier is 9 V. The gate is biased to 4 V by a pair of resistors. The amplifier is to operate at frequencies of 100 Hz and over. Calculate suitable values for the input coupling capacitor and resistors.

Multiple choice questions

1 A MOSFET is described as a unipolar transistor because:

    A current flows through it in only one direction.

    B it may explode when connected the wrong way round.

    C it has only one type of charge carrier.

    D the channel is made of n-type material

2 A follower amplifier is often used as a buffer between:

    A a high resistance output and a low resistance input

    B a low resistance output and a high resistance input

    C a high resistance output and a high resistance input

    D a low resistance output and a low resistance input.

3 A follower amplifier has a gain of:

    A approximately 100

    B exactly 1

    C slightly less than 1

    D slightly more than 1.

4 The total parallel resistance of resistors of 56 kΩ and 120 kΩ is:

    A 38.2 kΩ

    B 176 kΩ

    C 64 kΩ

    D 6720 kΩ.

5 The advantage of using a capacitor to couple an amplifier to a signal source is that:

    A it stabilises the gain

    B the amplifier is not affected by the quiescent output voltage of the source

    C the capacitor acts as a highpass filter

    D the capacitor filters out noise from the signal.

6 In p-channel MOSFETs the charge carriers flow from:

    A drain to source

    B source to gate

    C source to drain

    D negative to positive

7 For a MOSFET to conduct, its gate-source voltage must be:

    A negative.

    B above 3 V.

    C greater than the threshold.

    D 0.7 V.

8 A common-source amplifier:

    A is an inverting amplifier.

    B has unity gain.

    C has high output resistance.

    D has low current gain

9 The function of the drain resistor of a common-drain amplifier is to:

    A limit the output current.

    B amplify the signal.

    C convert the drain current to a voltage.

    D act as a low-pass filter.

10 The resistor and capacitor at the output of a common-drain amplifier act as a:

    A potential divider.

    B low-pass filter.

    C current to voltage converter.

    D high-pass filter.

Extension questions

1 Describe the structure and action of an n-channel enhancement MOSFET.

2 Design a common-source amplifier based on a p-channel enhancement MOSFET. Component values are not required.

3 Describe how conduction occurs in (a) n-type and (b) p-type semiconductors.

4 Why is there a forward voltage drop acoss a p-n junction?

9 BJT amplifiers

There are three basic ways in which a bipolar junction transistor may be used as an amplifier: common-emitter amplifier, common-collector amplifier and common-base amplifier.

These each have very different attributes and applications. We look at the circuits of the first two of these and also a few other types of BJT amplifier.

Common-emitter amplifier

The BJT amplifies current. In the simple common-emitter amplifier below, based on an npn BJT, a small current i_B flows through R1 into the base of Q1. Because of transistor action, this causes a much larger current i_C to flow in at the collector of Q1. The currents combine and flow out of Q1 at the emitter.

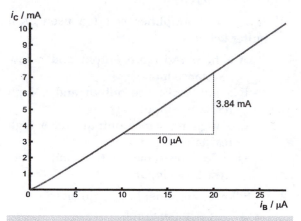

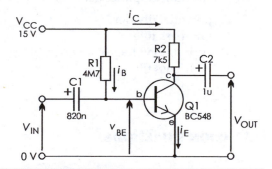

The graph of output current against input current of a BJT is almost a straight line. Compare this with the graph on p. 63, where the gradient of the curve increases with increasing input voltage.

In a common-emitter BJT amplifier the emitter terminal is common to both the input and output circuits. Compare with the MOSFET common-source amplifier on p. 63.

The emitter current i_E is:

$$i_E = i_B + i_C$$

These currents are shown on the circuit diagram above.

This one has two current meters, because both the input and output of a BJT are currents. Note that the meter for measuring i_B is a microammeter, while the meter for i_C is a milliammeter. This is because i_C is always much larger than i_B.

We can measure how much larger by plotting the graph of i_C against i_B This shows that, if i_B is increased by 10 μA, i_C increases by 3.84 mA.

Current gain

The relationship between i_B and i_C is known as the **forward transfer characteristic.** This graph is plotted by using a test circuit like that on the right. Compare this with the test circuit on p. 63, used for measuring the transconductance of MOSFETs.

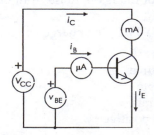

This is the kind of circuit used for measuring the forward transfer characteristic of a BJT.

The ratio between the two (or the slope of the curve) is known as the **small signal current gain**, symbol h_{fe}. In this example:

$$h_{fe} = 3.84/10 \ [\text{mA}/\mu\text{A}] = 384$$

The gain for the BC548 is often listed as 400 (but see below). Note that the graph is almost a perfect straight line.

Self Test

What is the h_{fe} of a BJT if i_C increases by 2.06 mA when i_B is increased by 5 μA?

Because the curve is almost a straight line we can measure gain in a slightly simpler way. Reading the value of i_C on the milliammeter when i_B = 20 μA, we find i_C = 7.68 mA. Now we calculate the **large signal current gain**, symbol h_{FE}:

$$h_{FE} = 7.68/20 \ [\text{mA}/\mu\text{A}] = 384$$

As is expected from the geometry of a straight-line curve, the gains are equal. The difference between h_{fe} and h_{FE} is theoretical, and rarely practical.

Voltage output

In the CE amplifier, the function of R2 is to convert the current i_C into a voltage, according to Ohm's Law. The values give a quiescent collector current i_C of 1 mA, which is a typical value for a CE amplifier. R2 is chosen so that the voltage drop across it is equal to half the supply voltage at this current. This allows the output voltage to rise and fall freely on either side of this half-way voltage without distortion.

Example

Current = 1 mA, voltage drop = 7.5 V, so:

$$R_1 = 7.5/0.001 = 7500$$

R_1 should be 7.5 kΩ.

Biasing

A suitable size for the collector current in a common-emitter amplifier is 1 mA. This minimises noise (p. 242). The value of R1 is often chosen to provide a base current sufficient to produce such a collector current.

Example

Current gain = 380, and i_C = 1 mA, so:

$$i_B = 1/380 \ [\text{mA}] = 2.63 \ \mu\text{A}$$

Assuming that the base-emitter voltage drop v_{BE} is 0.7 V, the voltage drop across R1 is 15 – 0.7 = 14.3 V. The resistance of R1 is:

$$R_1 = 14.3/2.63 \ [\text{V}/\mu\text{A}] = 5.44 \ \text{M}\Omega$$

We could use a 5.6 MΩ resistor. But to allow for the gain of the transistor to be a little less than 380, make i_B a little larger by making R1 a little less. We decide on 4.7 MΩ.

Base-emitter voltage

v_{BE} is usually between 0.6 V and 0.7 V for a silicon transistor.

This illustrates one of the problems with such a simple circuit. The correct value of R1 depends on the gain of the transistor. But transistors of the same type vary widely in their gain. Any given BC548 may have h_{fe} between 110 and 800. When building this circuit it is almost essential to match R1 to the particular transistor you are using.

Emitter resistance

From the circuit diagram, it looks as if this circuit has a very high input resistance, the resistance of R1. But, unlike the gate of a MOSFET, the base of a BJT is *not* insulated from the body of the transistor. Current i_B flows into the base, through the emitter layer and out by the emitter terminal. Along this path it encounters resistance, typically about 25 Ω. This is the emitter resistance r_e, which can be thought of as a resistor inside the transistor.

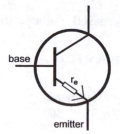

When current flows from the base of a BJT to the emitter, it is subject to the emitter resistance.

In the way it affects current entering the base, this resistance *appears* to be h_{fe} times its actual value.

Example

If h_{fe} is 400, r_e appears to be 400 × 25 = 10 kΩ. In comparison with this, the value of R1 in parallel with r_e can be ignored.

The amplifier has a low input resistance of only 10 kΩ.

Frequency response

As with the MOSFET amplifiers, C1 and R1 form a highpass filter. The value of C1 is chosen to pass all signals above 20 Hz. The values of R2 and C2 are chosen to pass the same frequencies on the output side.

Voltage gain

We will work through a rough calculation to find the voltage gain of the amplifier.

Example

If the input voltage v_{IN} is increased by a small amount, say 10 mV, this increases the base current i_B. If the input resistance is 10 kΩ, the increased current is:

$$i_b = 10/10 \ [\text{mV/k}\Omega] = 1 \ \mu\text{A}$$

Given that h_{fe} is 400, the increased current through R3 is:

$$i_c = 400 \times i_b = 400 \ \mu\text{A}$$

This produces an increased voltage drop across R3:

$$v_{out} = -400 \times 7.5 \ [\mu\text{A} \times \text{k}\Omega] = -3 \ \text{V}$$

The voltage gain of this amplifier is:

$$A_v = -3/10 \ [\text{V/mV}] = -300$$

The gain is 300, and the negative sign indicates that the signal is inverted.

There are several approximations in this calculation but the result agrees well with a practical test, which showed a voltage gain of −250.

Improving stability

The main disadvantage of the amplifier shown in p. 72 is its lack of stability. Its performance depends on the value of h_{fe}, which varies from transistor to transistor. Also, h_{fe} is affected by temperature. The amplifier may be operating in a particularly hot or cold environment which would make its performance even more variable. Even in a room at a comfortable temperature, the temperature of the amplifier increases after it has been run for a while.

The circuit is improved if we connect the positive end of R1 to the collector of Q1. The positive end of R1 is now held at *half* the supply voltage, so we halve its resistance to obtain the same i_B as before. This does not affect the amplification but has an effect on stability.

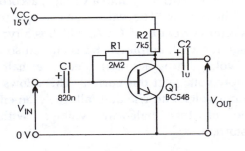

Stability is improved when R1 connects the collector to the base.

To see how this works, suppose that for some reason h_{fe} is higher than was allowed for in the design. Perhaps the transistor has a higher than average h_{fe} or perhaps h_{fe} has been raised by temperature. If h_{fe} is higher than usual, i_C is *larger* than usual. If i_C is larger, the voltage drop across R2 is greater. This lowers the voltage at the collector and so lowers the voltage at the positive end of R1. This reduces i_B, which in turn *reduces* i_C.

In summary, a change which increases i_C is countered by a reduction in i_C. The result is that i_C tends to remain unaffected by variations in h_{fe}. The gain of the amplifier is stable.

Further improvement in stability is obtained by biasing the base with two resistors, instead of only one:

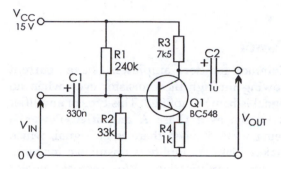

Biasing the base with a voltage divider helps stabilise the common-emitter amplifier.

R1 and R2 act as a potential divider to provide a fixed quiescent voltage, independent of h_{fe}. We have also introduced an additional improvement, the **emitter resistor**, R4. This is not to be confused with the emitter *resistance*, discussed on p. 73.

The emitter resistance and emitter resistor are in series but the emitter resistance is usually about 25 Ω and may be ignored when the emitter resistor has a value of 1 kΩ or more. If i_C is 1 mA, a convenient value for R4 is 1 kΩ, which produces a voltage drop of 1 V in the quiescent state. Now that the emitter voltage is 1 V, the base voltage must be 0.7 V more than this, to maintain the required v_{BE} of 0.7 V. The values of R1 and R2 are calculated so as to produce 1.7 V at the base.

Input resistance

The input resistance of the amplifier is the total of three resistances in parallel: R1, R2, and R4.

As with emitter resistance, a current flowing into the base is affected by h_{fe} times the resistance between the base and the 0V line. Now that there is a 1 kΩ resistor there, we can ignore r_e. The resistance of R4, as seen from the base, is h_{fe} times 1 kΩ, that is, 400 kΩ. The total of these three resistances in parallel is 27 kΩ. This gives the amplifier a moderate input resistance, but one which is far less than that of MOSFET amplifiers.

Given the new value of the input resistance, we can calculate a new value for C1.

Voltage gain

In this discussion, v_{in} means 'a small change in v_{IN}', and the same with other symbols.

In the amplifier on the left the signal passes more-or-less unchanged across C1:
$$v_{in} = v_b$$
where v_b is the base voltage. This stays a constant 0.7 V higher than v_E, the emitter voltage. We can write:
$$v_{in} = v_b = v_e$$

Applying Ohm's Law, we also know that:
$$i_e = v_e / R_4 \quad \text{and} \quad v_{out} = -i_c \bullet R_3$$

The negative sign indicates that we are dealing with a voltage *drop* across R_3.

From these equations we find that:
$$v_{out} = -i_c \bullet R_3 = -i_e \bullet R_3 = -v_e \bullet R_3 / R_4 = -v_{in} \bullet R_3 / R_4$$

From this we obtain:

$$\text{voltage gain} = v_{out} / v_{in} = -R_3 / R_4$$

The result shows that the voltage gain depends only on the values of the collector and emitter resistors. The voltage gain is independent of h_{fe}, so *any* transistor, either of the same type or of a different type, will produce the same voltage gain. Voltage gain is also virtually independent of temperature. The amplifier has high stability, but at the expense of low gain.

Example

In the amplifier with R_3 = 7.5 kΩ and R_4 = 1 kΩ, the expected voltage gain is:

$$-7.5/1 \ [kΩ/kΩ] = -7.5$$

A test on this circuit showed an actual voltage gain of –7.25.

Frequency response

With the capacitor values given in the diagram on p. 75, the frequency response of the circuit shows the lower cut-off point to be at 20 Hz, as calculated. In the higher frequencies, the cut-off point is approximately 11.7 MHz. This bandwidth of 11.7 MHz is satisfactory for many applications.

An upper cut-off point of 11.7 MHz may be essential for an amplifier in a radio receiver, but it is unnecessary for an audio amplifier. The human ear is not sensitive to frequencies higher than about 20 kHz. In Chapter 27, we show that there are disdvantages in amplifying frequencies outside the required range.

Bypass capacitor

The use of an emitter resistor (R4) gives improved stability but reduced gain. If required, the gain can be restored by wiring a high-value capacitor across R4. This holds the emitter voltage more or less constant, so the calculations on p. 75 do not apply. Without the capacitor, the voltage at the emitter rises and falls with the signal. This provides negative feedback.

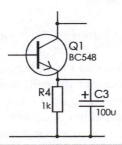

The addition of a bypass capacitor holds the emitter voltage steady and maintains voltage gain.

For example as i_B rises (tending to increase v_{BE}), i_C rises, and the emitter voltage rises. This tends to decrease v_{BE}, which decreases i_C and counters the rise in emitter voltage. Another way of looking at this is to say that the capacitor shunts the signal at the emitter through to the ground. For this reason C3 is called a **bypass capacitor**. With this capacitor in place, the voltage gain of the amplifier is about 280. The lower cut-off point is raised to 130 Hz, so bandwidth is slightly reduced.

Power

Common-emitter amplifiers have current flowing through the transistor even when no signal is being amplified. This type of amplifier is known as a **Class A amplifier**. Power is being wasted when there is no signal, which makes Class A amplifiers unsuitable for high-power amplification. We discuss power amplifiers in Chapter 15.

Activities – BJT amplifier

1 To measure the forward transfer characteristic of a BJT, set up the circuit on p. 58. V_{CC} is a fixed voltage source of 15 V DC, v_{BE} is a variable voltage source. Adjust the v_{BE} to produce a range of base currents, from zero to 25 μA in steps of 5 μA. Measure the corresponding values of i_C. Plot i_C against i_B and calculate the small signal current gain.

2 Investigate the action of a BJT common-emitter amplifier. Set up the circuit on p. 75. Connect a signal generator to its input and an oscilloscope to its output. Using a 100 mV sinusoidal signal, find the voltage gain of the amplifier at 1 kHz and other frequencies.

3 Vary the values of R3 and R4 to confirm that these determine the voltage gain of the amplifier. Try this with several different transistors of known h_{fe} to show that voltage gain is independent of h_{fe}.

4 Investigate the effect of the bypass capacitor.

Common-collector amplifier

The common-collector amplifier below has an emitter resistor but the collector is connected directly to the positive rail. The base is biased by two resistors. Assuming that the quiescent emitter current is 1 mA, the voltage across the emitter resistor R3 is 7.5 V, bringing the output to exactly half-way between the supply rails. To provide for a v_{BE} of 0.7 V, we need to hold the base at 8.2 V. The values of R1 and R2 are calculated to provide this.

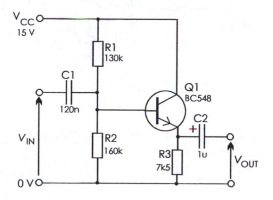

The voltage output of a common-collector amplifier is always approximately 0.7 V lower than its input.

Voltage gain

We calculate the voltage gain of the amplifier by considering what happens when the input voltage v_{IN} changes by a small amount v_{in}. The change in v_{IN} is carried across C1 to the base of Q1.

The change in base voltage is given by:

$$v_b = v_{in}$$

As the base voltage changes, the drop (v_{BE}) across the base-emitter junction remains constant. Therefore, the emitter voltage v_E changes by the same amount, which is v_b. This change is transferred across C2 to the output terminal.

For all such small changes in voltage we can say that:

$$v_{out} = v_e = v_b = v_{in}$$
$$\text{voltage gain} = v_{out}/v_{in} = 1$$

A gain of 1 is often referred to as **unity gain**.

The amplifier is a non-inverting amplifier with unity voltage gain. The output exactly follows all *changes* in the input, but is 0.7 V lower because of the drop across the base-emitter junction. If the signal is large enough, this drop can be ignored. Because the output follows the input, we refer to this kind of amplifier as a **voltage follower**. It is also known as an **emitter follower**.

The input resistance is high, being equal to the total resistance R_1, R_2, and $h_{fe} \times R_3$. With the values shown, the input resistance is 70 kΩ. The output resistance is the value of R3, which in this case is 7.5 kΩ. The output resistance is only about one tenth of the input resistance, making the amplifier suitable as a buffer (see p. 68).

Current and power gain

Because input current is small (due to the high input resistance) and output current is moderately large, this amplifier has high current gain and high power gain.

Frequency response

The frequency response is very good. In the case of the amplifier on the left, the value of C1 is chosen to produce a highpass filter with lower cut-off point at 20Hz. The upper cut-off point is at 128 GHz. The reason for this high cut-off point is that a transistor with its collector connected directly to the positive rail is not as subject to the effects of capacitance as one which is connected through a resistor. A common-collector amplifier can be said to have a wide bandwidth.

Darlington pair

A Darlington pair consists of two BJTs connected as below. The emitter current of Q1 becomes the base current of Q2. The current gain of the pair is equal to the product of the current gains of the individual transistors.

Example

If h_{fe} = 100 for each transistor in a Darlington pair, the gain of the pair is $100 \times 100 = 10\,000$.

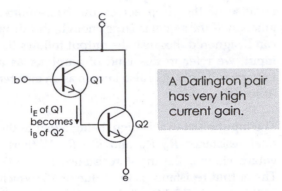

A Darlington pair has very high current gain.

A Darlington pair may be made up by wiring two individual transistors together. The two transistors may also be manufactured as a unit on a single silicon chip. This has three terminals, collector, base and emitter. The chip is enclosed in a normal three-wired package looking just like an ordinary transistor. When the pair is made from individual transistors, they need not be of the same type. Very often Q1 is a low-voltage low-current type while Q2 is a power transistor.

> **Self Test**
>
> A Darlington pair is made up of a BC548 with h_{fe} = 400 and a TIP31A (a power transistor) with h_{fe} = 25. What is the gain of the pair?

A Darlington may replace the single transistor the BJT common-emitter and common-collector amplifiers. This gives increased sensitivity to small base currents and, in the case of the common-emitter amplifier may result in increased gain.

Darlingtons may also be used as switching transistors, as in Chapter 1. They require only a very small current to trigger them.

The circuit below is a practical example of a Darlington pair being used as a switching transistor.

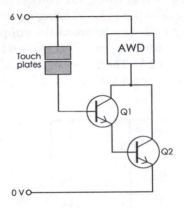

Using a Darlington pair in a touch-switch to operate a buzzer.

Two metal plates about 1 cm $\times$ 2 cm are mounted side by side with a 0.5 mm gap between them. When a finger touches the plates so as to bridge the gap, a minute current flows to the base of Q1. This current is only a few tens of microamperes, but the pair of transistors has a current gain of about 160 000. The resulting collector current through Q2 is sufficient to power the solid-state buzzer.

Activity — Darlington switch

Set up a Darlington pair as a switch controlling an audible warning device (see diagram above). A pair of BC548 transistors are suitable. Alternatively, use a Darlington pair transistor such as an MPSA13.

When the circuit is working, insert a microammeter and a milliammeter into it to measure the current through the finger and the current through the AWD.

What is the current gain of your circuit?

Differential amplifier

This differential amplifier (below) is also known as a **long-tailed pair**. A differential amplifier has two inputs v_{IN+} and v_{IN-}. The purpose of the amplifier is to amplify the voltage *difference* between its inputs. When it does this, we say that it is operating in the **differential mode**.

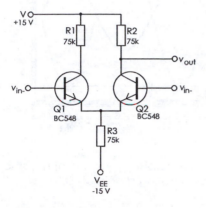

A differential amplifier has two inputs and one or two outputs. The second output (not shown here) is taken from the collector of Q1.

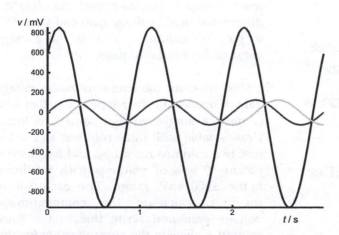

With the differential amplifier in differential mode, the output (black curve) is proportional to the difference between the non-inverting input (dark grey curve) and the inverting input (light grey curve). The inputs have 5 mV amplitude and are plotted on a 25 times scale.

To demonstrate this, two sinusoid signals that are 90° out of phase are fed to the two input terminals. The output signal is taken from the collector of Q2. At any instant:

$$v_{OUT} = (v_{IN+} - v_{IN-}) \times \text{differential-mode voltage gain}$$

Measurements on the graph (below, left) show that the differential voltage gain is approximately 129.

If we feed the *same* signal to both inputs, we are operating the amplifier in **common-mode**. We obtain the result shown on p. 80. The amplitude of the output signal is only half that of the input signal and it is *inverted*:

$$v_{OUT} = v_{IN} \times \text{common-mode gain}$$

The common-mode gain is −0.5.

The differential amplifier has high differential-mode gain but very low common-mode gain. It is suited to measuring voltage differences when the voltages themselves are subject to much larger changes affecting both voltages equally.

In medical applications we may need to measure very small differences of voltage between two probes attached to different parts of the body surface. At the same time both probes are subject to relatively large voltage changes resulting from electromagnetic fields from nearby mains supply cables and from other electrical equipment in the same room.

The differential amplifier is able to amplify the small voltage differences between probes but is much less affected by the larger voltage swings affecting both probes. It is able to *reject* the common-mode signals. We express its ability to do this by calculating the **common-mode rejection ratio**:

$$\text{CMRR} = \frac{\text{diffl. mode voltage gain}}{\text{common mode voltage gain}}$$

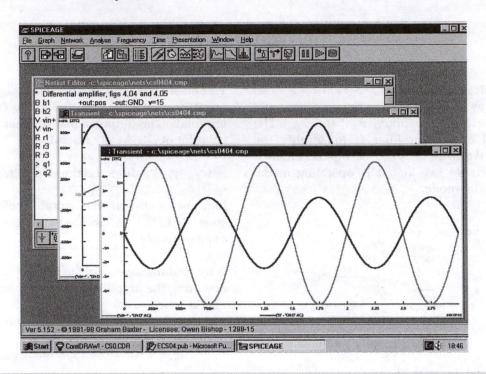

A simulation of the differential amplifier in common mode. The output (black curve) has smaller amplitude than the input (grey curve) and is inverted, demonstrating a gain of –0.5.

Example

We use the differential mode voltage gain as taken from the graph. We ignore negative signs when calculating the CMRR:

The CMRR is 129/0.5 = 258.

The CMRR can also be expressed in decibels:

$$CMRR = 20 \times \log_{10}(129/0.5) = 48.2 \text{ dB}$$

The common mode rejection ratio is 258, or 48.2 dB.

Activities — Differential amplifier

Build the differential amplifier (p. 79) on a breadboard or simulate it on a computer.

(a) First measure the **differential-mode voltage gain**. Connect a voltage source to each of the inputs and a voltmeter to the output. Draw a table with 5 columns. Head the first 3 columns v_{IN+}, v_{IN-} and v_{OUT}. Set the two voltage sources to two different voltages within the range ±100 mV.

Record the two voltages (v_{IN+} and v_{IN-}). Measure and record v_{OUT}. Repeat this for 10 pairs of input voltages. For each pair calculate ($v_{IN+} - v_{IN-}$) and write this in the fourth column. For each pair calculate the differential-mode voltage gain and write this in the fifth column. Calculate the average value of this for the 10 pairs.

(b) Next measure the **common-mode voltage gain**. Connect the two inputs together and connect a single voltage source to them. Draw a table with three columns and in the first two columns record v_{IN} and v_{OUT} as you obtain 10 sets of readings with the input in the ±100 mV range. For each set of readings, calculate the common-mode voltage gain and write this in the third column. Calculate the average value for this for the 10 sets.

(c) Finally, calculate the **common-mode rejection ratio**. Using the two gains found in (a) and (b), calculate the common-mode rejection ratio. Express this in decibels.

Tuned amplifier

Tuned amplifiers are used at radio frequencies. Their essential feature is a tuned network, usually consisting of a capacitor in parallel with an inductor (p. 57). There are several ways in which the LC network may be built into the amplifier, but one of the commonest ways is to use it to replace the resistor across which the voltage output signal is generated. In a BJT common-emitter amplifier it replaces the collector resistor. In an MOSFET common-source amplifier, it replaces the drain resistor. Below is a basic common-emitter amplifier (p. 77) with its collector resistor R_3 replaced by an LC network. The transistor has been replaced by a high-frequency type.

The resonant frequency of the network is given by:

$$f_0 = \frac{1}{2\pi\sqrt{LC}}$$

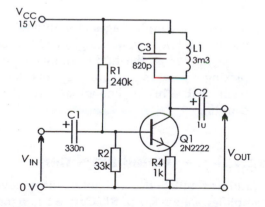

A tuned amplifier based on a simple common-emitter amplifier.

Example

The resonant frequency of the LC network is:

$$f_0 = \frac{1}{2\times\pi\times\sqrt{3.3\times820}} = 96.75 \text{ kHz}$$

$$[1/(\text{mH}\times\text{pF})]$$

The LC network is tuned to resonate at 96.75 kHz. If the amplifier is fed with a 10 mV signal in a range of frequencies from 95 kHz to 100 kHz, we can obtain the frequency response plot shown below. Looking at the curve nearest the front of the diagram, it can be seen that the output amplitude peaks sharply at just under 97 kHz, as predicted by the calculation above.

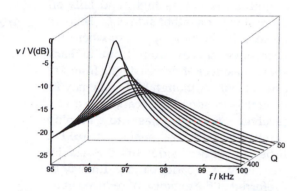

The frequency response of the tuned amplifier varies according to the Q of the inductor.

The reason for this peak is the fact that, at the resonant frequency, the capacitor-inductor network behaves as a pure resistance of high value. This is equivalent to replacing the network with a resistor equivalent to the one that was present as R3 in the amplifier on p. 77.

As the collector current flows through the network, a voltage is developed across it, and this is the output of the amplifier. At lower or higher frequencies the effective resistance of the network is not as great, either the capacitor or the inductor having low resistance. The voltage developed across it is less at lower or higher frequencies.

Because the resonating network has a high resistance (in theory, an infinite resistance, but less than this in practice due to losses in the capacitor and inductor) the voltage developed across it is high. In the plot above, the amplitude of the output signal at the resonant frequency is 15 V. This is a voltage gain of 1 500 times.

Quality factor

The output of the circuit varies mainly with the properties of the inductor, including the resistance of its coil. The **quality factor** of the inductor is given by $Q = 2\pi f L / r$, where r is the resistance of the coil. If the inductor has a high Q (say, 400), the amplifier output is high but falls off sharply on either side of f_0.

If Q is low (say, 50) the peak output amplitude is not as high, and falls off more slowly on either side of f_0. This can be seen in the plot opposite, where the successive curves from front to back show the effect of decreasing Q from 412 down to 50. Although high gain may be desirable in some applications, a radio receiver usually has to amplify sidebands with frequencies spaced a little to either side of the carrier frequency. A broader bandwidth may be preferred at the expense of reduced gain. This can be obtained by choosing an inductor with a suitable Q or by wiring a low-value resistor in series with the inductor.

As explained on p. 75, changing the value of the emitter resistor alters the voltage gain of the amplifier. This has an effect on frequency response. In the frequency responses plotted opposite we see the effect of varying R_3 over a range from $1\,\Omega$ to $1\,k\Omega$. Gain is highest when R_3 is $1\,\Omega$, with a broad bandwidth (13 kHz). At the other extreme, when R_3 is $1\,k\Omega$, gain is only 37 and the bandwidth is narrowed to 2 kHz.

Self Test

What are the reactances of C_3 and L_1 in the LC network when the frequency is (a) 96 kHz, (b) 96.75 kHz and (c) 98 kHz?

Advantages

The advantages of this amplifier for radio-frequency applications are:

- It has very **high gain**.

- Capacitance would reduce gain in a normal amplifier but, here, the stray capacitances in the transistor become, in effect, part of the network capacitance and help to bring about **resonance**. They are automatically allowed for.

- A tuned amplifier is **selective**. It is sensitive only to signals within its bandwidth.

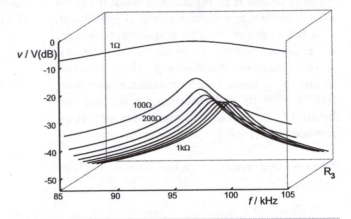

Changing the voltage gain of the amplifier by altering the emitter resistor has a marked effect on gain and bandwidth.

- It is easy to **couple** stages of a multi-stage tuned amplifier by winding a coil on the same former as the tuning inductor and picking up the signal for the next stage from that. This is known as inductive coupling.

Activities — Tuned amplifier

1 Build or simulate the common-emitter tuned amplifier shown on p. 81. Connect a signal generator to its input and an oscilloscope to its output. Set the signal amplitude to 10 mV. Vary the frequency over the range 10 kHz to 1 MHz and find the frequency (f_c) at which amplification is a maximum. Compare this with the resonant frequency of the loop formed by C3 and L1. Calculate Q for the inductor. Plot the frequency response for a range of frequencies on either side of the resonant frequency. Repeat, using a different capacitor and inductor for C3 and L1.

2) Use the tuned amplifier built for Activity (1) and find the bandwidth for 2 or 3 different values of R3. Find the resonant frequency (f_c) as above, and note the voltage amplitude of the output signal at that frequency.

Find the lower -3dB point. This is the frequency (lower than f_c) at which the amplitude is 0.7 times the amplitude at the resonant frequency.

Find the upper -3dB point. This is the frequency (higher than f_c) at which the amplitude is 0.7 times the amplitude at the resonant frequency.

The difference between the frequencies found in the previous two tests is the bandwidth.

Bipolar junction transistors

Extension Box 16

A bipolar transistor is a three-layer device consisting either of a layer of p-type sandwiched between two n-type layers or a layer of n-type between two p-type layers. These are referred to as **npn** and **pnp** transistors respectively.

The fact that conduction occurs through all three layers, which means that it involves both electrons (negative) and holes (positive) as charge carriers, is why these are sometimes called **bipolar transistors**. Their full name is **bipolar junction transistors (or BJT)** because their action depends on the properties of a pn junction, as is explained later. The diagram shows the 'sandwich' structure of an npn transistor, the most commonly used type.

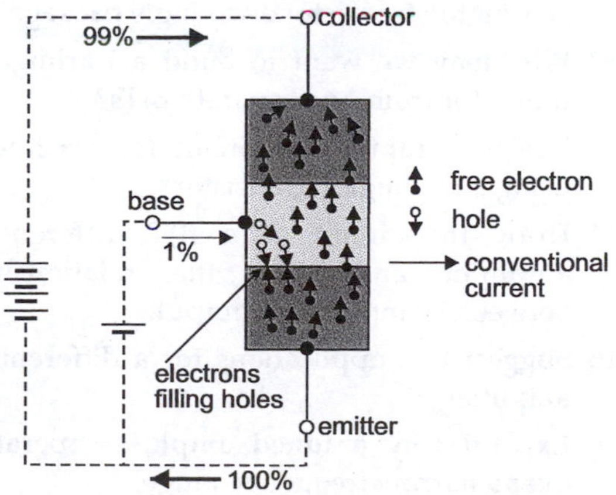

The three layers of the transistor are known as the **collector**, the **base** and the **emitter**. In effect, the transistor consists of two pn junctions (in other words, diodes) connected back-to-back. It would seem that it is impossible for current to flow from the collector to the emitter or from the emitter to the collector.

Whatever the direction of the pd, one or other of the p-n junctions is sure to be reverse-biased. This is where the features of the base layer are important:

● It is very thin (though not shown like that in the drawing).

● It is lightly doped, so it provides very few holes.

When the transistor is connected as in the diagram, the base-emitter junction is forward-biased. Provided that the base-emitter pd is greater than about 0.7 V, a base current flows from base to emitter. When describing the action in this way, we are describing it in terms of conventional current, as indicated by the *arrows* in the drawing.

What *actually* happens is that electrons enter the transistor by the emitter terminal and flow to the base-emitter junction. Then they combine with holes that have entered the transistor at the base terminal. As there are few holes in the base region, there are few holes for the electrons to fill. Typically, there is only one hole for every 100 electrons arriving at the base-emitter junction.

Conduction in a BJT is by electrons *and* holes.

[Continued on p. 84

BJTs — Extension Box 16 (cont)

The important result of the action is that the collector current is about 100 times greater than the base current. We say that there is a **current gain** of 100. The remaining 99 electrons, having been accelerated toward the junction by the field between the emitter and base, are able to pass straight through the thin base layer. They also pass through the depletion layer at the base-collector junction, which is reverse-biased. Now they come under the influence of the collector-emitter pd. The electrons flow on toward the collector terminal, attracted by the much stronger field between emitter and collector. They flow from the collector terminal, forming the collector current, and on toward the battery. In effect, the base-emitter p.d. starts the electrons off on their journey but, once they get to the base-emitter junction, most of them come under the influence of the emitter-collector pd.

If the base-emitter pd is less than 0.7 V, the base-emitter junction is reverse-biased too. The depletion region prevents electrons from reaching the junction. The action described above does not take place and there is no collector current. In this sense the transistor acts as a **switch** whereby a large (collector) current can be turned on or off by a much smaller (base) current. If the base-emitter p.d. is a little greater that 0.7 V, and a varying current is supplied to the base, a varying number of electrons arrive at the base-emitter junction. The size of the collector current varies in proportion to the variations in the base current. In this sense the transistor acts as a **current amplifier**. The size of a large current is controlled by the variations in the size of a much smaller current.

The structure and operation of pnp transistors is similar to that of npn transistors, but with polarities reversed. Holes flow through the emitter layer, to be filled at the base-emitter junction by electrons entering through the base.

Questions on BJT amplifiers

1 Explain how a small increase in the voltage input of a common-emitter amplifier produces a much larger increase in the output voltage.

2 What is the difficulty with using a single resistor to bias a common-emitter amplifier?

3 Explain what is meant by the small signal current gain of a BJT.

4 Describe the action of an emitter resistor in stabilising the common-emitter amplifier.

5 Briefly describe two ways in which negative feedback helps to stabilise the common-emitter amplifier.

6 Describe the action and applications of a common-collector amplifier.

7 Compare the common-emitter BJT amplifier with the common-source MOSFET amplifier.

8 Explain what is meant by negative feedback. Quote two examples of negative feedback in amplifiers.

9 A transistor has an h_{fe} of 110. By how much does the collector current increase when the base current is increased by 25 µA?

10 Describe a Darlington pair and state how to calculate its current gain.

11 What are the advantages of having a Darlington transistor as a single package?

12 Why may we want to build a Darlington transistor from two separate BJTs?

13 Design a rain-alarm circuit for washday, using a Darlington transistor.

14 Draw the circuit of a BJT differential amplifier and state the relationship between its inputs and output.

15 Suggest two applications for a differential amplifier.

16 Explain why a tuned amplifier operates over a narrow frequency range.

17 What are the advantages of a tuned amplifier for use at radio frequencies?

18 If the differential mode gain of a differential amplifier is 50 and the common mode gain is −0.25, calculate the CMRR.

Multiple choice questions

1 In a common-collector amplifier the output is taken from the:

    A collector
    B base
    C positive supply
    D emitter.

2 A transistor has an h_{fe} of 80. When the base current is increased by 15 μA, the collector current increases by:

    A 15 mA   C 1 mA
    B 1.2 mA   D 15 μA.

3 A common-emitter amplifier:

    A has a voltage gain of 1
    B has high output resistance
    C has low input resistance
    D in an inverting amplifier.

4 The voltage gain of a common-emitter amplifier with an emitter resistor depends on:

    A the value of the emitter resistor only
    B the value of the emitter resistance only
    C the values of the emitter and collector resistors
    D h_{fe}.

5 Compared with FET amplifiers, BJT amplifiers have:

    A higher input resistance
    B smaller voltage gain
    C better linearity
    D higher output resistance.

6 If the differential mode gain of a differential amplifier is 50 and the common mode gain is −0.25, the CMRR is:

    A 50.25   C 23 dB
    B 200   D 0.005.

7 The output of a tuned network shows a peak at the resonant frequency because:

    A the capacitor and inductor have equal reactance
    B the resonant network has high resistance
    C the inductor has high Q
    D resonance generates high voltages.

8 The transistors of a Darlington pair have current gains of 150 and 200. Their combined current gain is:

    A 30 000   C 400
    B 350   D 50.

9 The effects of stray capacitance are eliminated in a tuned amplifier because:

    A capacitance has no effect at high frequency
    B the capacitance acts as part of the resonant network
    C the capacitance is cancelled by the inductance
    A the transistor is connected so as to avoid capacitance effects.

10 The stabiliy of a common-emitter amplifier is improved if:

    A we use a high-gain transistor.
    B the base is biased with a resistor connected to the collector.
    C the base is biased with resistors connected to the positive supply and the 0 V line.
    D we use a high-power transistor.

11 A bypass capacitor connected in parallel with the emitter resistor of a common-emitter amplifier:

    A improves stability.
    B decreases gain.
    C increases bandwidth.
    D increases gain.

More on transistors

The Companion Site has more questions on MOSFETs and BJTs. There are Calculator windows to help you find the answers.

10 JFET Amplifiers

Junction field effect transistors have been replaced by MOSFETs as the most commonly used field effect transistors. If you are working for an examination, check the subject specification to find out whether or not JFETs are included in it.

Like MOSFETs, JFETS are unipolar transistors, made in n-channel and p-channel versions (see Extension Box 15, p. 69). In this chapter we describe an amplifier based on an n-channel JFET.

JFET amplifier

The JFET common-source amplifier shown below is very similar to the MOSFET amplifier described on p. 63 and operates in much the same way. The main difference is in the biasing of the gate.

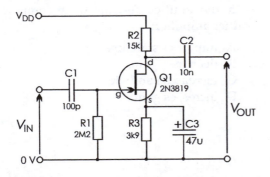

A JFET common-source amplifier is biased by a single pull-down resistor.

In JFETs the gate is insulated from the substrate of the transistor by being reverse-biased. It is held negative of the source, causing a depletion region to form between the gate and the substrate, in the same way that it is formed in a reverse-biased diode. The usual way of biasing the gate is to connect a single high-value resistor (R1) between the gate and the 0 V line. This holds the gate very close to 0 V.

We then make the source a few volts *positive* of the gate by wiring a resistor (R3) between the source and the 0 V line. Relative to the source, the gate is negative.

The threshold voltage of the 2N3819 is –4 V, so a suitable quiescent gate voltage is in the region of –2 V. The gate can be biased at any voltage between the threshold and about +0.5 V. Above 0.5 V, the junction at the gate becomes forward-biased, the depletion region breaks down and the gate is no longer insulated from the body of the transistor.

Example

In the circuit (left) the quiescent current through the transistor is 0.5 mA. The voltage drop across R3 is:

$$v_{GS} = -i_D R_3 = -0.0005 \times 3900 = -1.95 \text{ V}$$

In practice, the voltage across R3 varies with the signal, making v_{GS} vary too. As the signal increases, v_{GS} increases, which tends to partly turn off the transistor and so reduces the signal. This is **negative feedback**, which reduces the gain of the amplifier. A large-value capacitor C3 is used to stabilise the voltage at the source of Q1, so reducing the feedback and increasing the gain.

Looking at this another way, the capacitor passes the signal through from the source to the 0 V line, leaving only the quiescent voltage at the source. For this reason, the capacitor is called a **by-pass capacitor**.

Transconductance

The transconductance of JFETs is generally much lower than that of MOSFETs. For the 2N3819, the transconductance $g_m = 0.4$ mS. This compares with 150 mS for the ZVN3306A MOSFET (p. 13). We calculate the expected voltage gain as for the MOSFET common-source amplifier.

Example

If v_{IN} increases by an amount $v_{in} = 100$ mV, the corresponding increase in i_D is:

$$i_d = g_m \times v_{in} = 0.0004 \times 0.1 = 40 \text{ }\mu\text{A}$$

If the current through R2 increases by 40 μA, the voltage across R2 increases by:

$$v_{out} = i_d \times R_2 = 0.0004 \times 15\,000 = 0.6 \text{ V}$$

The voltage at the drain *falls* by 0.6 V.

$$\text{Voltage gain} = -0.6/0.1 = -6$$

The calculation assumes that there is no change in g_m due to variation in v_{GS}. As explained above, the gain is reduced by negative feedback if there is no by-pass capacitor.

Output resistance

The output resistance of a common-source amplifier depends on the value of the drain resistor, R2. Drain currents are usually small in JFET amplifiers. So we need a larger drain resistor to bring the quiescent output voltage to about half the supply voltage. The result is an output resistance (15 kΩ in this circuit) that is reasonably low, but not as low as that obtainable with a MOSFET amplifier.

Frequency response

The frequency response of a JFET amplifier is similar to that of a MOSFET amplifier, except that the amplitude begins to fall off above about 500 kHz instead of extending as far as 100 MHz. In other words, the JFET amplifier does not have such a high bandwidth as the MOSFET amplifier. This is because of the greater input capacitance of JFETs when compared with MOSFETs. The input capacitance of the 2N3819, for example, is 8 nF, compared with only 28 pF for the ZVN3306A.

Summing up, the JFET C-S amplifier shares with the MOSFET version the advantage of very high input resistance. Its disadvantages are that its output resistance is higher, its voltage gain is not as high, and it has a narrower bandwidth.

Activity — JFET amplifier

Using the diagram opposite as a guide, build or simulate a single-stage JFET common-source amplifier. Use the component values quoted in the diagram. Try varying the frequency of the input signal in steps over the range 10 Hz to 100 MHz. Keep the input amplitude constant at 100 mV. Measure the output amplitude at each frequency.

Plot a graph of output amplitude against frequency.

Try reducing the by-pass capacitor C3 to 10 μF and then to 1 μF and observe the effect on signal gain. Finally, remove C3 and test again.

Activity — Transconductance

Set up the circuit for measuring forward transfer characteristic (p. 63). Use this to measure the transconductance of the JFET that you used in the previous Activity. Use a power pack set to 15 V for V_{DD} and a variable power pack for v_{GS}. Connect the variable power pack with its negative terminal to the gate of the transistor and its positive terminal to the 0 V line (the line to which the source is connected).

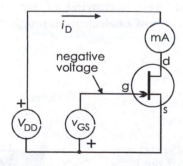

Vary v_{GS} over the range –10 V to 0 V.

Plot a graph similar to that on p. 47 and, from this, calculate two or three values of g_m at different negative values of v_{GS}.

Compare these values with those obtained for the MOSFET used in the Activity on pp. 66-67.

Junction field effect transistors

This name comes from the fact that its action depends on what happens at a p-n junction. The structure of a JFET is shown below. A bar of n-type semiconductor has metal contacts at each end. On either side of the bar is a layer of p-type material, the two layers being connected together by a fine wire.

If a pd is applied to the ends of the bar, a current flows along the bar. Because the bar is made from n-type material, the current is carried by electrons. We say that this is an **n-channel JFET**.

When referring to the potentials in this diagram we take the potential of the source to be zero. Electrons enter the bar at the **source** terminal (the reason for its name), and leave at the positive end of the bar, the **drain** terminal. The regions of p-type material are known as the **gate**. The gate regions form a pn junction with the n-type material of the bar. If the gate regions are at 0 V or a slightly more positive potential, they have no effect on the flow of electrons. With the gate negative of the bar, the junction is reverse biased and a depletion region (p. 6) is formed. The channel through which they flow is made narrower. In effect, the resistance of the bar is increased. The narrowing of the channel reduces the flow of electrons through the bar.

A JFET is always operated with the p-n junction reverse-biased, so current never flows from the gate into the bar. The current flowing into or out of the gate is needed only to change the potential of the gate. Since the gate is extremely small in volume, only a minute current (a few picoamps) is required.

A small change in potential of the gate, controls the much larger current flowing through the channel. This property of the transistor can be used in the design of amplifiers.

JFETs have many applications, particularly in the amplification of potentials produced by devices such as microphones which are capable of producing only very small currents. They are also useful in potential-measuring circuits such as are found in digital test-meters, since they draw virtually no current and therefore do not affect the potentials that they are measuring.

A JFET similar to that described above is manufactured from a bar of p-type material with n-type gate layers. This is a p-channel JFET. Current is conducted along the bar by holes and this is known as a **p-channel JFET**. In operation, the gate is made positive of the source to reverse-bias the p-n junction.

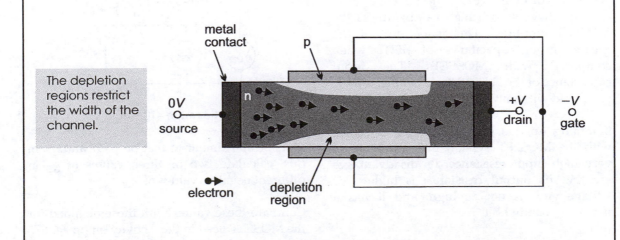

The depletion regions restrict the width of the channel.

metal contact

p

0V
source

n

+V
drain

−V
gate

electron

depletion region

Questions on JFETs

1 What are the main similarities and differences of the design and performance of MOSFET and JFET common-source amplifiers?

2 Why is it not possible to operate a JFET amplifier when the quiescent gate voltage is more than about 0.5 V?

3 Describe how to measure the transconductance of a JFET.

4 Explain the action of the by-pass capacitor in a JFET common-source amplifier.

5 Design a JFET circuit to switch on a buzzer when the light level falls below a set level. Use either an n-channel or a p-channel JFET.

6 Design a JFET common-source amplifier to operate on 9 V DC and to cut signals of frequency lower than 100 Hz.

Multiple choice questions

1 The operating gate voltage of an n-channel JFET is:

    A always positive
    B always negative
    C lower than –5 V
    D less than 0.5 V.

2 Compared with its MOSFET equivalent, the bandwidth of a JFET common-source amplifier is:

    A less
    B much wider
    C about the same
    D a little wider.

3 The output resistance of a common-source amplifier depends on:

    A the transconductance of the FET
    B the value of the output coupling capacitor
    C the value of the drain resistor
    D the gate-source voltage.

11 Operational amplifiers

Operational amplifiers are precision, high-gain, differential amplifiers. They were originally designed to perform mathematical *operations* in computers, but nowadays this function has been taken over by digital microprocessors. Operational amplifiers (or 'op amps' as they are usually called) are still widely used in many other applications. An op amp can be built from individual transistors and resistors but practically all op amps are manufactured as integrated circuits. Dozens of different types of op amp are available with various combinations of characteristics.

Terminals

All op amps have at least five terminals:

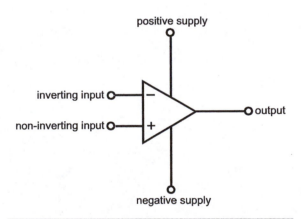

- **Positive and negative supply:** Most op amps run on a **dual supply (or split supply)**. Typical supply voltages are ±9 V, ±15 V or ±18 V. Some op amps run on low voltages such as ±1 V or ±2 V. Certain op amps that are capable of accepting input voltages close to the supply rails can also run on a single supply, such as 2 V, up to 36 V. The power supply terminals are often omitted from circuit diagrams to simplify the layout.

- **Inverting and non-inverting inputs:** These are the iputs to the first stage of the amplifier, which is a differential amplifier (see p. 65). In this book we refer to these inputs as the (–) and (+) inputs.

- **Output terminal**.

Amplifiers may also have two or three other terminals, including the **offset null** terminals.

Packages

Most op amps are available as a single amplifier in an 8-pin integrated circuit package, and practically all have the standard pinout shown below.

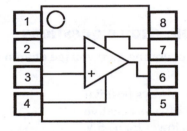

Op amp integrated circuits have these pin connections. A circular dimple indicates which is pin 1.

Pins 1 and 5 are used for other functions, depending on the type. Many op amps are also available with 2 or 4 amplifiers in a single 14-pin integrated circuit package. They share power supply pins and usually do not have terminals for special functions.

Differential output

Assuming that the op amp is working on a dual supply (the mid-rail voltage is 0 V), output is positive when the (+) input voltage exceeds the (-) input voltage. It is negative when the (-) input voltage exceeds the (+) input voltage. It is zero when inputs are equal.

Ideal op amp

An ideal op amp has the following features:

- Infinite voltage gain.
- Gain is independent of frequency.
- Infinitely high input resistance.
- Zero output resistance.
- Zero input voltage offset.
- Output can swing positive or negative to the same voltages as the supply rails.
- Output swings instantly to the correct value.

Practical op amps

All op amps fall short of these ideals. A more practical list is:

- Very high voltage gain. The gain without feedback (known as the **open-loop gain**) is of the order of 200 000.
- Gain falls with frequency. It is constant up to about 10 kHz then falls until it reaches 1 at the transition frequency, f_T. Typically, f_T is 1 MHz, but is much higher in some op amps.
- High input resistance. This is usually at least 2 MΩ, often much more.
- Low output resistance. Typically 75 Ω.
- Input voltage offset is a few millivolts (see below).
- The output voltage swings to within a few volts of the supply voltages (typically ±13 V for an amplifier run on ±15 V).
- Output takes a finite time to reach its correct value and may take additional time to settle to a steady value (see *Slew Rate,* below).

Input voltage offset

Because an op amp is a differential amplifier, its output should be 0 V when there is no difference between its inputs. In other words, when its inputs are at equal voltages.

In practical amplifiers, the output is 0 V when the inputs *differ* by a small amount known as the **input offset voltage**. Many op amps have a pair of **offset null** terminals. These are terminals 1 and 5 in the op amp shown below:

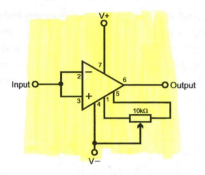

Offset null adjustment requires a variable resistor with its wiper connected to the negative supply or (with some op amps) to 0 V. Some op amps require a 100 kΩ variable resistor.

The offset null terminals allow the input offset voltage to be nulled, that is, made equal to zero. To do this, the same voltage is applied to *both inputs*, as in the diagram and the variable resistor is adjusted until the output is 0 V.

Slew rate

The maximum rate at which the output voltage can swing is called the **slew rate**. It is usually expressed in volts per microsecond. Slew rates range from a fraction of a volt to several hundred volts per microsecond, but a typical value is 10 V/μs. Op amps with the highest slew rates are the best ones for use at high frequencies.

When an op amp is amplifying a high frequency sinusoidal signal, the slew rate may be exceeded. For example, with 25 MHz signal, amplitude 100 mV, has an average slew rate of 10 V/μs. The effect is mainly due to capacitance within the op amp and the result is that the output voltage fails to keep up with the rapidly rising and falling input voltage.

The output waveform is then distorted into a triangular waveform. At the same time, its amplitude is considerably reduced.

Effect of frequency on gain

At DC and low frequencies the **open-loop gain** of an op amp is the value quoted in the data tables, often 200 000. However, the gain falls off at higher frequencies. This limits the gain attainable by closed-loop amplifier circuits at high frequencies.

The op amp is said to operate at 'full power' at all frequencies from 0 Hz (DC) up to the frequency at which the power output is half its DC power. This is the frequency at which output is 3 dB down on the open-loop maximum. From 0 Hz up to this frequency is the **full-power bandwidth**.

Beyond this frequency the gain falls at a rate such that the product of the gain and the frequency is constant. This constant is the **gain-bandwidth product**.

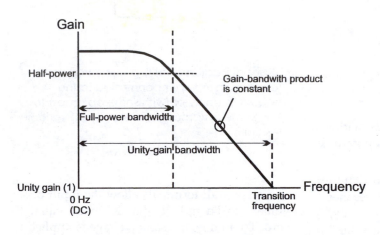

The gain of an op amp falls off steadily at frequencies beyond the full-power bandwidth. The graph is plotted on logarithmic scales for both frequency and gain.

Example:

If the open-loop gain is 20 at 100 kHz, it is 10 at 200 kHz, 5 at 400 kHz, and so on to a gain of 1 at 2 MHz. In every case:

Gain-bandwidth product = gain × frequency = 2 MHz

The frequency at which gain becomes 1 is known as the **transition frequency**, and is numerically equal to the gain-bandwidth product. The data table on p. 99 lists the full-power bandwidth and gain-bandwidth product of a selection of op amps.

Voltage comparator

This is the only application in which we make direct use of the op amp's very high voltage gain. The object of the circuit (right) is not to *measure* the difference between the two input voltages, but simply to indicate which is the higher voltage:

- Output swings **negative** if v_1 is more positive than v_2.

- Output swings **positive** if v_2 is more positive than v_1.

Output is **zero** or close to zero if the inputs differ by only a very small amount.

Using an op amp as a differential amplifier to compare two input voltages.

Because of the high open loop voltage gain, the difference between input voltages is usually large enough to make the output swing as far as it can go, in one direction or the other.

Example

A TL081C op amp has an open-loop gain of 200 000. If the amplifier is operating on a power supply of ±15 V, its output can swing to within 1.5 V of either rail. Its swing is ±13.5 V. To produce a swing of this amount, the inputs must differ by 13.5/200 000 = 67.5 µV. As long as the differences are 67.5 µV or more, the amplifier saturates and its output swings to ±13.5 V.

> **Gain in decibels**
>
> Data sheets often quote the open-loop gain in decibels. A gain of 200 000 is equivalent to 106 dB.

In this application it is essential to choose an op amp with a small input offset voltage. If this is not done, input offset could make the output swing the wrong way when inputs are close.

Example

The TL081C has a typical input offset voltage of 5 mV. Inputs must differ by at least 5 mV to obtain a reliable output swing. Putting it the other way round, this op amp is unsuitable for use as a comparator if the voltages that we want to compare are likely to differ by less than 5 mV. If differences are likely to be 5 mV or smaller, it is essential to use an op amp with small offset. Bipolar op amps are best in this respect. For example, the OP27 has an input offset voltage of only 30 µV.

> **Self Test**
>
> Which way does the output of an op amp swing when:
>
> 1 The (+) input is at +2.2 V and the (-) input is at +1.5 V?
>
> 2 The (+) input is at –4.6V and the (-) input is at + 0.5 V?

Activity — Voltage comparator

Demonstrate the action of an op amp as a voltage comparator using the circuit below. The meters must be able to read both positive and negative voltages. Either use digital multimeters with automatic polarity, or centre-reading moving-coil meters.

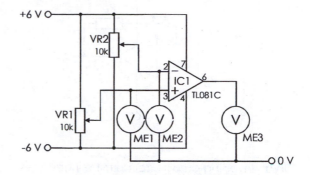

Measuring the input offset voltage of an op amp.

Keep the (+) input constant (adjust VR1 to bring ME1 to +1V, for example) and then adjust VR2 to set the (–) input at a series of values ranging from –6V to +6V.

Record the readings of the three meters in a table. Repeat with a different voltage on ME1, or hold the (–) input constant and vary the (+) input.

Write a summary interpreting your results. What is the reading on ME3 when the input voltages are equal? If ME3 does not show a reading of 0V, can you explain why? Does the output voltage ever swing fully to +6V or –6 V? How far does it swing in each direction? Repeat for another type of op amp.

Inverting amplifier

Of the many applications of op amps, the two most commonly found are the inverting amplifier and the non-inverting amplifier. They have very different properties.

The inverting amplifier (p. 94) has part of the output signal fed back to the inverting input. This is **negative feedback,** and the circuit has three interesting properties as a result.

Equal input voltages

The (+) input has high input resistance. Only a minute current flows into it and so only a minute current flows through R_B. As a result, there is almost no voltage drop across R_B and the (+) input is practically at 0 V.

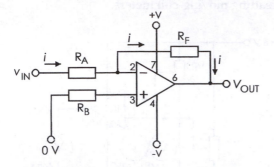

In an inverting amplifier circuit, the input goes to the (−) terminal and there is negative feedback to that terminal.

If v_{IN} is made positive v_{OUT} becomes negative. There is a fall in voltage across the resistor chain R_A and R_F. At the v_{IN} end it is positive and at the v_{OUT} end it is negative. Somewhere along the chain the voltage is zero. As v_{OUT} falls, it pulls down the voltage at the (−) input until this is the point at which the voltage is zero.

If v_{OUT} falls any further, the (−) voltage becomes *less* than the (+) voltage and v_{OUT} begins to rise. If the voltage is exactly zero (taking this to be an ideal amplifier in which we can ignore input offset voltage), v_{OUT} neither rises or falls. Feedback holds it stable. The important thing is that the (−) input has been brought to the *same voltage* as the (+) input.

The same argument applies if v_{IN} is made negative. It also applies if the (+) input is held at any voltage above or below zero. This allows us to state a rule:

An inverting amplifier comes to a stable state in which the two inputs are at equal voltage.

Gain set by resistors

The diagram below shows typical resistor values. The voltages are shown for when v_{IN} is a constant 100 mV and the circuit has reached the stable state referred to above. Because of the high input impedance of the (−) input, no current flows into it (ideally). All the current flowing along R_A flows on along R_F. Call this current i.

Because the (−) input is at 0 V, the voltage drop across R_A is:

$$v_{IN} = iR_A$$

Similarly, the voltage drop across R_F is:

$$v_{OUT} = iR_F$$

Combining these two equations, we obtain:

$$\text{Voltage gain, } A_V = \frac{v_{OUT}}{v_{IN}} = \frac{iR_F}{iR_A} = \frac{R_F}{R_A}$$

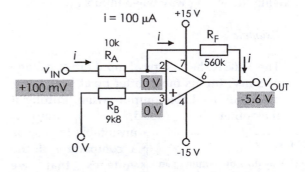

The output voltage stabilises when the voltages at the (+) and (−) inputs are equal.

This leads to another rule for op amps:

The voltage gain of the circuit is set only by the values of the two resistors.

Example

An inverting op amp amplifier circuit has an input resistor R_A of 10 kΩ, and a feedback resistor R_F of 560 kΩ. Its voltage gain is:

$$A_V = \frac{R_F}{R_A} = \frac{560\ 000}{10\ 000} = 56$$

Although the op amp may have an open-loop gain of 200 000, the amplifier *circuit* has a gain of only 56. This is known as the **closed-loop gain**, and must always be substantially less than the open-loop gain of the op amp.

Self test

An inverting amplifier has R_A = 2.2 kΩ and R_F = 820 kΩ. Calculate its closed-loop gain.

The open-loop gain is subject to tolerance errors resulting from differences arising during manufacture. But the closed-loop gain depends only on the precise values of the resistors. If 1% tolerance resistors are used the gain is precise to 1%.

The closed-loop gain obtained from a given combination of R_A and R_F can never be as great as the open-loop gain of the op amp.

Virtual earth

If the (+) input is at 0 V, the circuit stabilises with (−) at 0 V too. The input current flows through R_A toward the (−) input and an *equal* current flows on through R_F. Although no current flows into the (−) input, the action is the same as if there is a direct path from the (−) input to the 0 V rail. The rule is:

> The (−) input of an op amp connected as an inverting amplifier acts as a virtual earth.

Virtual earth is a helpful concept when analysing certain op amp circuits.

Input resistance

As a consequence of the virtual earth, a signal applied to the input has only to pass through R_A to reach a point at 0 V. Therefore, the input resistance of the amplifier in this example is only 10 kΩ. Although the input resistance *of the op amp* may be at least 2 MΩ and possibly as high as 10^{12} Ω, the input resistance *of the amplifier circuit* is nearly always much less.

Self Test

What is the input resistance of the amplifier described in the Self Test above?

If resistor values are being chosen for high gain, R_F is made large and R_A is made small. As a result, the input resistance of a high-gain inverting amplifier is nearly always a few tens of kilohms. For higher input resistance we use an additional op amp as a voltage follower, as explained later in this chapter.

In the diagram on p. 94, we show zero voltage at both ends of R_B. Ideally no current flows through it. In that case, omitting R_B altogether should make no difference to the operation of the circuit. If some loss of precision is acceptable, R_B can be left out of this circuit and the other circuits in which it appears. This saves costs, makes circuit-board layout simpler, and reduces assembly time.

The conclusion reached in the previous paragraph relies on the 'ideal' assumption that no current flows into the inputs. In practice, a very small current known as the **input bias current** flows into both inputs. In bipolar op amps, which have BJT input transistors, this is the base current that is necessary to make the transistor operate, and is around 100 nA.

Op amps with FET inputs do not require base current but there is leakage current of a few picoamperes. This leakage current can generally be ignored but the base current of bipolar inputs can not. Although it is small, it flows through a high input resistance (around 2 MΩ) which leads to a voltage of a hundred or more millivolts at the inputs. The input bias currents of the two inputs differ due to manufacturing differences in the internal circuitry, so the voltages differ, leading to an input voltage offset that is additional to the one already described.

Calculating R_B

In practical circuits it is possible to minimise the voltage offset by considering the paths to the 0 V line through the input resistors. If we make these paths to 0 V equal in resistance, the voltages developed across them are equal and the voltage offset is zero. A worked example appears overleaf.

Take the case in which v_{IN} is zero. In the absence of voltage offset, v_{OUT} is also zero. Now consider the base currents flowing to the inputs. A small base current flows through R_B to the (+) input, so there is a small voltage drop across it and the (+) input is at a small negative voltage. Bias current for the (−) input comes along R_A and R_F from two points both at 0 V. If R_B has the same resistance as R_A and R_F in parallel, the voltage drop is the same for both inputs. The voltage of the (−) input is equal to that at the (+) input. This minimises the effect of input bias currents.

Example

In the example given (p. 80), R_B needs to be equal 10 kΩ and 560 kΩ in parallel:

$$R_B = \frac{10 \times 560}{10 + 560} = \frac{5600}{570} = 9.8$$

The calculation is in kilohms, so the value of R_B is 9.8 kΩ.

> **Self Test**
>
> Calculate the best value for R_B for an inverting amplifier in which R_A = 2.2 kΩ and R_F = 820 kΩ.

Activity — Input voltage offset

Set up this circuit for measuring input voltage offset:

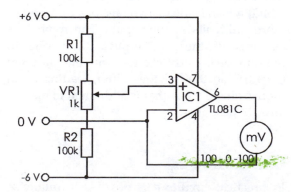

> Compare this drawing with those on pp. 93-94 and note that the op amp is drawn with its input terminals exchanged. This is done to make the diagram simpler. Remember always to check this point when setting up circuits.

The circut requires a split power supply of ±6 V. The (–) input terminal is connected directly to 0 V. R1, R2 and VR1 form a potential divider for adjusting the voltage at the (+) input to about 100 mV on either side of zero. VR1 should preferably be a multi-turn variable resistor to allow the voltage to be adjusted precisely. One with 18 or 22 turns is suitable.

Adjust VR1 until the output of the amplifier rests at 0 V. Then use a digital multimeter to measure the voltages at the inputs. Subtract one from the other to obtain the input offset voltage.

The TL081C may be expected to have an offset of several millivolts. Repeat the measurements with other FET-input op amps and also with some bipolar-input op amps (see data on p. 99). Good ones to try are the popular 741 and the more recent OP-177GP, which has an offset so small that you may not be able to measure it.

Activity — Frequency response

Set up an inverting amplifier circuit (p. 94) with a closed loop gain of 100. Connect its input to a signal generator delivering a 100 mV sinusoid at 100 Hz. Observe the output on an oscilloscope.

Proceed by running the circuit at frequencies from 100 Hz to 10 MHz, in steps of ten times. At each frequency, measure and record the output amplitude.

The first frequency of importance is that at which the output amplitude is 3 dB down on the input. For an input of 100 mV, and a gain reckoned at 100 times, the –3 dB point is 100 × 100 × 0.7071 = 707 mV. Continue in this way, increasing the frequency until the amplifier has unity gain.

Explain your results by referring to the values of R_A, R_B, and R_F. Also show how the performance of the amplifier circuit is related to the full-power bandwidth and the transition frequency of the amplifier IC.

Non-inverting amplifier

This amplifier uses negative feedback taken from a potential divider connected between the output and the 0 V rail. The input signal is fed to the non-inverting input.

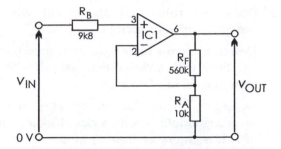

The positive and negative power rails and their connections to the op amp are omitted from this diagram of a non-inverting amplifier and in all op amp circuits after this one.

The voltage v_F at the junction of R_A and R_F is given by the usual equation for a potential divider:

$$v_F = v_{OUT} \times \frac{R_A}{R_A + R_F}$$

This is the voltage at the inverting input. Assuming that the current flowing through R_B is so small that it can be ignored, the voltage at the non-inverting input equals v_{IN}. Applying the same reasoning as for the inverting amplifier, we find that the circuit becomes stable when the input voltages are equal:

$$v_{IN} = v_F = v_{OUT} \times \frac{R_A}{R_A + R_F}$$

Rearranging the equation:

$$A_v = \frac{v_{OUT}}{v_{IN}} = \frac{R_A + R_F}{R_A}$$

The rule we derive from this is the same as for the inverting amplifier:

The voltage gain of the circuit is set only by the values of the two resistors.

Example

For comparison, we take resistors of the same values as in the inverting amplifier. Working in kilohms:

$$A_V = \frac{10 + 560}{10} = 57$$

The voltage gain is 57 times. This is very slightly greater than that of the inverting amplifier.

For the same reasons as in the inverting amplifier, the value of R_B is equal to the value of R_A and R_F in parallel. If the op amp has FET inputs, the current flowing through R_B is so small that R_B can be omitted.

The input resistance of this amplifier is the input resistance of the op amp itself, which is very high. With typical bipolar op amps it is 2 MΩ, and with FET op amps it is up to 10^{12} Ω, or 1 teraohm. A non-inverting amplifier is often used when high input resistance is important.

Voltage follower

A particular instance of the non-inverting amplifier is illustrated overleaf. There is no feedback resistor R_F, and R_A is omitted. The full v_{OUT} is fed back to the inverting input. As before, the circuit settles so that the inputs are at equal voltage. The non-inverting input is at v_{IN}, so the inverting input must be at v_{IN} too.

Self Test

A non-inverting amplifier based on a TL081C op amp has R_A = 33 kΩ and R_F = 270kΩ.

1 What is its voltage gain?

2 What is its input resistance?

3 What is a suitable value for R_B?

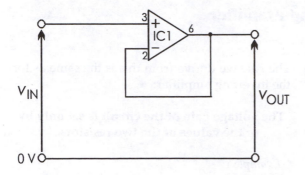

A voltage follower is useful as a buffer between circuits.

But the inverting input is at v_{OUT}, being connected directly to the output. As a result of this:

$$v_{OUT} = v_{IN}$$

Because the output voltage equals the input voltage, this is called a **unity-gain** amplifier. It is also known as a **voltage follower**. It has the same applications as the emitter follower and source follower transistor amplifiers, to act as a buffer between a circuit with high output resistance and a circuit with low input resistance.

The op amp is well suited as a buffer because the input resistance can be up to 1 TΩ, and its output resistance is only 75 Ω. The op amp follower is preferred to the transistor followers because its output follows input exactly, apart from a small error due to input offset voltage. In contrast, the transistor followers always have a much larger offset (0.7 V in the case of the emitter follower) and their gain is not *exactly* 1 (usually nearer to 0.9 in the source follower).

Questions on op amps

1 Describe the appearance of a typical operational amplifier packaged as an 8-pin integrated circuit.

2 How would you measure the open-loop gain of an op amp? What result would you expect to get?

3 Describe what is meant by slew rate. Why may we sometimes need to employ an op amp with high slew rate?

4 List the terminals present on all op amps and state their functions.

5 Describe how the open-loop gain of an op amp varies over the frequency range from 0 Hz (DC) to 100 MHz.

6 State two rules that are useful when designing op amp circuits.

7 Design a circuit for an op amp inverting amplifier with a closed-loop gain of −200, assuming an ideal op amp.

8 Design a circuit for an op amp non-inverting amplifier with a closed-loop gain of 101, assuming an ideal op amp.

9 Explain why op amp inverting amplifiers have a virtual earth.

10 What is input offset voltage, what is its effect, and how is it nulled?

11 Explain the meaning of (a) gain-bandwidth product, and (b) transition frequency.

12 Describe how an op amp voltage follower works and give an example of its applications.

13 Select an op amp from the table on p. 99, suitable for a battery-powered temperature alarm. The op amp compares a standard voltage with one produced by a temperature sensor and triggers the alarm when the selected temperature is reached.

14 An inverting amplifier based on a 355 op amp has $R_F = 100$ kΩ and $R_A = 5.6$ kΩ. What is a suitable value of R_B? What is its input resistance? What is the open-loop gain of the 355? Estimate its closed-loop gain in this circuit at 1 kHz, 1 MHz, and 2.5 MHz.

15 Select an op amp from the table on p. 99 for use in a battery-powered domestic audio intercom system, explaining the reasons for your choice.

16 For what kind of application would you use an op amp as a non-inverting amplifier rather than as an inverting amplifier?

17 Describe the part played by the TL081 in the delay circuit on p. 41.

DATA — Operational amplifiers

Op amps are listed by type number under the three kinds of input stage. BJT input op amps usually have the lowest input offset voltages. Bifet and MOSFET types have very high input resistance, and very low input bias current. Many types of MOSFET allow the output to swing to one or both supply rail voltages, but have lower slew rates than Bifet types. The op amps marked * in the list are also available in dual or quadruple versions. These versions save board space and simplify the pcb layout. Op amps marked † are suitable for running from a single supply voltage.

Type	Supply voltage range (V)	Open loop gain (dB)	Input resistance (Ω)	Input offset (mV)	Slew rate (V/µs)	Full-power bandwidth (kHz)	Transition frequency (MHz)	Output swing (V)	Applications
BJT inputs									
741*	±5 to ±18	106	2 M	1	0.5	10	1	±13	General purpose
OP27	±4 to ±18	123	4 M	0.03	2.8	-	5	±11	Low input offset, low noise, precision
5539	±8 to ±12	52	100 k	2	600	48000	1200	+2.3 / -2.7	High slew rate and full-power bandwidth
Bifet inputs (JFET + BJT)									
LF355	±4 to ±18	106	1 T	3	5	60	2.5	±13	General purpose
TL071*	±2 to ±18	106	1 T	3	13	150	3	±13.5	Low supply voltage, low noise
TL081*	±2 to ±18	106	1 T	5	13	150	3	±13.5	Low supply voltage, general purpose
MOSFET inputs									
CA3130E †	±3 to ±8	110	1.5 T	8	10	120	4	O/P swings to both rails.	wide bandwidth
CA3140E* †	±2 to ±18	100	1 T	5	9	110	4.5	O/P swings to neg. rail	general purpose

Selecting op amps

When selecting an op amp, input offset voltage, slew rate, and output swing are the usually the most important factors to consider. The open loop gain is nearly always far greater than the designer needs.

Sometimes high input resistance is important, but remember that the input resistance of many op amp circuits equals that of the input resistor. It is not dependent on the input resistances of the op amp IC itself.

Full-power bandwidth is important for radio-frequency circuits and high-speed digital signals. The transition frequency (p. 92) is a limitation that should be considered for signals of such high frequency.

Low noise op amps, such as the TL071, are often essential in applications where a sensor produces a signal of very low voltage, or when the signal consists of small variations in voltage. Normally the closed-loop gain is high and any noise in the signal is amplified along with the signal itself.

Multiple choice questions

1   The input resistance of a bipolar op amp is typically:

    A 75 Ω   C 200 kΩ
    B 75 kΩ   D 2 MΩ.

2   The output resistance of an op amp is typically:

    A 75 Ω   C 200 kΩ
    B 75 kΩ   D 2 MΩ.

3   The open-loop gain of an op amp falls to 1 at:

    A 1 MHz
    B the transition frequency
    C 10 kHz
    D 75 kHz.

4   The input resistor of an inverting op amp is 12 kΩ and the feedback resistor is 180 kΩ. The input voltage is −0.6 V. The output voltage is:

    A 1.5 V
    B 9 V
    C −9.5 V
    D −16 V

5   In the amplifier described in Question 4, the value of the resistor between the (+) input and 0V should be:

    A 192 kΩ
    B 11.25 kΩ
    C 180 kΩ
    D 2 MΩ

6   The input resistor of a non-inverting op amp is 39 kΩ and the feedback resistor is 1.2 MΩ. The output voltage is −0.5 V. The input voltage is:

    A −15.7 mV
    B −31.8 mV
    C 16.3 mV
    D −16.3 mV.

7   An op amp has an open-loop gain of 10 at 250 kHz. Its gain-bandwidth product is:

    A 250 kHz
    B 25 kHz
    C 1 MHz
    D 2.5 MHz

8   At 400 kHz the open-loop gain of the op amp described in Question 7 is:

    A 160
    B 1
    C 6.25
    D none of these.

12 Applications of op amps

Operational amplifiers are such a useful and generally inexpensive building block that they appear in a wide range of applications.

These include adders, difference amplifiers, integrators, ramp generators, triggering circuits and active filters.

Adder

The amplifier shown below is an extension of the inverting amplifier and depends on the virtual earth at the inverting input of the op amp. The version shown has four inputs, but there may be any practicable number of inputs, each with its input resistor. In the figure, all resistors have the same value and a suitable value is 10 kΩ.

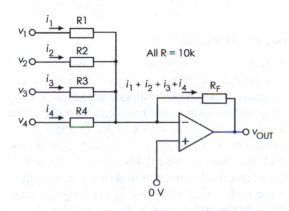

An adder or summer circuit with equal input resistors produces an output voltage equal to the negative of the sum of the input voltages.

When voltages (the same or different) are applied to the inputs, currents flow through each input resistor to the (–) input of the op amp.

Because of the virtual earth, each input resistor has its input voltage at one end and 0 V at the other. The currents through each input resistor are independent of each other and their values are determined by Ohm's Law. The currents through the resistors are obtained by using the Ohm's Law equation.

For the four input currents:

$$i_1 = \frac{v_1}{R_1} \qquad i_2 = \frac{v_2}{R_2}$$

$$i_3 = \frac{v_3}{R_3} \qquad i_4 = \frac{v_4}{R_4}$$

In fact, the currents flow on past the (–) input terminal and join togther. R_F carries the *sum* (*i*) of the currents to the output:

$$i = i_1 + i_2 + i_3 + i_4$$

Substituting from the four current equations and with that all resistors having resistance R:

$$i = \frac{1}{R}(v_1 + v_2 + v_3 + v_4)$$

But considering the current that flows on through R_F, Ohm's law tells us that:

$$i = -\frac{v_{OUT}}{R}$$

Combining the last two equations:

$$v_{OUT} = -iR = -(v_1 + v_2 + v_3 + v_4)$$

The output voltage is the negative of the sum of the input voltages. A circuit such as this can be used as a mixer for audio signals. If the input resistors are variable, the signals can be mixed in varying proportions.

This leads to the concept of **weighted inputs**, practical examples of which are described in Chapters 24 and 40 .

Difference amplifier

One form of the difference amplifier or *subtractor* is shown below. All its resistors are equal in value. Consider the resistor chain R3/R4 . The (+) input of the op amp has high input resistance, so no current flows out of the chain through this. One end of the chain is at 0 V and the other end is at v_2. There is a voltage drop of v_2 along the chain and, because R3 and R4 are equal, the voltage at the (+) input is $v_2/2$.

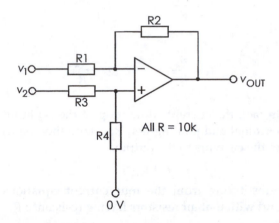

All R = 10k

0 V

A subtractor circuit produces an output equal to the difference between its inputs.

The voltage at the (−) input is $v_2/2$. For the same reasons as for the other amplifiers, the circuit comes to a stable state when the voltages at the (+) and (−) inputs are equal.

Consider the resistor chain R1/R2. This has a voltage v_1 at one end, v_{OUT} at the other end and $v_2/2$ in the centre between the two resistors. No current leaves the chain to flow into the (−) input. Because the resistors are equal in value there are equal voltage drops across them. The voltage drops are:

$$v_1 - v_2/2$$

and $$v_2/2 - v_{OUT}$$

These are equal:

$$v_1 - v_2/2 = v_2/2 - v_{OUT}$$

Rearranging gives:

$$v_{OUT} = v_2 - v_1$$

The output voltage equals the difference between the input voltages.

The difference is amplified if R1 = R3 and R2 = R4, with R3 and R4 being larger than R1 and R3. Then:

$$v_{OUT} = (v_2 - v_1) \times \frac{R_2}{R_1}$$

Alternatively, the difference may be reduced in scale by making R_2 and R_4 smaller than R_1 and R_3.

Input resistance

A disadvantage of this circuit is that the input resistances of the two input terminals are unequal. When input voltages are equal, the (−) input of the amp is a virtual earth so the input resistance at the v_1 input is R_1. But the input resistance at the v_2 input is R_3 and R_4 in series. If all resistors are equal, the input resistance of the v_2 terminal is double that of the v_1 terminal. If the resistances of R_2 and R_4 are large because the voltage difference is to be amplified, the input resistance of the v_2 terminal is much greater than that of the v_1 terminal.

The effect of v_1 having a smaller input resistance is to pull down all voltages applied to that terminal, making v_{OUT} smaller than it should be. In some applications, both input signals include certain amounts of mains hum or other interference picked up equally on the input cables (see next section). A differential amplifier has common-mode rejection (p. 79) so this normally does not matter. But, if the inputs have unequal input resistances, as in the difference amplifier circuit, the interference has a stronger effect on the input with the higher input resistance, and is not rejected.

Medical instrumentation

Difference amplifiers are often used for measuring the small electrical potentials generated within the human body. The activity of the heart is often explored by measuring the potentials generated by its beating muscle and displaying the results as an **electro-cardiograph**. Silver or platinum probes are attached to the wrists and leg. Difference amplifiers measure the potential differences between them.

Another common application is the measurement of potentials produced by the brain. The plotted results is an **electro-encephalogram**. High-gain circuits are needed because the brain is less active than the heart and produces potentials of only a hundred microvolts or so.

The difficulty in measuring potentials within the human body is not that they are small but that the potentials of interest are swamped by other potentials generated in the body and in the equipment.

The body is composed largely of conductive fluids. Electromagnetic fields such as those surrounding electrical equipment and wiring can induce electric currents within the fluids. Typically, these currents alternate at the mains frequency, 50 Hz. If you have ever touched a finger against an input terminal of an audio amplifier, you will probaby have heard a loud 50 Hz buzz. This is the kind of background noise from which we must extract the far weaker signals originating from the heart or brain.

Potentials coming from the heart or brain affect one of the inputs of the amplifier more than the other. They are differential-mode signals and are amplified. Potentials arising from electromagnetic interference affect both inputs equally. They are also equally likely to be picked up by the leads connecting the probes to the amplifier. They are common-mode signals, and are largely rejected, depending on the common mode rejection ratio (CMRR) of the amplifier.

Most op amps have a high CMRR. Of the types listed on p. 99, most have a CMRR in the region of 90-100 dB. This means that the differential-mode gain is 31 060 to 100 000 times greater than the common-mode gain. This allows the differential mode signal to be picked out clearly against a background of common-mode interference and other noise.

Some op amps, are less suitable for this application as they have lower CMRRs. An example is the TL081 with a CMRR of 76 dB.

Bridge measurements

A number of measuring circuits are based on the Wheatstone Bridge. The bridge is made up of four resistances:

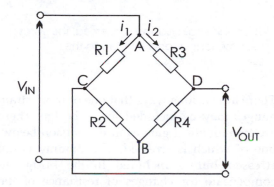

The basic Wheatsone bridge produces a voltage difference, V_{OUT}.

Think of the bridge as two potential dividers ACB and ADB, side by side. Currents i_1 and i_2 flow through the two dividers.

One or more of the resistances are adjustable, as will be explained later. Adjustments are made until the bridge is **balanced**. In this state, points C and D are at the same voltage, and $v_{OUT} = 0$ V.

Therefore:

$$v_{AC} = v_{AD} \quad \text{and} \quad v_{CB} = v_{DB}$$

and, by Ohm's Law:

$$i_1 R_1 = i_2 R_3 \quad \text{and} \quad i_1 R_2 = i_2 R_4$$

Dividing gives:

$$\frac{R_1}{R_2} = \frac{R_3}{R_4}$$

If the bridge is balanced and we know three of the resistances, we can calculate the fourth.

An example of the use of this technique is the measuring of mechanical strain by means of a **strain gauge**. The gauge is a thin metal foil etched to form a number of fine wires. This is embedded in a plastic film and cemented to the object (such as a steel girder) which is to be stressed.

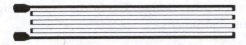

When this strain gauge is stressed, the wires become longer and thinner, so their resistance increases.

There are various ways that one or more strain gauges may be included into a bridge. There are two identical gauges in the version below, one of which is stressed. The dummy is not stressed but is included in the bridge to compensate for changes of resistance of the gauge and its leads due to temperature.

In this version of the Wheatstone bridge, R2 consists of a fixed resistor and a variable resistor to allow the total resistance of R2 to be adjusted.

R1 and the fixed part of R2 consist of high-precision resistors. The variable part of R2 is a high-precision variable resistor with a calibrated knob.

Traditionally, points C and D are joined by a centre-reading microammeter. This displays a zero reading when the bridge is balanced. Instead, we can use an op amp connected as a difference amplifier. The variable resistor of R2 is adjusted until the output of the amplifier is zero. The high CMRR of the op amp makes it possible to detect small differences of potential between C and D, in spite of common-mode voltage signals picked up in the leads and other parts of the circuit.

To make a measurement, we subject the test piece (for example, the girder) to stress. We then adjust the variable resistor until the output of the amplifier is zero. We already know the resistance of R1 and the fixed portion of R2. We now read off the resistance of the variable portion of R2. The resistance of the dummy at the ambient temperature is taken from data tables. From these values we calculate the resistance of the strain gauge. Such equipment could be used for measuring forces, including weighing large loads.

The resistance of the strain gauge can then be read off in terms of mechanical strain by consulting data tables. Alternatively, it is possible to calibrate the equipment by subjecting the test piece to a range of known stresses and plotting a graph to relate strain to the setting of R2.

The difference amplifier is used for other measurements in which small variations of resistance of voltage must be determined against a background of common-mode interference. The platinum resistance thermometer depends on the increase with temperature in the resistance of a coil of platinum wire. These thermometers have high precision and can be used for temperatures ranging from −100°C to over 1500°C. However, the temperature coefficient of metals is very small, so a bridge method is used to obtain the required precision.

Integrator

The best way to understand what the circuit below does is to see how it behaves when a given input is applied to it. The (+) input is at 0 V, and the effect of negative feedback is to make the (−) input come to 0 V. It is a virtual earth, as described in Chapter 9. If there is a constant input v_{IN}, current flows through R to the virtual earth. The current is $i = v_{IN}/R$, because the virtual earth causes the voltage across R to be equal to v_{IN}. In fact, this current does not flow into the (−) input but flows toward the capacitor.

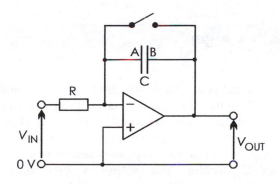

In an integrator, the output falls as charge flows toward plate A of the capacitor.

Putting this into equations, we can say that if current i flows for t seconds, the charge Q flowing toward plate A is (using the relationship between Q, i, and t as in the definition of the coulomb):

$$Q = it$$

If this charge were to accumulate on Plate A, it would produce a voltage v on the plate which is (by definition of the farad):

$$v = Q/C$$

Combining these two equations:

$$v = it/C$$

Substituting the value of i:

$$v = v_{IN} \times t/RC$$

This assumes that the charge accumulates and that the voltage increases. But negative feedback holds the input at 0 V. To make this happen, the output v_{OUT} falls, pulling down the voltage at plate B, which in turn pulls down the voltage at plate A, and holding it at 0 V. The fall in output exactly compensates for the charge that accumulates on plate A:

$$v_{OUT} = -v$$

$$v_{OUT} = \frac{-1}{RC} \times v_{IN}t$$

If v_{IN} is constant, v_{OUT} equals v_{IN} multiplied by time t, multiplied by a scaling factor $-1/RC$. As time passes, v_{OUT} decreases at steady rate proportionate to $1/RC$. The larger the values of R and C, the more slowly it decreases.

> **Self test**
>
> If v_{IN} is constant at 20 mV, R is 10 kΩ and C = 470 nF, what is v_{OUT} after 25 ms?

This discussion assumes that the capacitor is uncharged to begin with, that is, when t = 0. This is the purpose of the switch, which is closed to discharge the capacitor at the beginning of each run. The result of the *Self Test* problem (above) illustrates the fact that, with components of typical values, the output voltage rapidly falls as far as it can in the negative direction and the op amp then becomes saturated.

It also reminds us that quite a small input offset voltage can cause the circuit to swing negative in a second or so, even when v_{IN} is zero. Consequently it is important to use an op amp with small input offset voltage, and to use the offset null adjustment, as shown on p. 91.

If the input voltage is varying, we can work out what will happen by thinking of the time as being divided into many very short time intervals of length δt. We can consider that the voltage is constant during each of these, but changes to a new value at the start of the next interval.

During each interval, charge accumulates according to the constant-voltage equation. The total charge, and also the fall in output voltage at any given time is the sum of the charges and voltage drops that have occurred during all previous intervals. For each interval, the voltage drop is $v_{IN}\delta t$ multiplied by $-1/RC$. For all the intervals between zero time and time t, we obtain the total voltage drop by calculating:

$$v_{OUT} = \frac{-1}{RC} \int_0^t v_{IN} \mathrm{d}t$$

This is the *integral* of v_{IN} with respect to time, from time zero to time t, multiplied by the factor $-1/RC$.

If we divide v_{OUT} by the factor $-1/RC$ we obtain the value of the integral, and this is the average value of v_{IN} during the period from 0 to t. In this way, the circuit is a useful one for averaging a voltage over a period of time.

Ramp generator

The circuit has another important application, as a **ramp generator**. If we apply a constant positive or negative voltage to an integrator, v_{OUT} ramps down or up at a steady rate. Ramping voltages are often used in timing circuits. The output voltage is proportional to the time elapsed since the beginning of the ramp.

Inverting Schmitt trigger

This circuit below has the same kind of action as the trigger circuit on p. 17. Note that the op amp uses a single supply in this application, so the type chosen must be suitable for use in this way. Preferably, the output should be able to swing to equal either supply voltage, which is why a CA3130E has been selected.

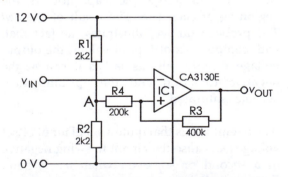

An inverting Schmitt trigger uses positive feedback to give it its 'snap' action.

R1 and R2 form a potential divider and, because they are equal in value, the voltage at point A is 6 V, half-way between the two supply rails. Low-value resistors are selected for R1 and R2 so that the voltage at A is not seriously affected by current flowing into or out of the divider.

A second potential divider is formed by R3 and R4. This has one end connected to the op amp output and the other end at point A. These resistors have relatively high values so that they do not draw excessive current from the R1-R2 divider. The output of this second divider goes to the (+) input of the op amp. Since the op amp input resistance is 1 TΩ, this draws virtually no current from the second divider and the voltage at the (+) terminal will be as calculated.

Schmitt action

The graph opposite (taken from a simulation) displays the action of the circuit when the input is a sawtooth waveform, amplitude 4 V, centred on + 6V.

Starting with v_{IN} (light grey) at its lowest peak (2 V), the op amp is saturated, and its output (black) is high. Feedback through R3 pulls the (+) input voltage (dark grey) above the halfway value to 8 V. This is the **upper threshold voltage** (UTV). Consequently, v_{IN} has to rise to 8 V before the (−) input exceeds the (+) input and the output of the amplifier swings low.

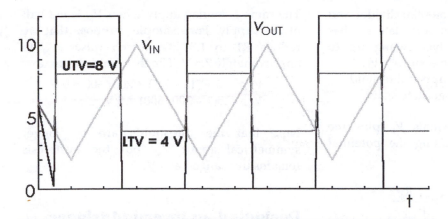

As the input (light grey) to the trigger circuit ramps slowly up and down, the output (black) alters abruptly. Changes in the threshold are plotted in dark grey.

Schmitt triggers are often used for squaring irregular waveforms. Square waves, pulses and digital signals become distorted in shape after transmission over a long distance. They may also pick up interference from electromagnetic and other sources. Passing the signal through a Schmitt trigger removes irregularities and restores the square shape.

As soon as v_{IN} exceeds 8 V, the voltage at the (+) input is greater than that at the (–) input. The output begins to swing low, toward 0 V. As it swings, it pulls the voltage at the (–) input down below 8 V, so increasing the differential between the two inputs. This accelerates the downward swing and produces the characteristic 'snap' action of the Schmitt trigger. Note the inverting action.

The other result is that, once the swing has begun, any slight decrease in v_{IN} can have no effect. Once begun, the change in state can not easily be reversed.

Once the output has fallen to 0 V, the potential divider R3-R4 has 0 V at one end and the halfway voltage (6V) at the other. The voltage at the (+) input is now below the half-way value, at 4 V. This is the **lower threshold voltage** (LTV). Consequently, as v_{IN} falls from its positive peak, it has to fall below 4 V before the trigger changes back to its original state. Then v_{IN} is less than the LTV and the output goes high.

With a sawtooth input (or any other periodic waveform such as a sinusoid), the action continues as above, the output falling every time input exceeds the UTV and rising every time input drops below the LTV. The difference between UTV and LTV is the **hysteresis** of the trigger, 4 V in this example.

Calculating thresholds

As an example, we will calculate the thresholds of the trigger circuit just described. The first step is to calculate the voltage at point A. This is easy in this example because R1 = R2. The voltage at A, which we will call V_M, is half-way between 0 V and 12 V, so $V_M = 6$ V. If R1 and R2 are unequal, and the supply voltage is V_S, the formula is:

$$V_M = V_S \times \frac{R_2}{R_1 + R_2}$$

This is derived from the usual potential divider formula. Looking at the second potential divider (R3-R4) we see that, when v_{OUT} is high, the end connected to A is at V_M, while the other end is at V_S. The voltage drop across the divider is $V_S - V_M$. This is divided into two parts depending on the ratio between R_3 and R_4. The upper threshold voltage is the voltage V_U at the point between R3 and R4.

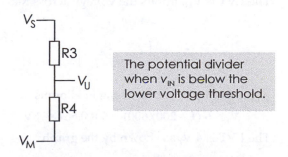

The potential divider when v_{IN} is below the lower voltage threshold.

The diagram shows this potential divider and the voltages concerned. In this state v_{IN} has fallen below LTV, v_{OUT} has swung up to become equal to the positive supply V_S. Now v_{IN} has to rise above the upper threshold V_U before the trigger will change state again.

From the diagram, V_U equals V_M plus the voltage rise across R4. Using the potential divider formula:

$$V_U = V_M + (V_S - V_M) \times \frac{R_4}{R_3 + R_4}$$

In the example the formula can be simplified, because $V_M = V_S/2$:

$$V_U = V_M \left(1 + \frac{R_4}{R_3 + R_4}\right)$$

Substituting the values used in the example, and working in kilohms.

$$V_U = 6(1 + 200/600) = 6 \times 1.333 = 8 \text{ V}$$

The UTV is 8 V, as shown by the graph.

When v_{IN} rises above the UVT, v_{OUT} swings down to the negative supply, which is 0 V in this single-supply circuit.

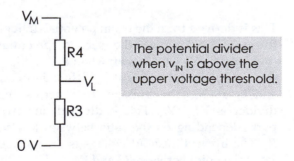

The potential divider when v_{IN} is above the upper voltage threshold.

The LVT is V_M, minus the voltage across R4:

$$V_L = V_M \left(1 - \frac{R_4}{R_3 - R_4}\right)$$

In the example, and working in kilohms:

$$V_L = 6 (1 - 200/600) = 6 \times 0.667 = 4 \text{ V}$$

The LVT is 4 V, as shown by the graph.

The same formulae apply when V_M is *not* half of the supply. For example, suppose that we reduce R1 to 1.3 kΩ, which raises V_M to approximately 7.5 V. The thresholds become:

$$V_U = 7.5 + (12 - 7.5) \times 200/600 = 9 \text{ V}$$
$$V_L = 7.5(1 - 200/600) = 5 \text{ V}$$

Note that the thresholds are no longer symmetrical about V_M, but the hysteresis remains unchanged at 4 V.

Designing an inverting trigger

It can be seen from the discussion above that the size of the hysteresis is decided by the values of R3 and R4. The location of the hysteresis band is decided by R1 and R2, but it is not necessarily symmetrical about V_M.

The calculation assumes that we can ignore the effects of current drawn from the R1-R2 divider to supply the R3-R4 divider. It also assumes that the output swings fully to the positive and negative supply rails. Many op amps do not allow the output to swing as far as the rails (see table, p. 99), and the extent of swing may vary in different op amps. If such op amps are used, it is preferable to measure the maximum swings before calculating the final values of resistors.

The design process begins with the selection of the threshold values V_U and V_L. The other known value is the supply voltage V_S (alternatively, the limits to which v_{OUT} swings). In order to calculate resistor values we first need to calculate V_M.

Referring to the voltage diagram on p. 107, and representing $R_4/(R_3 + R_4)$ by R to simplify the equations:

$$V_U = V_M + R(V_S - V_M) \qquad \text{[Equation 1}$$

Referring to the voltage diagram on this page:

$$V_L = V_M - RV_M \qquad \text{[Equation 2]}$$

Subtracting Equation 2 from Equation 1:

$$V_U - V_L = RV_S$$

From this we obtain:

$$R = \frac{V_U - V_L}{V_S}$$

This gives us the data we need for calculating V_M. Equation 2 above can be re-written:

$$V_L = V_M(1 - R)$$

From which we obtain:

$$V_M = \frac{V_L}{1 - R}$$

Based on the equations above, the routine for designing an inverting trigger is as follows:

1 Decide on values required for V_U and V_L.

2 Calculate R, using:

$$R = \frac{V_U - V_L}{V_S}$$

3 Calculate V_M, using:

$$V_M = \frac{V_L}{1 - R}$$

4 Find two resistors R1 and R2, which total a maximum of 5 kΩ, to give the required value of V_M. This is most easily done by estimating a value for R2, then calculating R1:

$$R_1 = R_2\left(\frac{V_S}{V_M} - 1\right)$$

5 Find two resistors R3 and R4, which total a minimum of 50 kΩ, to give the required value of R. This is most easily done by estimating a value for R4, then calculating R3:

$$R_3 = \frac{R_4(1 - R)}{R}$$

Example

Design an inverting Schmitt trigger, using the circuit on p. 91. The supply voltage is 9 V, LVT = 3 V, UVT = 4 V.

We follow steps 2 to 5 above:

2 $R = (4 - 3)/9 = 1/9$

3 $V_M = 3/(8/9) = 3.375$ V

4 Let $R_2 = 2.2$ kΩ, then, working in kilohms:

$R_1 = 2.2(9/3.375 - 1) = 3.667$ kΩ.

5 Let $R_4 = 200$ kΩ, then, working in kilohms:

$R_3 = 200 (1 - 1/9)/(1/9) = 1600$ kΩ = 1.6 MΩ.

As a check on the calculations, the graph below was plotted for a simulated trigger circuit based on these results.

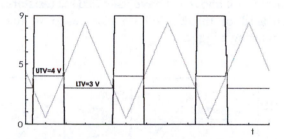

The values calculated for the four resistors produce a trigger action with UTV = 4 V and LTV = 3 V.

The method above can be used for any op amp that swings its output fully to the supply rails. It yields approximate results for op amps which do not swing to the rails, provided that the operating voltage is near the maximum for the op amp used.

Self test

Design an inverting trigger for an op amp running on a 15 V single supply, so that UTV = 10 V and LTV = 7.5 V.

Non-inverting Schmitt trigger

This has a circuit similar to the inverting trigger, except that the connections to the op amp inputs are reversed. The input signal v_{IN} goes to the non-inverting (+) input, by way of the R3-R4 potential divider.

The point A, at V_M, is connected to the inverting input. The effect of these changes is to make v_{OUT} swing high as v_{IN} rises above the upper threshold. It swings low as v_{IN} falls below the lower threshold.

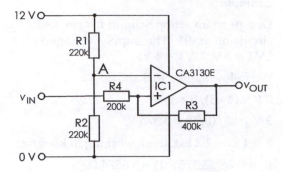

A non-inverting Schmitt trigger also uses positive feedback.

The R3-R4 divider has one end at v_{IN} and the other end at 0 V. As v_{IN} rises, the voltage at the (+) input of the op amp (dark grey, call this v_+) rises too. It is lower than v_{IN} because v_{OUT} is low and pulls it down. As v_{IN} rises, v_+ stays below it, until v_+ reaches v_- the voltage of the R1-R2 divider. The level of v_{IN} at this stage is the upper threshold voltage (UTV).

As v_{IN} exceeds UTV, v_{OUT} swings sharply to the positive supply (V_S). Instead of being pulled down, v_+ is now pulled up, clearly above UTV. Small reversals in the rise of v_{IN} can not affect the circuit after this. It is triggered.

When v_{IN} falls, $v+$ falls too but stays a little higher than v_{IN}. This is because v_{OUT} is pulling it up. Eventually, v_{IN} falls to the lower threshold level (LVT). At this stage, v_+ falls below v_-. The circuit is triggered back to its original state, with v_{OUT} low.

In this circuit the R1-R2 potential divider is connected only to the (–) input of the op amp. This draws a very small current, so large-values resistors can be used for R1 and R2. In the circuit above we have used 220 kΩ resistors instead of the 2.2 kΩ resistors used in the inverting trigger.

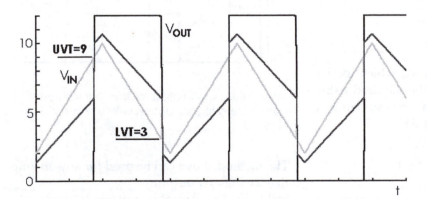

The (+) input to the op amp (dark grey line) is a sawtooth like v_{IN} (light grey) but is shifted up or down by 4.2 V every time it reaches 6 V. This happens because of v_{OUT} pulling up or down every time the op amp changes state.

Thresholds

As before, it is helpful to sketch the R3-R4 divider and mark the voltages on it. As before, we simplify the equations by putting $R = R_4/(R_3+R_4)$. The result is shown below and we can see that:

$$v_+ = v_{IN} - v_{IN} \times R$$

Collecting terms gives:

$$v_+ = v_{IN}(1 - R)$$

It is obvious from the graph that the action of this circuit differs in several ways from that of the inverting trigger (see graph, p. 117). The graph shows the action of the circuit when the input is a sawtooth waveform, amplitude 4 V, centred on +6 V.

Starting with v_{IN} (light grey) at its lowest peak, the op amp is saturated and its output (black) is low.

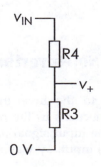

The potential divider when v_{IN} is below the upper threshold voltage.

v_{IN} rises until v_+ equals v_-, the voltage at the (–) input of the op amp. At that point:

$$v_{IN} = \frac{v_-}{1-R}$$

This is the upper threshold voltage, at which v_{OUT} swings from low to high.

The drawing below shows the divider after v_{IN} has passed through the UTV.

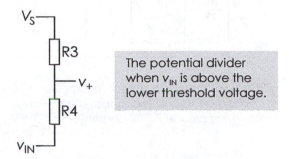

The potential divider when v_{IN} is above the lower threshold voltage.

Adding voltage drops:

$$v_+ = v_{IN} + (V_S - v_{IN})R$$

From which we deduce that, at the LVT:

$$v_{IN} = \frac{v_- - V_S R}{1 - R}$$

In the example, first calculate R, working in hundreds of kilohms:

$$R = 2/6 = 0.333$$

At the UVT, given that $V_S = 12$ V, $v_- = 6$ V, and working in hundreds of kilohms:

$$v_{IN} = \frac{6}{1 - 0.333} = 9 \text{ V}$$

This agrees with the value obtained in the simulation. At the LVT:

$$v_{IN} = \frac{6 - 12 \times 0.333}{1 - 0.333} = 3 \text{ V}$$

The thresholds are 9 V and 3 V, as shown in the diagram, and the hysteresis is 6 V.

Designing a non-inverting trigger

The hysteresis is decided by the values of R3 and R4. The location of the hysteresis band is determined by R1 and R2, but it is not necessarily symmetrical about v_-.

The design process begins with the selection of the threshold values V_U and V_L. The supply voltage V_S is known. The quantity R is defined as on p. 109.

Referring to the voltage diagram opposite:

$$V_U = v_+ + Rv_U \qquad \text{[Equation 1]}$$

Referring to the diagram on this page:

$$V_L = v_+ - R(V_S - v_L) \qquad \text{[Equation 2]}$$

Subtracting terms in Equation 1 from corresponding terms in Equation 2:

$$V_U - V_L = R(V_U - V_L + V_S)$$

From this we obtain:

$$R = \frac{V_U - V_L}{V_U - V_L + V_S}$$

This gives us the data we need for calculating v_+ (and hence v_-). From Equation 1:

$$v_+ = V_U(1 - R)$$

Based on the equations above, the routine for designing a non-inverting trigger is as follows:

1 Decide on values required for V_U and V_L.

2 Calculate R, using:

$$R = \frac{V_U - V_L}{V_U - V_L + V_S}$$

3 Calculate v+ using:

$$v_+ = V_U(1 - R)$$

This is also the value of v_-.

4 Find two resistors R1 and R2, to give the required value of v-. This is most easily done by estimating a value for R2, then calculating R1:

$$R_1 = R_2 \left(\frac{V_S - v_+}{v_+} \right)$$

5 Find two resistors R3 and R4 to give the required value of R. This is most easily done by estimating a value for R4, then calculating R3:

$$R_3 = R_4\left(\frac{1-R}{R}\right)$$

Example

Design a non-inverting Schmitt trigger operating on a 15 V supply, with LVT = 6 V and UVT = 11 V.

Taking the calculation from step 2, onward:

2 $R = (11-6)/(11-6+15) = 0.25$

3 $v_+ = 11(1-0.25) = 8.25$

4 Let $R_2 = 200$ kΩ, then, working in kilohms:

$R_1 = 200(15-8.25)/8.25 = 164$ kΩ

5 Let $R_4 = 200$ kΩ, then, working in kilohms:

$R_2 = 200(1-0.25)/0.25 = 600$ kΩ

As a check on the calculations, the graph below was plotted for a simulated trigger circuit based on these results.

Self Test

Design a non-inverting trigger for an op amp running on a 6 V single supply, so that UTV = 4.5 V and LTV = 2 V.

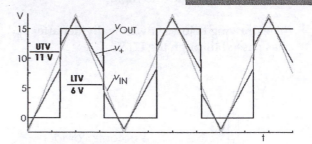

The input signal is a sawtooth, amplitude 10 V, centred on +7.5 V. The output is a square wave, amplitude 7.5 V, centered on + 7.5 V.

Ramp and square-wave generator

This is an application of the inverting Schmitt trigger. The block diagram below shows its main features.

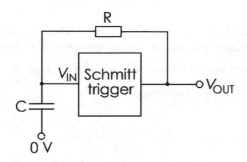

A resistor-capacitor network connected to an inverting Schmitt trigger forms a simple square-wave generator.

The circuit repeatedly charges and discharges a capacitor through a resistor. When v_{OUT} is high, current flows through R to charge C. The voltage across C, and hence the input v_{IN} to the trigger, rises until it reaches the UTV of the trigger. Then the output swing sharply low, to 0 V.

With v_{OUT} at 0 V, the capacitor discharges through R. Discharge continues until the voltage across C, and the voltage at v_{IN} has fallen to the LTV. At this point v_{OUT} swings high again.

The result of this action is that v_{OUT} swings repeatedly between 0 V and the supply voltage. It produces a continuous square wave. The graph below shows the output from such a circuit, based on the trigger circuit on p. 106. The additional components are the resistor R, value 1 MΩ and the capacitor C, value 1 μF.

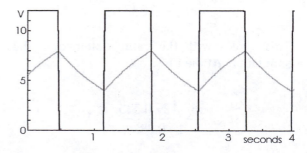

The curve for v_{OUT} is a square wave. The curve for v_{IN} is close to a sawtooth, but the segments are slightly curved.

The diagram below shows the circuit in more detail:

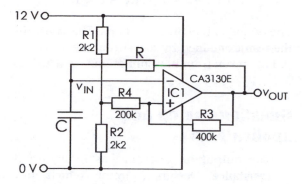

> This trigger circuit is exactly the same as that on p. 106 with the addition of the timing components R and C.

Ramp generator

Instead of taking the output of this circuit from the op amp output (v_{OUT}), we can take it from the point marked v_{IN}. This is plotted in grey on the graph opposite. The voltage ramps up and down, producing a sawtooth wave. However, the segments of the waveform are really segments of an exponential curve (p. 36). Provided that the difference between UVT and LVT is not too large, the segments are close enough to straight lines for most applications.

Taking the output from the point labelled v_{IN} may draw an appreciable current from this circuit. This will make the charging times too long and the discharging times too short. In practice, the input terminal of a unity gain voltage follower (p. 97) is connected to this point. This provides an ample source of current for feeding the sawtooth signal to another circuit.

Timing

Charging and discharging a capacitor takes time, as explained on pp. 36-37. The greater the capacitance of C and the larger the resistance of R, the longer it takes to charge and discharge the capacitor.

The frequency of the waveforms thus depends on the values of R and C. It also depends on the hysteresis of the trigger. The greater the hysteresis, the larger the rise and fall of v_{IN} and the longer it takes.

On pp. 36-37 it is shown that the times taken to charge to or discharge from certain key points on the exponential curves depends solely on the time constant, t, which equals RC. Timing is independent of the supply voltage. It follows that, given the threshold levels set by resistors R1 to R4, the frequency of the output signals depends only on the values of R and C. This is a useful feature of this circuit, for its frequency does not vary with changes in supply voltage.

Measurements on the plot opposite show that the period of the signals is almost exactly 1.4 s. The time constant of the circuit is:

$$t = RC = 1 \times .1 = 1 \text{ s} \qquad [\text{M}\Omega \times \text{F}]$$

From this we deduce that the period is 1.4 time constants. Testing the circuit with other values of R and C shows that this is holds true. However, period depends also on the thresholds of the trigger, a topic which we study in the next section.

Thresholds and period

It was shown on p. 97 that the thresholds for trigger circuits with the resistor values shown in the diagram are:

$$UVT = 8 \text{ V}$$
$$LVT = 4 \text{ V}$$

These values apply when $V_S = 12$ V. Putting the thresholds as fractions of the supply voltage we have:

$$UVT = 0.667$$
$$LVT = 0.333$$

During the charging phase we need to calculate the time required to charge C from $0.333V_S$ to $0.667V_S.$

The Extension Box 6 on p. 38 shows how to calculate times when charging from 0 V up to to a chosen fraction of the supply voltage. Here we are charging from one fraction to another.

During charging, t_p is the time taken to charge from 0 V to pV_S, p having any value between 0 and 1.

Therefore, from the equation on p. 38:

$$1 - e^{\frac{-t_p}{RC}} = p$$

and therefore:

$$\frac{-t_p}{RC} = \ln(1-p)$$

$$-t_p = \ln(1-p)RC$$

During charging, the time to charge from 0 V to the LVT, where p = 0.333 is:

$$t_{0.333} = \ln(1 - 0.333)RC = 0.405RC$$

The time taken to charge from 0 V to the UVT, where p = 0.667 is:

$$t_{0.667} = \ln(1 - 0.667)RC = 1.100RC$$

The time taken to charge from LVT to UVT is the difference between these two times, so:

Charging time = $(1.100 - 0.405)RC = 0.695RC$

Now to consider discharging, for which the equation is:

$$-t_p = \ln(p)RC$$

The time to discharge from V_S to the UVT, where p = 0.667 is:

$$t_{0.667} = -\ln(0.667)RC = 0.405RC$$

The time taken to charge from V_S to the LVT, where p = 0.333 is:

$$t_{0.333} = -\ln(0.333)RC = 1.100RC$$

The time taken to discharge from UVT to LVT is the difference between these two times, so:

Discharging time = $(1.100 - 0.405)RC$
$$= 0.695RC$$

Charging and discharging times are equal because, in this trigger, UVT and LVT are symmetrical about $V_S/2$. This does not necessarily apply to other trigger circuits.

The total length of one cycle is equal to the charging time plus the discharging time:

Period = $2 \times 0.695RC = 1.39\,RC$

The period is 1.39 time constants. This confirms the result obtained by measuring the graph on p. 112, taken from the simulation.

Questions on op amp applications

1 The output of a sensor has high output resistance. Design a circuit (which may have more than one op amp in it) to take the output from the sensor, subtract 1.5 V from it and display the result on a multi-meter.

2 Design a circuit to mix three audio signals, two at equal amplitude and the third signal at half its original amplitude.

3 Explain why an op amp difference amplifier is used for measuring a small voltage signal against a background of interference. Illustrate your answer by quoting two practical examples.

4 Describe a technique for using a Wheatstone bridge to measure an unknown resistance.

5 Explain how a bridge may be used to measure temperature, using a platinum resistance thermometer as sensor.

6 Explain how an op amp integrator works when the input voltage is constant.

7 How can an op amp integrator be used as a ramp generator?

8 An op amp integrator circuit has R = 22 kΩ and C = 560 nF. Its output rises from 0 V to 125 mV in 40 s. What is the average input during this period?

9 Explain how an op amp integrator is used to measure the average value of the input voltage over a period of time.

10 Explain the action of an op amp inverting Schmitt trigger. Describe two examples of applications of this type of trigger circuit.

11 Explain the working of an op amp non-inverting Schmitt trigger circuit.

12 Calculate the upper and lower threshold voltages of this Schmitt trigger circuit:

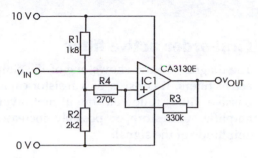

How much is the hysteresis of this trigger?

13 Design an op amp inverting Schmitt trigger that has its upper threshold voltage at 10 V and its lower threshold at 4 V. The supply voltage is 15 V.

14 Calculate the upper and lower threshold voltages of this Schmitt trigger circuit:

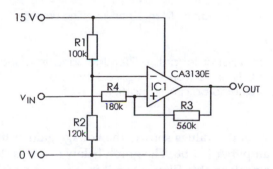

In what way does the action of this trigger differ from that in Q. 12?

15 Design an op amp non-inverting Schmitt trigger that has its upper threshold voltage at 7.5 V and its lower threshold at 3.5 V. The supply voltage is 9 V.

16 Calculate the upper and lower threshold voltages of this trigger circuit:

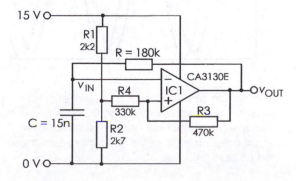

17 Calculate the time constant of the circuit of Q. 16. How long is its period, in time constants? What is its period in seconds and its frequency in hertz?

18 Draw a circuit diagram to show how to use this circuit as a ramp generator.

Multiple choice questions

1 An op amp used in an integrator circuit should have:

    A low input offset voltage
    B high open-loop gain
    C low slew rate
    D high input resistance.

2 An op amp integrator receives a sinusoidal input, amplitude 50 mV. Assuming the integrator does not have time for its output to swing fully in either direction, its output after three cycles of the sinusoid is:

    A 50 mV
    B 150 mV
    C 0 V
    D −25 mV.

3 An inverting Schmitt trigger circuit changes state when:

    A a rising input goes just above the upper threshold
    B a rising input goes just below the lower threshold
    C a falling input goes just below the upper threshold
    D the upper and lower thresholds are equal.

4 An op amp used in a Schmitt trigger should have:

    A low input offset voltage
    B high input resistance
    C low output resistance
    D high common-mode rejection ratio.

5 The difference between the upper and lower thresholds of a Schmitt trigger is known as:

    A feedback
    B input offset voltage
    C input bias current
    D hysteresis.

13 Active filters

Active filters are another important application of operational amplifiers.

When a signal of mixed frequencies is passed through a filter circuit, some of the frequencies are reduced in amplitude. Some may be removed altogether. The simplest filter consists of a resistor and a capacitor (p. 44). As it is built from two passive components it is known as a **passive filter**.

Passive component

One that does not use an external power source. Examples are resistors, capacitors, inductors, and diodes.

Other passive filters may consist of a network of several resistors and capacitors, or may be combinations of resistors and inductors, or capacitors and inductors.

A passive filter must have at least one reactive component (a capacitor or an inductor) in order to make its action frequency-dependent.

Passive filters are often used, but a large number of individual filters must be joined together to effectively cut off unwanted frequencies. Such filters take up a lot of space on the circuit board and can be expensive to build. When filtering low frequencies the inductors are necessarily large, heavy and expensive.

Another serious disadvantage of passive filters is that passive components can only *reduce* the amplitude of a signal. A signal that has been passed through a multi-stage passive filter is usually very weak. For these reasons, we prefer **active filters**, based on active components, especially op amps. When these are used we can obtain much sharper cut-off with relatively few filtering stages, there is no need to use inductors, and the filtered signal may even show amplitude gain.

Active component

One that uses an external power source. Examples include transistors, and operational amplifiers.

First-order active filter

The diagram below shows one of the simplest active filters. It consists of a resistor-capacitor passive filter followed by a non-inverting amplifier to restore or possibly increase the amplitude of the signal.

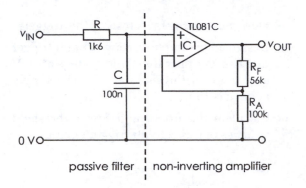

passive filter | non-inverting amplifier

In a first-order active filter, the signal from the RC filter network is amplified by a non-inverting op amp amplifier.

With the values shown, the voltage gain of the amplifier is 1.56. The graph below displays the signals in this filter when it is supplied with a 1 V sinusoid at 1 kHz.

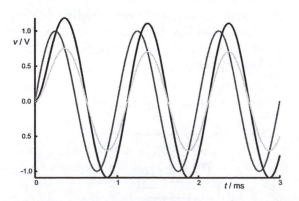

The action of a first-order active filter on a 1kHz sinusoid (dark grey). The RC filter network reduces its amplitude and delays its phase (light grey). The amplifier then increases the amplitude (black).

The RC network filters the input sinusoid (dark grey line). At the output of the network (light grey) the filtered signal has an amplitude of only 0.7 V. This is not the only change. The output reaches its peaks 125 µs behind the input. This is a delay of 1/8 of the period of the signal. In terms of angle, this is 360/8 = 45°. Because it is a delay, it has negative value. The output is *out of phase* with, or *lags* the input, by −45°. The black curve represents the output of the op amp.

> **Gain**
>
> Gain of non-inverting amplifier =
>
> $(R_A + R_F)/R_A$

With the values shown for R_F and R_A, the gain of the amplifier is 1.56. The output from the RC network has amplitude 0.7 V, so the overall gain of the active filter is 0.7 × 1.56 = 1.1. The amplitude of the output is 1.1 V.

The amplifier does not change the phase any further, so the output lags the input by 45°.

Frequency response

The frequency response (see p. 46) is plotted below.

The plot (black) shows that amplitude is the full 1.56 V up to about 300 Hz. From then on amplitude begins to fall off slightly, reaching 0.7 V (equivalent to -3dB, or half power) at 1 kHz. The bandwidth of the filter is 1 kHz.

The frequency of f_C, the -3 dB point or cut-off point, depends on the values of the resistor and capacitor.

Example

With the values of R and C shown on p. 116, the cut-off point is:

$$f_C = \frac{1}{2\pi RC}$$

$$f_C = 1/(2\pi \times 1.6 \times 100) = 1 \text{ kHz} \quad [1/(k\Omega \times nF)]$$

As frequency is increased beyond 1 kHz, the amplitude rolls off even faster and eventually the curve plunges downward as a straight line. If we read off frequency and amplitude at two points on this line we find that for every doubling of frequency, the amplitude falls by 6 dB. We say that the roll-off is –6 dB per octave.

> **Octave**
>
> An interval over which frequency doubles or halves.

Another way of expressing this is to measure how much the amplitude falls for a ten-times increase in frequency, or decade. In the plot, the roll-off is 20 dB per decade.

Note that these roll-off rates are characteristic of single-stage passive or active filters and do not depend on the values of the capacitor and resistor.

The plot also shows how the phase change of the output varies with frequency. The scale shows phase angle on the right margin of the figure. We can see that the phase change at 1 kHz is −45°, confirming what we found in the output plot opposite.

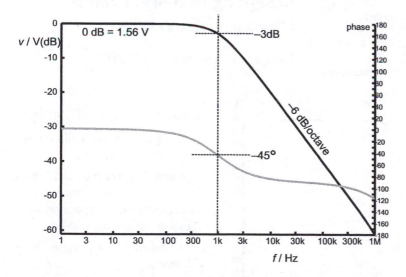

The Bode plot of the frequency response of the first-order lowpass filter shows that it falls to its half-power level (amplitude 0.7 V) and has phase lag of 45° at the cut-off.

A phase lag of –45° at the cut-off point is another characteristic of single-stage passive or active filters. The phase lag is less at lower frequencies. At 0 Hz (DC) the phase change is zero. It increases at frequencies above the cut-off point.

Highpass first-order active filter

The lowpass filter is converted to a highpass filter by transposing the resistor and capacitor.

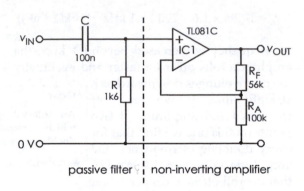

passive filter | non-inverting amplifier

A first order highpass active filter consists of a highpass passive filter followed by a non-inverting amplifier.

When this is done, the frequency response has the appearance shown below:

The amplitude curve rises up from 0 Hz at +6 dB per octave, or +20 dB per decade. The equation for calculating the cut-off point is the same for both lowpass and highpass filters, so this is 1 kHz as before. Beyond the cut-off point, amplitude rises to 0 dB above about 4 kHz.

The phase response curve shows that the filter has a 45° *lead* at its cut-off frequency. This is the opposite of the phase change found in the lowpass filter.

Looking more closely at the frequency response, we note that the amplitude starts to fall again from about 400 kHz upward. This is not the effect of the RC filter but is due to the fall in gain of the op amp at high frequencies. The transition frequency of the TL081C is 3 MHz (see p. 85). If we were to plot the response at higher frequencies, we should see a steep roll-off.

In practice, if we extend the operation of this filter into higher frequencies it acts as a bandpass filter.

Designing first-order filters

1 Decide on a capacitor value.

2 Apply the equation: $R = 1/2\pi f_C C$.

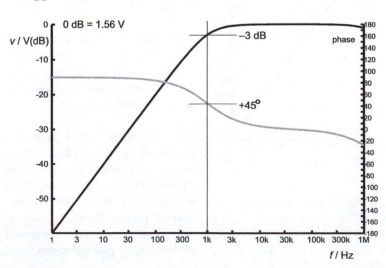

Example

To design a lowpass filter with cut-off point at 40 kHz.

1 For stability, try the calculation for a 220 pF capacitor, so that a ceramic plate capacitor with zero tempco can be used.

2 Applying the equation, $R = 18086 \, \Omega$.

Use an 18 kΩ, 1% tolerance, metal film resistor. For a highpass filter, exchange the resistor and capacitor.

Second-order active filter

Adding an extra stage to the first-order lowpass filter allows feedback to be introduced. The circuit below has two RC filters, one of which is connected to 0 V, as in the first-order lowpass filter. The other is part of a feedback loop from the filter output.

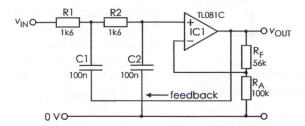

The second-order active filter has positive feedback from the output to the capacitor of the first stage.

The feedback action makes the filter resonate at frequencies around the cut-off frequency. Resonance is not strong, but there is enough extra feedback around this frequency to push up the 'knee' of the frequency response curve.

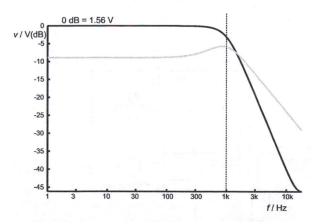

The Bode plot of the response of the second-order filter (black) also shows the amplitude of the feedback signal (grey), which pushes up output at the knee of the curve. The doubled filter network produces a steep roll-off at high frequencies.

The grey curve is a plot of the voltage across C1, and it can be seen that the strongest feedback occurs around 1 kHz. The response of the filter turns sharply down above 1 kHz. From then on, there is a steady and more rapid fall in response, the rate depending on the number of stages. With two stages in the filter, the roll-off is –12 dB per octave or –40 dB per decade.

This sharpening of the 'knee' is accomplished in a passive filter by using an inductor-capacitor network (p. 57) to produce the resonance. Using an op amp instead provides the same effect but without the disadvantages of inductors.

A highpass version of the second order filter is built by substituting resistors for capacitors and capacitors for resistors. Higher-order filters with even steeper roll-off often consist of two or more second-order filters and possibly a first-order filter cascaded together.

Bandpass filter

A bandpass filter may be constructed by cascading a lowpass filter and a highpass filter (see upper circuit, opposite). The cut-off frequency of the lowpass filter is made *higher* than that of the highpass filter so that there is a region of overlap in which both filters pass frequencies at full strength. This area of overlap can be made narrow or broad by choosing suitable cut-off frequencies. Both lower and upper cut-off frequencies are calculated using the equation $f = 1/2\pi RC$.

In the example opposite, the cut-off frequency of the lowpass filter is 4.8 kHz, while that of the highpass filter is 1 kHz. The frequency response shows a broad passband from 645 Hz to 7440 Hz, with a central frequency of 2.2 kHz. The bandwidth of the filter is 6795 Hz.

Another approach to bandpass filtering is the **multiple feedback filter** (see lower circuit, overleaf). This has two loops, each providing *negative* feedback.

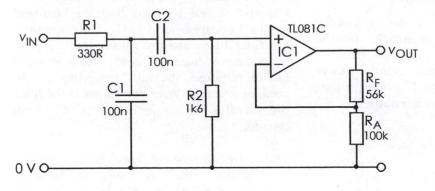

The result (see below) is a passband of narrow bandwidth. The centre frequency is 1.8 kHz. The lower cutoff frequency is 1 kHz and the upper cutoff frequency is 3 kHz. This gives a bandwidth of 2 kHz. The centre frequency (but not the bandwidth or the amplitude) may be tuned by varying the value of R3.

A broad-band bandpass filter can be made by cascading a lowpass filter and a highpass filter. This is followed by a non-inverting amplifier to restore the amplitude to a reasonable level.

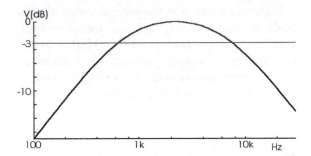

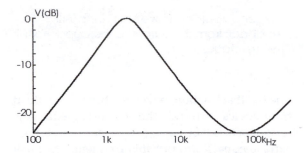

The bandpass filter above produces a broad passband, with response falling off at 6 dB per octave on either side.

The bandpass filter below produces a narrow passband, with response falling off at 6 dB per octave on either side.

- R1 and C1 form a highpass filter; high frequency signals passing through this are fed back negatively, partly cancelling out the high frequencies in the original signal.

- R2 and C2 form a lowpass filter; low frequency signals passing through this are fed back negatively, partly cancelling out the low frequencies in the original signal.

Only signals in the medium range of frequency are able to pass through the filter at full strength.

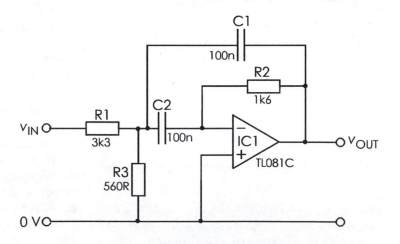

A multiple feedback filter cancels out both high and low frequencies by feeding them back negatively, leaving the middle frequencies unaffected.

The centre frequency of the filter is calculated from the equation:

$$f_o = \frac{1}{2\pi C} \sqrt{\frac{R_1 + R_3}{R_1 R_2 R_3}}$$

The response curve shows an upturn in the higher frequency range. The is the result of the transition frequency of the op amp. The gain of the op amp is reduced at higher frequencies and therefore the higher frequencies in the signal are not completely cancelled out by the negative feedback.

Bandstop filters

Bandstop filters, often known as **notch filters**, are used to remove a limited range of frequencies from a signal. For example, such a filter may be used to remove a 50 Hz mains 'hum' that has been picked up by electromagnetic interference. This may be necessary when measuring small voltage signals from the human body.

There are two approaches to bandstop filtering, both of which show interesing applications of electronic principles. The first circuit (below) depends on one of the basic applications of op amps.

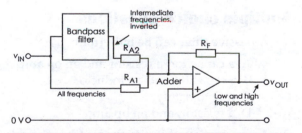

This bandstop filter uses two op amps, one as part of an active filter, the other as an adder.

The signal is sent to an op amp adder, both directly and indirectly through a bandpass filter. The filter could be a multiple feedback filter (see opposite page). This not only filters the signal, removing low and high frequencies, but also inverts it.

The second op amp mixes the unfiltered signal with the inverted filtered signal. The result is to subtract the medium frequencies from the original signal, creating a notch in the frequency response curve. The depth of the notch depends on the relative values of R_{A1} and R_{A2}.

Another type of bandstop filter depends on an LC resonant network (p. 57). This is combined with a resistor and an op amp wired as an inverting amplifier.

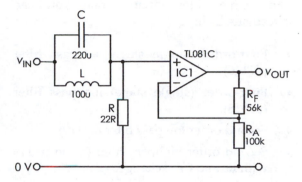

This bandstop filter uses an LC resonant network to produce a very narrow notch in its frequency response (below).

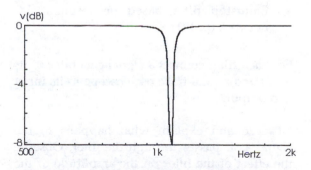

The LC bandstop filter has a steep-sided notch at 1.07 kHz.

At the resonant frequency, the capacitor and inductor have equal impedance. Their impedance in parallel is a maximum. This reduces the amplitude of signals of that frequency, producing the narrow notch in the frequency response.

To calculate the notch frequency we use the equation on p. 57. The depth of the notch is determined by the value of R. The smaller the value of R, the deeper the notch.

Activities — Active filters

It is easier to work this activity on a circuit simulator, and this has the advantage that an analysis of frequency response can be done very quickly. But you should build and test at least two filter circuits using a breadboard.

The active filter circuits that you can investigate include:

- First-order (single stage) lowpass filter (p. 116).
- First-order (single stage) highpass filter (p. 118).
- Second order lowpass filter (p. 119).
- Second order highpass filter (as on p. 119 with Rs and Cs exchanged).
- Bandpass filter consisting of a lowpass filter cascaded with a highpass filter (cutoff point for the lowpass filter must be *higher* than that for the highpass filter).
- Multiple feedback bandpass filter (p. 120)
- Bandstop filter based on resonant LC network (p. 121).

For each filter connect a signal generator to its input and a dual-trace oscilloscope to its input and output.

Observe and explain what happens to the signal as it passes through the filter. Observe the effect of the filter on the amplitude of the signal and on the shape of the waveform, trying this with both sinusoidal and square waves, and at different frequencies.

For example, when testing a lowpass filter, use a signal with a frequency well below the cut-off point, try another signal at the cut-off point and a try a third signal well above the cut-off point. Summarise your results in a table or by sketching frequency response curves.

Remember that the transition frequency of the op amp itself may be about 1 MHz, so do not use frequencies higher than 100 kHz in these investigations.

Questions on active filters

1 List and discuss the advantages of op amp active filters, compared with passive filters.

2 Design a first-order lowpass filter with a cut-off frequency of 15 kHz.

3 Design a first-order highpass filter with a cut-off frequency of 20 Hz.

4 Explain the action of a second-order op amp lowpass filter and state why its performance is superior to that of a first-order filter.

5 Design a bandpass filter consisting of a lowpass and a highpass first-order filter cascaded. The specification is: low cut-off point = 2 kHz; upper cut-off point = 5 kHz; gain at the centre frequency = 1.5.

6 Explain the action of a multiple feedback bandpass filter. In what way is its frequency response different from that of cascaded lowpass and highpass filters?

7 Outline the structure of three different types of active filter. Describe two examples of how each type can be used.

Multiple choice questions

1 A passive filter can be built from:

    A a capacitor, a resistor and an op amp
    B a capacitor and an inductor
    C resistors only
    D a resistor, and an op amp.

2 An op amp used in a highpass filter should have:

    A a high gain-bandwidth product
    B low slew rate
    C low input bias current
    D high output resistance.

3 At the cut-off point of a lowpass filter the phase change is:

    A +90°   C −45°
    B 0°   D −90°.

4 The cut-off frequency of a first-order active high pass filter is 5 kHz. It has a 1 nF capacitor. The nearest E12 value for the resistor is:

A 200 kΩ
B 31.8 Ω
C 33 kΩ
D 220 kΩ.

5 An active filter must have:

A a resistor
B a diode
C no more than two passive components
D a least one reactive component.

6 The roll-off of a 2nd-order highpass filter is:

A +6 dB per octave
B +40 dB per decade
C −12 dB per decade
D −20 dB per octave.

7 An active bandpass filter is built from a lowpass filter cascaded with a highpass filter. It has a lower cutoff frequency of 200 Hz, and a bandwidth of 800 Hz. The cutoff points of its two filters are:

A lowpass = 200 Hz, highpass = 1 kHz.
B lowpass = 1 kHz, highpass = 200 Hz.
C lowpass = 200 Hz, highpass = 800 Hz.
D lowpass = 800 Hz, highpass = 1 kHz.

8 In a multiple feedback bandpass filter, there is:

A negative feedback of low frequencies
B negative feedback of middle frequencies
C positive feedback of middle frequencies
D negative feedback of low and high frequencies.

9 A bandstop active filter produces a notch in the frequency response because:

A of the inverting action of the op amp
B the reactances of the capacitor and inductor are equal at the notch frequency
C the reactance of the capacitor and inductor in parallel is a maximum at the notch frequency
D the LC network resonates strongly at the centre frequency.

10 A bandstop active filter has a resonant network in which $C = 10$ nF, and $L = 27\,\mu H$. The centre frequency of the notch is at:

A 589 kHz
B 306 kHz
C 962 Hz
D 1.85 MHz.

14 Oscillators

An oscillator produces an alternating output voltage of fixed or variable frequency.

The waveform of many types of oscillator is a sinusoid (p. 44) but other waveforms may be generated such as triangular, sawtooth, and square waves. Circuits that produce square waves, often called astable multivibrators, are described on pp. 43 and 190.

Oscillators may be based on a single transistor but an op amp is preferred for greater reliability. Each of the three oscillators described in this Topic operates on a different principle but are all op amp oscillators.

The main requirement is that the op amp should have a full-power bandwidth that includes the intended operating frequency.

Phase shift oscillator

All three oscillators rely on **positive feedback** to keep them oscillating. Part of the output of the op amp is fed back to its non-inverting input to maintain the oscillations.

The network consists of three high-pass filters in series. On p. 48, it is explained that the output from a low-pass filter lags behind the input by up to 90°. Conversely the output of a high-pass filter *leads* input by up to 90°.

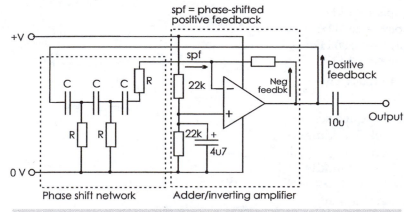

The phase shift oscillator consists of two sub-circuits: an op amp adder/inverting amplifier and a phase shift network. The gain of the op amp is set to about 30.

Three filters could produce a phase lead of up to 270°. At a particular frequency the lead will be 180°. The signal passing through the phase-shift network is fully *out of phase* with the op amp output. This signal is now amplified and inverted by the op amp, so it is *in phase* with the negative feedback signal. The output from the circuit is unstable and oscillates strongly.

The voltage at the positive terminal of the op amp is held constant at half the supply, and stabilised by the capacitor.

A rising voltage at the output is fed back to the inverting input (–) , causing a fall of output. This is negative feedback and the op amp would have a stable output. There is positive feedback too, through the phase shift network.

The output waveform is a sinusoid with frequency $f = 1/15.39RC$.

The circuit is a simple one that is best used as a fixed-frequency oscillator. It would be too difficult to have variable resistors or capacitors in the network and to tune them all at the same time.

Colpitts Oscillator

This type of oscillator depends on a resonant network consisting of two capacitors (total series capacitance = C) and an inductor (L) in parallel with them. As shown on p. 81, this L-C network resonates at a frequency, $f = 1/2\,\pi\sqrt{(LC)}$.

The op amp is wired as an inverting amplifier with a gain of about 30. Its non-inverting (+) input is held at half the supply voltage (+V/2) by the two 22 kΩ resistors acting as a postential divider.

The LC network is placed in the positive feedback loop of the op amp. At the resonant frequency the output from the op amp makes the network resonate. The tapped point between the capacitors is at +V/2, but the part of the oscillating signal across C2 is fed to the inverting amplifier. It is amplified and keeps the network oscillating strongly.

When the circuit is first switched on there is no signal, but small random currents in the circuit are amplified. Those at the resonant frequency are amplifed more than others and gradually build up in amplitude until the circuit is oscillating strongly.

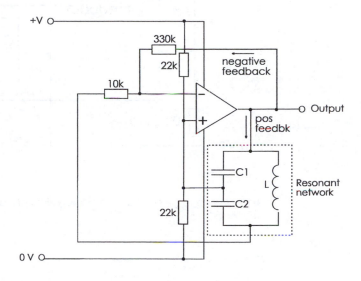

If you are trying out this circuit, the capacitors can be 100 nF each. For the inductor, wind two or three turns of ordinary single-stranded connecting wire about 10 mm in diameter. Neither core nor former is needed.

Oscillator in action

An animated diagram of a Colpitts oscillator is present on the Companion Site.

Wien Bridge oscillator

This oscillator (diagram overleaf) relies on a pair of LR resonamt networks to shift the output signal by exactly 180° and one particular frequency.

The pahse-shifted signal is then fed back to the non-inverting (+) input of the op amp. It is amplified and reinforces the output signal. A strong sinusoidal waveform is generated.

In the Wien network, both capacitors and both resistors are equal in value. The signal frequency is $f = 1/2\pi RC$.

The amplifier has the advantage that it is easily made tunable by using a dual ganged variable resisistor for the two resistors in the Wien network.

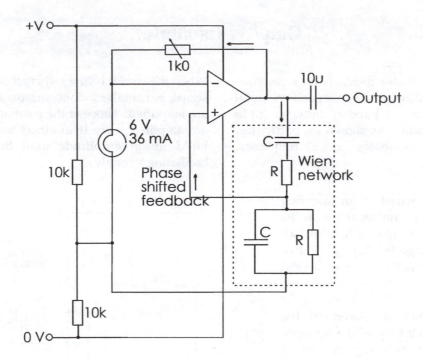

The Wien bridge oscillator produces low-distortion sine waves in the audio range.

The gain of the op amp needs to be held very close to 3. Oscillations die away if gain is less than 3, and the op amp becomes saturated if it is more. Gain is set by the variable feedback resistor and the resistance of the filament lamp.

The resistance of the lamp depends on the average current flowing through it, which depends on the amplitude of the signal. With high gain and high current the filament soon heats up and its resistance increases. This increase of resistance decreases the gain. If gain falls, current falls and the filament cools. The resistance of the lamp decreases and gain is reduced. This is a good example of negative feedback leading to stability.

The point at which the gain stabilises is set by the variable resistor. This is adjusted until a strong permanent signal is produced at the output terminal. The amplifier then has the required gain, close to 3.

Things to do

Try breadboarding these oscillators. View their output signals with an oscilloscope, or measure them using a testmeter with a frequency measuring facility.

On the Web

There are thousands of sites describing oscillator circuits. Search with the keyword 'Colpitts oscillator' and visit the sites that describe the many versions of this type of oscillator. You need a design for a practical project, the Web is the place to find it.

Try the same thing with other types of oscillator.

15 Power amplifiers

A power amplifier is used to produce a major effect on the surroundings. Examples are power audio amplifiers, producing sound at high volume, or motor control circuits actuating the arm of an industrial robot or aligning the dish of a radio telescope. Other examples include the circuits that produce the dramatic effects of disco lighting. The power of these devices is rated in tens or hundreds of watts, sometimes more. Even a pocket-sized CD player can have a 700 mW output, yet the laser light signal that its head is picking up from the disc is far less powerful than that.

The electrical signals that initiate any of these actions may be of extremely low power. For example, the power output of a microphone or many other types of sensor, or the control outputs from a microcontroller, are usually rated at a few milliwatts. The aim is to amplify the power of the signals from these devices so that they can drive powerful speakers, motors or lamps.

The power at which a device is operating depends on only two quantities, the *voltage* across the device and the *current* flowing through it. The relationship is simple:

$$P = IV$$

The amplifiers that we have described in Chapters 8 to 12 have mainly been voltage amplifiers. Some, such as the FET amplifiers, actually produce a current that is proportional to their input voltage but, even then, we usually convert this current to an output voltage by passing it through a resistor. Thus the early stages of amplification are usually voltage amplification. Since P is proportional to V, amplifying the voltage amplitude of a signal amplifies its power by the same amount.

Descriptions of voltage amplifiers often refer to the fact that the currents in amplifiers are small. In many instances the collector or drain current is only 1 mA. There is good reason for this.

As explained in Chapter 27, large *currents* through a semiconductor device generate noise. This type of noise is a random signal, which shows up as background hissing in an audio circuit. If the audio signal is weak, it may not be possible to pick it out against a noisy background. In other kinds of circuit it may become evident as unpredictable behaviour, making the circuit unreliable.

When a signal has been through several stages of amplification it may have reached an amplitude of a few volts or perhaps a few tens of volts. But, for the reason given above, its current is rated as no more than a few milliamps, perhaps less. Consequently, its power level is low. The next and final stage is to amplify the current.

Current amplifiers

The two current amplifiers most often used are the common-drain amplifier and the common-collector amplifier. Both of these are voltage follower amplifiers. Their voltage gain is just less than 1 and they both have high current gain. In addition, they have low output resistance, a useful feature when driving high-power devices.

The voltage output of these amplifiers is generated by passing a variable current through a resistor, either the drain resistor or the collector resistor. The value of the resistor is chosen so that, when there is no signal, the output voltage sits at half way between the 0 V line and the supply voltage. This gives the output voltage room to swing to maximum extent in either direction without clipping or bottoming.

An amplifier of this type is known as a Class A amplifier. We examine this type in more detail in the next section.

Self Test

What are the alternative names for CD and CC amplifiers?

Class A amplifiers

The main disadvantage of Class A amplifiers is that current is flowing through the output transistor and its resistor even when there is no signal. Power is being used but no sound or other form of output activity occurs. Such amplifiers are inefficient because they waste 50% of the energy supplied to them. If an amplifier is to produce enough output power to drive a motor or high-wattage speaker, we must design the output stage of the circuit to avoid such waste.

Class B amplifiers

The **complementary push-pull amplifier** (below) is called 'complementary' because it consists of two amplifiers, one based on an npn transistor (Q1) and the other based on a pnp transistor (Q2). It is called 'push-pull' because one amplifier pushes the output in one direction and the other pulls it in the opposite direction. The amplifiers are both common-collector amplifiers. The circuit is intended as an audio amplifier, with the speaker taking the place of the emitter resistor in both amplifiers.

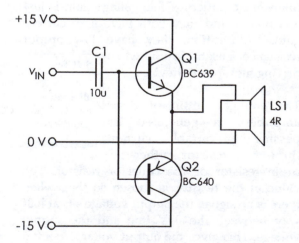

In a Class B amplifier the transistors are biased so that they are off when there is no input signal.

Q1 and Q2 are chosen to have similar characteristics except for their polarity. The amplifier runs on a split supply. When there is no signal the input is at 0 V. This means that both amplifiers are off, and no power is being wasted. This is the distinguishing feature of Class B amplifiers.

When the input goes positive, the pnp amplifier remains off but the npn amplifier is turned on. The reverse happens when the signal goes negative. The resistance of the speaker coil is only a few ohms so a high current passes through it.

Example

A push-pull amplifier operates on a ±12 V supply. The impedance of the speaker coil is 4 Ω. What is the maximum current through the speaker?

$$\text{Current} = 12/4 = 3 \text{ A}$$

This result is only an approximation, since it ignores the base-emitter voltage drop of 0.7 V. Also the impedance of the coil varies with frequency and is greater at high frequencies.

We can take the example further. If the current through the speaker coil is 3 A and the voltage across it is 12 V, the power dissipated is :

$$P = IV = 3 \times 12 = 36 \text{ W}$$

When the input to the amplifier goes negative, the npn transistor goes off. Then the pnp transistor turns on and current flows through the speaker in the opposite direction.

The advantage of a Class B amplifier such as this is that it consumes power only when there is a signal to be amplified. It is much more efficient than Class A. Its greatest drawback is illustrated in the graphs opposite. An npn transistor does not conduct until the base-emitter voltage is greater than about 0.7 V. A pnp transistor does not conduct until its base is about 0.7 V below the emitter. For this reason, signals of amplitude less than 0.7 V have no effect on the amplifier. The output is zero.

With signals greater than 0.7 V in amplitude there is a period of no output every time v_{IN} swings across from positive to negative or from negative to positive.

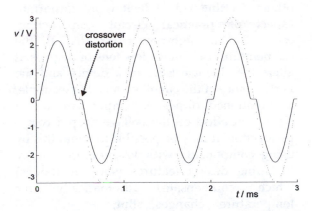

The amplifier opposite shows severe crossover distortion as the signal (grey) swings between positive and negative.

This introduces considerable distortion, known as **crossover distortion**. It is proportionately greater in small signals. The graphs above also illustrate the fact that the voltage gain of each of the follower amplifiers is less than 1. In this case it is approximately 0.8.

Eliminating crossover distortion

A common technique for eliminating crossover distortion is to bias the transistors so that they are just on the point of conducting. With BJTs, this means biasing them so that v_{BE} is equal to 0.7 V when no signal is present. The most convenient way to do this is to make use of the 0.7 V voltage drop across a forward-biased diode.

In the circuit on the right, current flows through the chain consisting of R1, D1, D2 and R2. There are voltage drops of 0.7 V across the base-emitter junctions of both transistors. These raise the base voltage of Q1 to 0.7 V above the value of v_{IN}. When there is no signal, v_{IN} is zero so the base of Q1 is at +0.7 V. It conducts as soon as v_{IN} increases above zero.

Similarly, the base of Q2 is at –0.7 V and Q2 is ready to conduct as soon as v_{IN} falls below zero. In short, as soon as v_{IN} begins to increase or decrease, one or other of the transistors is ready to conduct.

There is very little crossover distortion, if any. This modification of the Class B amplifier is referred to as a Class AB amplifier.

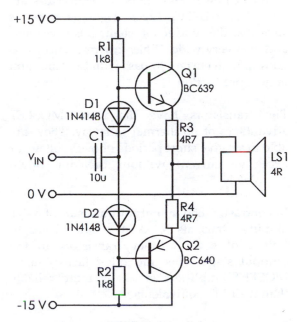

The diodes in this version of the Class B amplifier bias the transistors so that they are ready to conduct as soon as the signal rises above or falls below 0 V.

The diodes have an additional advantage. Consider Q1 and its associated diode D1. The diode, like the base-emitter junction of Q1, is a forward biased pn junction. As temperature changes, the forward voltage drop (v_{BE}) across the base-emitter junction changes. But the drop across the diode changes too, and compensates for changes in the transistor. Whatever the temperature, the voltage drops of D1 and Q1 are equal and Q1 remains biased, ready to begin conducting with any increase of v_{IN}. The same applies to Q2 and D2. In this way, the diodes improve the temperature stability of the amplifier.

MOSFET power amplifiers

A similar Class B amplifier can be based on power MOSFETs. A complementary pair is used, consisting of an n-channel MOSFET and a matching p-channel MOSFET. A potential divider running between the power rails holds the gates at their threshold voltages when there is no signal.

Often in power amplifiers the transistors are VMOS or HEXFET types. Their geometry is such that the conduction channel is very short and also very wide. It has low resistance. This allow a high current to pass with the minimum of heat production.

The transistors have the usual MOSFET advantages of no thermal runaway. They are able to survive short periods of over-voltage or excessive current without breaking down completely.

In contrast, when an output transistor of a BJT amplifier fails, its breakdown often leads to failure of several other transistors in the outoput stage. This and other factors make MOSFET amplifiers generally more reliable than their BJT equivalents.

In power amplifiers, MOSFET transistors have the advantage of being very fast-acting. They can vary currents of several amperes in a few nanoseconds. They are about 30 to 100 times faster than BJTs. The low input resistance of the gate means that low currents can be used in the drive circuits.

When a BJT is nearing saturation, the output waveform becomes severely and abruptly clipped. In an audio amplifier, this results in an unpleasant distortion of the sound. Because of the non-linear transfer characteristic (p. 63) of MOSFETs, the clipping is more gradual, producing a more rounded waveform. This is still a distortion but it sounds less unpleasant.

Valve amplifiers have a similar characteristic, producing a more mellow and 'warmer' sound than BJT amplifiers, and this accounts for the continuing popularity of valve amplifiers.

Heating problems

Heat is generated whenever current flows through a resistance, so heat is generated in all components, though more in some than in others. Getting rid of heat is an important aspect of practical circuit construction, especially in high-power circuits. If the temperature rises only a few tens of degrees, it alters the characteristics of a device and the performance of the circuit suffers. For example, as mentioned opposite, temperature has a significant effect on the voltage drop across a pn junction. It may be possible to avoid this by using components with low tempcos or by including design features (such as diodes) which compensate automatically for temperature changes. But, when the power dissipation of the circuit is rated in watts, temperatures may quickly rise above 150°C, at which temperature semiconductor devices are completely destroyed.

> **Tempco**
> Short for temperature coefficient.

The temperature that matters most is the temperature of the active regions of the device, the semiconductor junctions. This is where the maximum heat is generated and most damage is done. The task is to remove this heat and to tranfer it to the air around the circuit. We must assume that the temperature of the environment, the ambient temperature t_A, is sufficiently low and that the air is able to circulate freely to carry the heat away.

Heat sinks

The usual way of increasing the transfer of excess heat is to attach a heat sink to the device. This is made of metal, most frequently aluminium, and is provided with fins to allow maximum contact with the air.

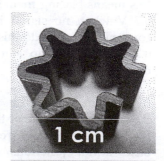

This heat sink clips on to a TO-39 cylindrical metal can. Its thermal resistance is 30°C/W.

The surface of the heat sink is usually anodised black, which, compared with polished aluminium, increases heat loss by radiation by as much as ten times.

This bolt-on heat sink has a thermal resistance of 19°C/W.

Small heat sinks may clip on to the transistor but the larger ones are bolted on. They often have pegs for soldering the heat sink securely to the circuit board.

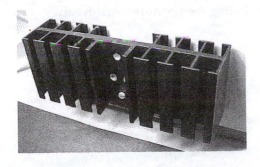

The thermal resistance of this high-power heat sink is 2°C/W. The heat sink is mounted on the outside of the enclosure to allow free circulation of air.

The largest heat sinks of all are bolted to the circuit chassis or the enclosure. Then one or more transistors are bolted to the heat sink. Components that are likely to become very hot in operation should be located near the edge of the circuit board, or even on a heat sink mounted on the outside of the enclosure.

Heat sinks are sometimes made of copper. This metal has higher thermal conductivity than aluminium, so copper heat sinks are used if there is a large amount of heat to be conducted away. The disadvantage is that copper sinks are heavier than aluminium sinks of the same size. Also copper is more expensive. If there is space on the circuit board, a heat sink can be made by leaving a large area of the copper layer unetched. Transistors are bent over on their leads and bolted to this copper area.

A heat sink must be mounted with its fins vertical to allow convection currents to flow freely. Cooling is reduced by around 70% if the fins are horizontal. For the same reason, it is helpful to mount circuit boards vertically if there are many of them in a small enclosure.

It is of course essential that the circuit enclosure should have adequate ventilation. This may be aided by a fan if the circuit dissipates large amounts of power, for example, a microcomputer circuit.

Thermal resistance

Thermal resistance (R_θ) is the ability of something to resist the transfer of heat. It is measured in degrees Celsius per watt (°C/W). If an object is being heated on one side by a source of heat rated at P watts, the hot object is t°C, and the ambient temperature is t_A°C, the thermal resistance of that object is given by:

$$R_\theta = \frac{t - t_A}{P}$$

The situation is illustrated by the diagram overleaf.

Examples

1 The metal tag of a power transistor operating at 2.2 W is at 90°C. It is bolted to a heat sink and the ambient temperature (inside the circuit enclosure) is 35°C. What is the thermal resistance of the heat sink?

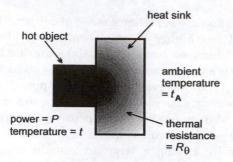

A hot object, such as an operating transistor, is placed in contact with a material that conducts heat away from it and passes it to the air around.

For a relatively small object such as a heat sink made from a metal of good thermal conductivity and with free air circulation around it, we can assume that all parts of its surface, except that in contact with the transistor tag, are at ambient temperature.

Temperature difference = 90 – 35 = 55°C

Thermal resistance = 55/2.2 = 25°C/W

2 For the same transistor, the manufacturers give the thermal resistance between the tag and the actual junction on the chip as 0.9°C per watt. What is the junction temperature and is it dangerously high?

Rearranging the equation gives:

$$t - t_A = R_\theta \times P$$
$$= 0.9 \times 2.2$$
$$= 1.98°C$$

The junction is approximately 2 degrees hotter than the tag, which is at 90°C, so:

Junction temperature = 92°C

Temperatures up to 150°C are usually regarded as safe.

In these examples we have ignored the fact that parts of the transistor case and the tag are in direct contact with the air. Some heat is lost from these surfaces by convection, but usually the amount is relatively small and can be disregarded.

The examples take three thermal resistances into account:

- the resistance between the chip and the outside of the case, usually a metal tag for power devices.

- the resistance between the area where the heat sink is bolted to the tag and the surfaces of the fins in contact with the air.

It is important for the tag to be in good thermal contact with the heat sink. Their surfaces must be as flat as possible. Often we coat the surfaces with silicone grease or special heat transfer compound to fill in the small gaps caused by slight unevenness of the surfaces.

Greasing the surfaces can reduce the thermal resistance of the surface-to-surface interface by about a half. If the tag is bolted to the heat sink to keep them firmly in contact (below) , the nut must not be tightened too much, for this may distort or even crack the transistor inside and lead to early failure. The effect can be mimimised by using a bolt that is just able to pass through the holes, as shown in the sectional diagram below. A washer helps by distributing the pressure more evenly.

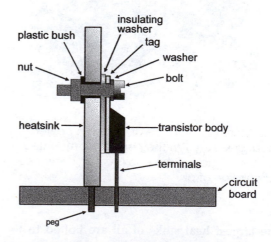

Showing how a transistor in a TO220 or similar package is attached to a heat sink. The heat sink may be larger than shown here. The heatsink has two projecting pegs, that are soldered to the circuit board to give firm support.

With certain types of transistor or other device the tag is in electrical contact with one of the terminals of the device. An example is the 2N3055 transistor in its metal TO66 case. It has two wire terminals connecting to the base and emitter. The connection to the collector is the case itself.

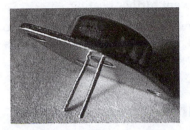

The TO3 power transistor case, showing the two wire terminals (base, emitter) and the case (collector).

With metal cased devices such as these, it is usually necessary to mount an insulating washer between the device and the heat sink. This introduces another thermal resistance.

Washers made of mica or plastic have thermal resistances of about 2–3°C/W, or about half of this if they are greased on both sides. There are also silicone rubber washers, which have even lower thermal resistance and are used without greasing. Washers may be needed to provide electrical insulation when two or more devices are mounted on the same heat sink.

Mica washers are used to electrically insulate the heat sink from the tag of the transistor. Also shown are three plastic bushes.

As well as a washer, it is usual to employ a plastic bush around the bolt to prevent it from making an electrical connection between the tag and the heat sink.

Example

A transistor running at 40 W, has a junction-to-tag thermal resistance of 0.8°C/W. It is bolted to a heat sink with resistance 2.4°C/W, and is insulated by a greased mica washer with resistance 0.1°C/W.

The heat sink is mounted outside the enclosure in an ambient temperature of 27°C. What is the junction temperature and is it safe?

The resistances are in series, so the solution is:
Junction temp. = power × sum of resistances + ambient temp.

$$= 40 (0.8 + 0.1 + 2.4) + 27$$
$$= 159°C$$

This is higher than the maximum safe operating temperature of 150°C. Use a heat sink with lower thermal resistance, such as 1.2°C/W. This reduces the junction temperature to 40 × 2.1 + 27 = 111°C, allowing the ambient temperature to rise higher with safety.

Integrated circuits

Power amplifiers have so many applications, especially in the audio and TV fields, that it is economic to produce them in large numbers as integrated circuits. Only a few external components are required. An integrated circuit amplifier simplifies wiring, saves circuit-board space, and is generally cheaper. Having all the active components on the same chip, so that they are all at the same temperature, leads to greater reliability and stability. Typical of the IC power amplifiers is the TDA2040, which is packaged as a 5-pin device with a metal tag for bolting to a heat sink. The circuit overleaf shows how the TDA2040 is connected to build a Class AB audio amplifier capable of driving a 4 Ω speaker at 22 W, when operating on a ±16 V supply.

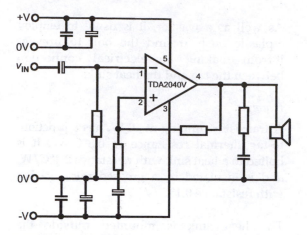

A power amplifier in integrated circuit form may need only a few external resistors and capacitors to complete it.

The circuit has many of the features of an inverting op amp circuit. There are two inputs, (+) and (–), with feedback from the output to the (–) input. The input stage of the IC is a differential amplifier. Some of the external resistors and capacitors make up filters. Others are there to hold the supply voltage steady against feedback through the speaker to the 0 V supply line.

The description in the data sheet makes it clear that this is not just a simple op amp. Like most IC audio amplifiers, it has several additional features, including:

The LM386 is one of the smaller IC power amplifiers, housed in an 8-pin case. In spite of its small size, it is capable of producing up to 780 mW of power, and has a voltage gain of up to 200. It requires five external capacitors and four resistors, one of which is a volume control.

- Short circuit protection. If the output terminal is short-circuited, the output current is automatically limited to prevent the transistors being destroyed by over-heating.

- Thermal shutdown. If the device becomes too hot for any reason, it is automatically shut down to limit current and thus prevent further heating.

- The tag is connected internally to the 0 V rail, so no insulating washer is needed between it and the heat sink.

Activity — Power amplifier

Build and test a Class B amplifier based on a complementary pair of BJTs, using the basic circuit shown on p. 128. Then modify the circuit as in the diagram on p. 129. In each case, use an oscilloscope to observe the waveform of the output when the input signal is a 500 Hz sinusoid, amplitude 1 V.

Questions on power amplifiers

1   Describe an example of a Class A amplifier and explain why it is inefficient.

2   Describe a complementary push-pull power amplifier and how it works. In what way is it more efficient than a Class A amplifier?

3   What is meant by *crossover distortion* and what can be done to avoid it?

4   Describe the precautions you would take to prevent the transistors of a power amplifier from overheating.

5   Using information from the manufacturer's data sheet, design an audio power amplifier based on an integrated circuit.

6   A transistor in a power amplifier has 12 V across it and passes 4.4 A. It is greased and then mounted on a heat sink with a thermal resistance 1.2°C/W. The junction-to-tag resistance is 1.1°C/W and the tag-to-heat-sink resistance of the grease is 0.1°C/W. What is the maximum ambient temperature under which the amplifier can safely operate?

7   Explain why it may be necessary to place an insulating washer between the tag of the transistor and the heat sink. In what way may the thermal resistance between the tag and the heat sink be made as small as possible?

8   With the amplifier described in Question 6 explain, giving reasons, what measures you could take to make the amplifier suitable for operating at higher ambient temperatures.

9   Explain the advantages of MOSFETs for use in the output stage of Class B amplifiers.

Multiple choice questions

1   A basic type of BJT amplifier used in power amplification is the:

   A common-base amplifier
   B common-emitter amplifier
   C common-collector amplifier
   D common-drain amplifier.

2   The voltage gain of a typical Class B power amplifier is:

   A 100
   B less than 1
   C more than 1
   D 20.

3   If the tag of a transistor is at 60°C, t_A is 25°C, and the transistor is running at 3.5 W, the thermal resistance of the heat sink is:

   A 10°C/W
   B 35°C
   C 10 W/°C
   D 0.1°C/W.

4   The maximum allowable temperature of a transistor junction is:

   A the ambient temperature
   B 150°C above the ambient temperature
   C 100°C
   D 150°C.

16 Thyristors and triacs

A thyristor is also known as **silicon controlled switch (SCS)** and as a **silicon controlled rectifier (SCR)**. The word *switch* tells us that it can be turned on and off. The word *rectifier* indicates that current flows through it in only one direction.

A thyristor consists of four semiconductor layers (below) with connections to three of them.

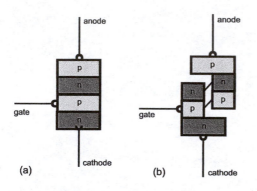

(a)   (b)

anode   anode

gate   gate

cathode   cathode

A thyristor is a four-layered device, but can be thought of as a pnp transistor joined to an npn transistor.

Its structure is easier to understand if we think of the middle layers being split into two, though still electrically connected. Then we have two three-layer devices, which are recognisable as a pnp transistor and an npn transistor.

The base of the pnp transistor (n-type) is connected to the collector of the npn transistor (also n-type). The collector of the pnp transistor (p-type) is connected to the base of the npn transistor (also p-type). These connections show more clearly in the drawing at top right, where the two transistors are represented by their usual symbols. In this figure the terminals have been given the names that apply to a thyristor.

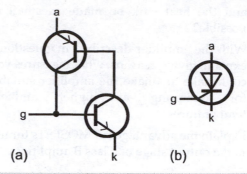

(a)   k   (b)

a   a   g   k   g

It is easier to understand thyristor action if we think of it as a pnp transistor and an npn transistor combined. Its symbol shows that it has the one-way conduction property of a diode.

The three terminals are:

- **Anode**: the emitter of the pnp transistor; current enters the thyristor here.

- **Cathode**: the emitter of the npn transistor; current leaves the thyristor here.

- **Gate**: the base of the npn transistor.

The names *anode* and *cathode* are the same as those given to the terminals of a diode. This shows that the thyristor has the same kind of one-way action as a diode and, like a diode, may be used as a rectifier. We will look more closely at this action to see how the voltage present at the gate terminal triggers it.

Thyristor action

When the gate is at the same voltage as the cathode, the npn transistor is off. No current can flow through the thyristor. If the gate voltage is made 0.7 V higher than that of the cathode, the npn transistor begins to turn on. Current begins to flow through the npn transistor. This current is drawn from the base of the pnp transistor.

Base-emitter voltage, v_{BE}

An npn transistor begins to turn on when this exceeds about 0.7 V.

Drawing current from its base begins to turn the pnp transistor on. Current begins to flow through it. This current goes to the base of the npn transistor, adding to the current already flowing there and turning the npn transistor more strongly on. Its collector current increases, drawing more from the base of the pnp transistor and turning it more strongly on. This is positive feedback and results in both transistors being turned on very quickly.

> **Note**
> This kind of action is described as *regenerative*.

The whole process, from the initial small gate current (base current) to the transistors being fully on, takes only about 1 microsecond.

Once current has begun to flow, it continues to flow even if the original positive voltage at the gate is removed. The current from the pnp transistor is enough to keep the npn transistor switched on. This means that the gate need receive only a very short positive pulse to switch the thyristor on. Typically, a gate voltage of about 1 V is enough to start turning the npn transistor on, and the current required may be only a fraction of a milliampere.

Once the thyristor has been triggered, it continues to pass current for as long as the supply voltage is across it. If the supply voltage is removed, current ceases to flow and it requires another pulse on the gate to start it again. There is a minimum current, known as the **holding current**, which must be maintained to keep the thyristor in the conducting state. If the supply voltage is reduced so that the current through it becomes less than the holding current, conduction ceases and a new gate pulse is needed to re-trigger it. Holding current is typically 5 mA but may be up to 40 mA or more with certain types of thyristor.

The voltage drop across a conducting thyristor is only about 1 V, though may be up to 2 V, depending on the type. Low voltage results in relatively low power, even if current is high. This means that the thyristor may pass a very high current without dissipating much energy and becoming excessively hot.

Most thyristors can pass currents measured in amperes or tens of amperes. Yet a relatively minute gate pulse triggers such currents. It is the high gain of thyristors, combined with their extremely rapid switch-on which makes them popular for power control circuits.

> **Note**
> The amount of heat generated depends on how many watts of power are dissipated in the thyristor. This depends on the voltage between anode and cathode and the current through the thyristor. $P = IV$.

A simple npn or pnp power transistor is much less satisfactory for this purpose. If it is to have sufficient gain to switch a large current, its base layer needs to be very thin. This makes it unable to withstand high reverse voltages. By comparison, thyristors can withstand a reverse voltage of several hundred volts without breaking down. Many thyristors have a reverse **breakover voltage** of 400 V, and some can withstand a reverse voltage of more than 1000 V. For this reason, thyristors are widely used for switching mains or higher voltages.

Switching DC

A thyristor may be used to turn on a high-power device when only a small triggering current is available. For example, the light-dependent resistor (LDR) in the circuit below triggers a thyristor to switch an alarm.

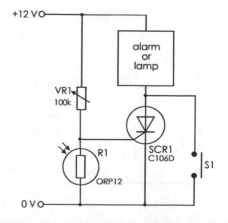

Making use of the thyristor's high gain and rapid action in an alarm circuit.

The resistance of an LDR decreases as the amount of light falling on it decreases. When the light level falls below the level set by adjusting VR1, the LDR resistance increases, voltage at the gate rises and triggers the thyristor. Once triggered, the thyristor conducts for an unlimited period, even if light increases again. The circuit is sensitive to very short interruptions in the light, such as might be caused by the shadow of a passing intruder falling on the LDR.

This kind of switching action is mainly limited to alarm-type circuits with a once-for-all action. The alarm sounds until the reset button S1 is pressed or the power is switched off. A lamp can replace the alarm so that the thyristor turns on the lamp when light level falls. With some circuits not based on a thyristor, the light coming on could cause negative feedback and turn the lamp off again. The lamp would flicker on and off. There is no such problem in this circuit because of the once-for-all action.

Switching AC

Most thyristor applications are concerned with AC circuits. The circuit below supplies power to a device represented by a 100 Ω resistance, R_L. The supply comes from the mains at 320 V AC (= 230 V rms, p. 363) and 50 Hz.

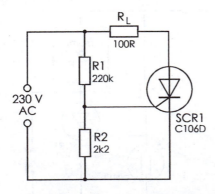

A basic half-wave controlled rectifier, using a thyristor.

Operating voltage

Most of the thyristor and triac circuits in this chapter are shown supplied from the mains. This is because the control of mains-powered devices is the most important application of these devices. To investigate the action of these circuits practically but safely, **use a supply of lower voltage**, such as **12 V** AC or **20 V** AC taken from a bench power unit. Or run the circuits on a simulator, when the *simulated* supply may safely be set to 322 V (the actual mains voltage, 230 V, is its rms voltage (p. 365)..

On the positive half-cycle there is a positive voltage across R2, which acts to turn on the thyristor. It conducts for as long as the voltage across it is in the positive direction, from anode to cathode. The thyristor is turned off at the end of the half-cycle. On the negative half-cycle the voltage across R2 is negative, so the thyristor is not turned on. The result is that the thyristor conducts only during the positive half-cycle. It behaves just like a diode, and gives half-wave rectified current.

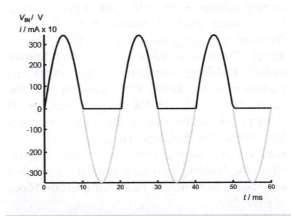

The output (black) of the half-wave rectifier follows the input (grey) only during the positive half-cycles. There is a very slight delay at the beginning of every cycle.

In the figure we see the curve for current (black) following the input voltage curve (grey) during the positive half-cycle only.

In this circuit, the thyristor is turned on at the beginning of every half-cycle. The resistor network provides the necessary pulses, which are synchronised with the alternating supply. The pulse occurs at the same point or phase early in every cycle. Because the timing of the pulses is related to the phase of the cycle, we call it **phase control**.

There can be no pulse at the very beginning of the cycle because the supply voltage is zero at that time. But very soon the input voltage has reached a level which produces a voltage across R2 that is sufficient to trigger the thyristor. The phase at which the pulse is generated is the **firing angle**. In the plot opposite, the pulse begins about 0.3 ms after the cycle begins, and the length of the cycle is 20 ms, so the firing angle is $360 \times 0.3/20 = 5.4°$. Current flows for the remainder of the half-cycle so the **conduction angle** is $180 - 5.4 = 174.6°$.

If R2 is a variable resistor it is possible to reduce its value and so reduce the amplitude of the triggering pulses. It then takes longer for the pulse to reach a value that will trigger the thyristor. Firing is delayed until later in the cycle. The plot below shows what happens if R2 is reduced to 560 Ω. The delay at the start of the cycle is now 2.2 ms. This gives a firing angle of $360 \times 2.2/20 = 40°$. The conduction angle is $180 - 40 = 140°$.

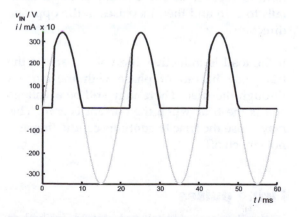

Reducing the value of R2 increases the delay at the beginning of each cycle, so that the conduction angle is smaller, and the amount of power supplied to the load is reduced.

The value of R2 controls the delay at the beginning of each cycle, which controls the firing angle and conduction angle. The conduction angle determines how much power is delivered to the load during each cycle. The circuit is a **half-wave controlled rectifier** which can be used for controlling the brightness of mains-powered lamps and the speed of mains-powered motors. There is also the fact that the thyristor rectifies the alternating current to produce pulsed direct current, which can be used for driving DC motors. But this is a half-wave circuit so, at the most, the lamp or motor is driven at only half power. This is a basic circuit and several improvements are possible, as will be explained later.

Full-wave controlled rectifier

There are various ways of using thyristors in full-wave controlled rectifiers. One of these has two thyristors in parallel but with opposite polarity so that one conducts on the positive half cycle and the other on the negative half cycle. They each have a pulse generator synchronised to switch on the thyristor during the appropriate half-cycle. A different circuit, which requires only one thyristor, is illustrated below.

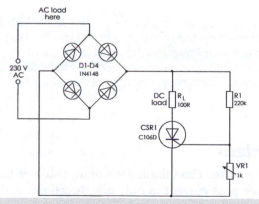

A full-wave controlled rectifier has the same circuit as opposite, but the power is first rectified by a full-wave diode bridge.

The Companion website has animated diagrams that show how a rectifier operates.

The current is rectified by a conventional diode bridge, and the thyristor acts to turn the current on and off during both positive and negative half-cycles. The thyristor is able to act during both half-cycles because it is located in the rectified part of the circuit so that both half-cycles are positive. This circuit is suitable only for loads that operate on DC.

As indicated in the circuit diagram, the load R_L can be placed in the mains side of the diode bridge instead. The current still has to flow through the thyristor to complete its path through the circuit. The thyristor turns the current on and off during both positive and negative half-cycles, as shown below.

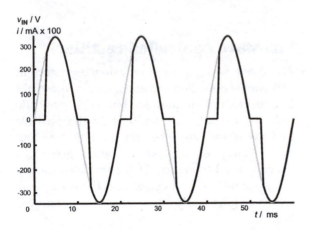

If the load on a full-wave controlled rectifier is placed on the mains side of the diode bridge, the current through the load (black) is alternating and is controlled during both positive and negative half-cycles.

Triacs

One of the chief limitations of thyristors is that they pass current in only one direction. At the most, they can operate only at half-power.

If we require power during both half-cycles we need to wire two thyristors in parallel with opposite polarity, or rectify the supply. The triac is a device in which two thyristors are combined in parallel but with opposite polarity in a single block of silicon.

Gating is provided for one of the thyristors, but when the thyristors are combined in this way, the gate switches either one of them, whichever one is capable of conducting at that instant.

Another useful feature is that either positive or negative pulses trigger the device. Like a thyristor, the device switches off whenever the supply voltage falls to zero, and then requires re-triggering. It also switches off if the current through it falls below the minimum holding current.

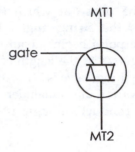

Because the triac conducts in both directions, its main terminals are simply named MT1, MT2. The third terminal is the gate.

Although triacs have advantages over thyristors, they have a disadvantage too. Since they conduct on both cycles they have little time to recover as the current through them falls to zero and then increases in the opposite direction.

If the load is inductive, the voltage across the triac may be out of phase with the current through the triac. There may still be a voltage across the triac when the current is zero. This may cause the triac to continue conducting and not switch off.

Firing pulses

In the thyristor and triac circuits described so far, we have used a pair of resistors as a potential divider to derive the firing pulse from the mains voltage.

The simple two-resistor system ensures that the pulse occurs at a fixed firing angle. The pulse can be made to occur at any time during the first 90° of the half-cycle (as voltage increases) but the conduction angle must always be between 90° and 180°. If we want a conduction angle between 0° and 90°, we use a capacitor to delay the pulse for more than 90°.

The lamp-dimming circuit below illustrates this technique. The triggering network consists of a resistor and capacitor. If VR1 is increased in value, the charge across the capacitor rises to its maximum later in the cycle.

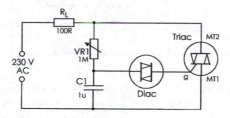

The plot below shows firing at 126° and 306°. The triac is triggered by a positive pulse during the positive half-cycle and by a negative pulse during the negative half-cycle.

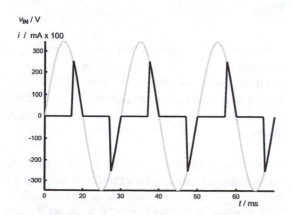

Although the resistors and capacitors in triggering networks pass very small currents, and so are not wasteful of power, there is always the risk of their eventually breaking down. They are directly connected to wires at mains voltage and there is danger in this. In commercially produced, manually operated circuits the control knobs are sufficiently well insulated to prevent the operator from receiving an electric shock, but when a circuit is automatically operated, perhaps by a computer, we need to take precautions to prevent high voltages reaching circuits where they can do expensive damage. Automatic thyristor and triac circuits often employ one of the following devices to isolate the control circuit from the thyristor or triac:

- **Pulse transformer:** This is usually a 1:1 transformer so that the pulse generated in the control circuit is transferred virtually unchanged to the gate of the thyristor or triac. The insulation between the primary and secondary windings withstands voltages of 2000 V or more.

- **Opto isolator:** Opto isolators are often used when it is necessary to isolate two circuits, one at low voltage and one at high voltage, yet to pass a signal from one to the other. A typical opto-isolator consists of an integrated circuit package containing an infrared LED and a phototransistor. A pulse is passed through the LED causing it to emit light.

> **Infrared LED**
>
> A light-emitting diode that emits infrared radiation (invisible to the eye) instead of visible red or light of other colours.

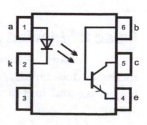

In a circuit controlled by a computer or microcontroller, the LED could be lit by a pulse from the output of a logic gate. The phototransistor receives the light energy and is switched on. Current flows through the transistor. The diagram below shows how this current is used to generate a pulse at the gate of a triac. The opto-isolator is constructed so that its insulation can withstand voltages of 1000 V or more between the diode and the transistor.

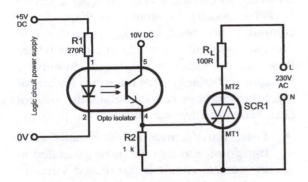

The light from the LED provides sufficient energy to turn on the phototransistor. The current through R2 then generates a positive pulse to turn on the triac.

- **Opto triac isolator:** This comprises an LED and a triac with a photosensitive gate. A pulse of current applied to the LED causes the triac to be triggered into conduction.

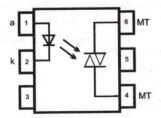

In an opto triac isolator, the light from the LED triggers the triac directly.

False triggering

If the voltage across a thyristor or triac increases too rapidly, this may lead to false triggering and loss of control. This may occur when:

- power is first applied to the circuit.

- there is an inductive load, such as a motor, causing voltage and current to be out of phase.

- there are spikes on the supply lines.

The cure for this is to wire a resistor-capacitor network across the device, to prevent the voltage across it from rising too rapidly. This is known as a **snubber** network and is illustrated by R1 and C2 below. This network could be added to any of the circuits described in this chapter.

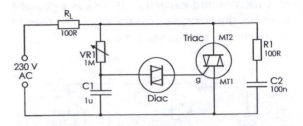

A snubber network (R1 and C2), when added to the triac circuit, prevents the triac from being triggered by a rapid rise in the voltage across it.

A relatively new development is the **snubberless triac**. This can be subjected to voltage change rate as high as 750 V/ms without causing false triggering. A snubber network is not required with this device.

Diac

A triac can be triggered directly by applying a positive or negative pulse to its gate, as in circuits such as that on the left, above. Triggering is made more reliable by including a diac in the trigger circuit, as on the right, above.

A diac has a structure similar to that of a triac but has no gate terminal. It conducts only when the voltage across it exceeds a certain value. This is usually in the range 20 V to 30 V. The action is like the avalanche breakdown of a Zener diode (p. 4).

Once the breakdown voltage is reached, current suddenly flows freely through the diac. The sudden increase in current generates a sharp triggering pulse at an accurately fixed point in the cycle.

RFI

When a thyristor or triac switches on or off during a half-cycle, there is a rapid change in current. Rapid changes in current are equivalent to a high-frequency signal. The result is the generation of **radio frequency interference**.

You can demonstrate this by holding a portable AM radio receiver close to a domestic lamp dimmer switch. RFI is radiated from the circuit and also appears on the mains supply lines which are providing power to it. It is conducted along the mains lines to other equipment, where it may interfere with their operation. The larger the currents being switched by the thyristor or triac, the stronger the interference.

To reduce RFI it is usual to place an RF filter on the supply lines, as shown in the circuit below. This is wired as close as possible to the thyristor or triac so as to reduce the amount of RFI being radiated from the lines joining the thyristor or triac to the load. L1 and C3 form a lowpass filter with the cutoff frequency in the RF band.

Integral cycle control

All the circuits described so far in this chapter are examples of **phase control**. The thyristor or triac is switched on at a particular phase of the half-cycle and remains on until the half-cycle ends. The amount of power supplied to the load depends on for how large a fraction of the half-cycle the thyristor or triac is switched on.

In integral cycle control the device is switched on for *complete* half-cycles. The amount of power supplied to the load depends on for how many half-cycles it is switched on and for how many half-cycles it is switched off. The advantage of this technique is that switching occurs when the supply voltage is at or close to zero. There is no large increase or decrease of current through the load, so RFI is reduced to the minimum. Integral cycle control is sometimes referred to as **burst fire** control

As an example, we could supply power to a motor for five half-cycles, then cut off the power for the same number of half-cycles. This supplies power for half the time. Power is interrupted for five half-cycles (for 0.1 s on a 50 Hz mains supply) but the inertia of the motor and the mechanism it is driving keeps it rotating during the off periods. An electric heater may also be controlled by this technique because interruptions of a fraction of a second or even several seconds makes little difference to the temperature of the heating element.

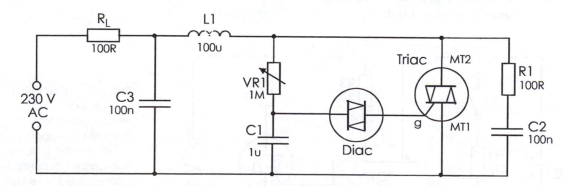

A lowpass RF filter (L1 and C3), when added to the circuit, prevents radiofrequency interference produced by the triac from spreading to the supply mains.

On the other hand, this form of control is not suitable for controlling the brightness of lamps because it makes them flicker noticeably.

Integral cycle control relies on being able to detect the instant at which the alternating supply has zero value and is changing direction.

This is the zero-crossing point at the beginning of every half-cycle. A zero-crossing detecting circuit can be built from discrete components but special integrated circuits are more often used. Some triac opto-couplers have a built-in zero-crossing detector that allows the triac to be switched on only at the beginning of the half-cycle.

A typical example of a zero-crossing switch IC is the CA3059, which can be used to control either thyristors or triacs. The diagram shows it being used to provide on/off control of a triac. The integrated circuit contains four main sub-circuits:

- A regulated 6.5 V DC power supply for the internal circuits. This has an outlet at pin 2 for supplying external control circuits.

- Zero-crossing detector, taking its input from the mains supply at pin 5.

- High-gain differential amplifier, with its inputs at pins 9 and 13. Depending on the application, this can accept inputs from:

  ◊ a switch (as in the diagram), for simple on/off control.

  ◊ a sensing circuit based on a device such as a thermistor or light-dependent resistor, for automatic control of a heater or lamp.

  ◊ a transistor or opto-isolator, for control by automatic circuits, including computers and microcontrollers.

  ◊ an astable for burst-fire control.

- Triac gate drive with output at pin 4. It produces a pulse when the mains voltage crosses zero volts. Pulses are inhibited if the pin 9 input of the differential amplifier is positive of the other input at pin 13. Pulses may also be inhibited by a signal applied to the 'inhibit' input at pin 1 and the 'fail-safe' input at pin 14. The drive can be triggered by an external input at pin 6.

As can be seen from the diagram, the CA3059 needs very few external components. These comprise:

- The triac, with its load R_L.

- A snubber network connected across the terminals of the triac (not shown in the figure).

- A dropper resistor R_3, which should be a high-wattage type (about 5 W).

- Smoothing capacitor C_1 for the 6.5 V supply.

- A control circuit.

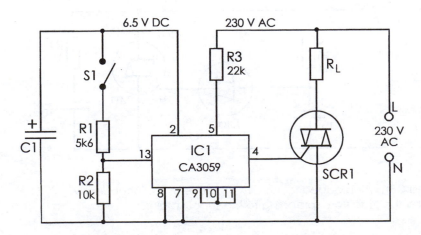

The integrated circuit is a zero-crossing switch, used to minimise RFI and to interface the triac to the control circuitry, represented here by a simple on/off switch.

In the figure, the control circuit consists of S1 and a pair of resistors. Pin 9 is connected to pins 10 and 11 and this puts it at mid-rail voltage (approx 3 V). When S1 is open, R2 pulls pin 13 down toward 0 V. The gate drive circuit is inhibited because pin 13 is negative of pin 9 and the triac does not fire.

When S1 is closed, pin 13 is raised to just over 4 V.

This makes it more positive than pin 9 and enabling the gate drive to produce pulses. The triac is fired at the beginning of every half-cycle and full power is applied to the load.

Activity — thyristor action

Assemble the circuit shown below. Apply the 12 V DC supply from a bench PSU.

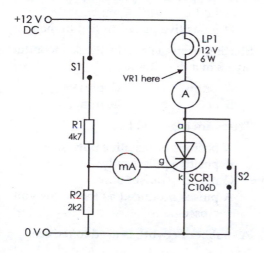

This circuit is used for investigating the action of a thyristor. Note that the supply is 12 V DC.

Record and explain what happens when:
- power is first applied.
- S1 is pressed and held.
- S1 is released.
- S2 is pressed and held.
- S2 is released.

Measure the current flowing to the gate when S1 is pressed and held. Measure the current flowing through the lamp when it is lit. Add a 22 kΩ variable resistor VR1 to the circuit. Begin with VR1 set to its minimum resistance so that maximum current flows. The lamp lights when S1 is pressed. Measure the current through it.

Repeat with VR1 set to various higher resistances and find the minimum holding current of the thyristor. The lamp may not light when current is small but the meter shows if a current is flowing.

Activity — Controlled rectifier

Set up the half-wave and full-wave circuits illustrated on pp. 138-140. Note the remarks in the box headed 'Operating voltage' (p. 138) and use a low-voltage AC supply, NOT the mains. Use an oscilloscope to monitor the input and output waveforms.

Phototransistors	Box 18

When light falls on the base of a transistor, some of the electrons in the lattice gain energy and escape. This provides electron-hole pairs, and is equivalent to supplying a base current to the transistor.

If the collector is positive of the emitter, the usual transistor action then occurs and a relatively large collector current flows through the transistor.

Transistors are usually enclosed in light-proof packages so that this effect can not occur, but a phototransistor is enclosed in a transparent plastic package or in a metal package with a transparent window.

Many types of phototransistor have no base terminal, as the device will work without it.

Opto devices   Box 19

Opto devices have an LED and an amplifying semiconductor device enclosed in a light-proof package. Their function is to pass a signal from one circuit to another. Opto **couplers** are used when, for *electronic* reasons, it is not suitable to connect the circuits directly (perhaps voltage levels are not compatible, or we want the two 'ground' rails to be isolated from each other to prevent interference being carried across). Opto **isolators** are specially designed to withstand a large voltage difference (usually 1000 V or more) between the two sides and are used for *safety* reasons, when a high-power circuit is to be controlled by a low-power circuit.

Problems on thyristors and triacs

1 A thyristor has its anode several volts positive of its cathode. A positive pulse is applied to its gate. Describe what happens next.

2 Once a thyristor is conducting, what can we do to stop it?

3 Describe how a thermistor and a thyristor may be used to control an electric heating element powered by the mains.

4 Explain why it is preferable to use a thyristor rather than a transistor for controlling large currents.

5 Explain what is meant by *firing angle* and *conduction angle*.

6 Draw a circuit diagram of a thyristor half-wave controlled rectifier and explain its action.

7 Draw a diagram of a simple phase-control triac circuit that could be used in a mains-powered lamp-dimming circuit.

8 List the methods used to isolate a triac circuit from its control circuit.

9 Explain the difference between an opto isolator and an opto coupler.

10 Explain the possible causes of false triggering in a triac power circuit and what can be done to prevent it.

11 Explain the differences between phase control and integral cycle control, and their advantages and disadvantages.

12 Describe the action of the CA3059 zero crossing switch IC and some of its applications.

Multiple choice questions

1 A thyristor is triggered by:

  A a negative pulse applied to the gate.
  B a negative pulse applied to the anode.
  C a positive pulse applied to the cathode.
  D a positive pulse applied to the gate.

2 Starting from the anode, the semiconductor layers in a thyristor are:

  A pnpn.   C pnp.
  B npn.   D npnp.

3 Triacs are triggered by:

  A positive or negative pulses.
  B positive pulses only.
  C negative pulses only.
  A pulses produced as the mains voltage crosses zero.

4 A snubber network is used to reduce:

  A overheating of the triac.
  B RFI.
  C the length of the trigger pulse.
  D false triggering.

5 Once a thyristor has been turned on, it conducts until:

  A the triggering pulse ends.
  B the current through it falls below the holding current.
  C the anode is made positive of the cathode.
  D the gate is made negative of the cathode.

17 Power supplies

Electronic circuits normally operate at low voltages and on direct current (DC). A typical circuit operates on 6 V or 9 V DC, but some can work on as little as 1 V. Often this low voltage supply is provided by a battery. A battery is ideal for portable equipment such as a mobile phone or a digital camera. But batteries are fairly expensive to replace, especially if the equipment is to be operated for long periods or requires a large current. In such cases, we prefer to use a power supply that takes power from the mains and converts it to a low voltage DC supply suitable for electronic circuits. DC power supplies are found in many different kinds of electronic equipment from TV receivers and computers to washing machines and cash registers.

This chapter illustrates three ways of building a power supply unit (PSU) suitable for running electronic circuits from the mains. This topic describes the simplest of the circuits, which has three stages:

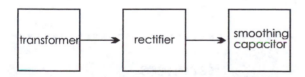

The functions of these stages are:

- **Transformer:** transforms mains power (230 V AC) into a lower AC voltage.
- **Rectifier:** converts low AC voltage into low DC voltage.
- **Smoothing capacitor:** smooths the DC output of the rectifier.

The **transformer** (T1) is of the step-down type wound on a soft-iron laminated core. The primary winding is rated to be connected to the live and neutral lines of the 230 V AC mains. The secondary winding has fewer turns than the primary winding, giving a turns ratio of less than 1. The amplitude of the transformed AC is typically as low as 12 V AC. The wire of the secondary winding should be rated to supply the current that it is intended to draw from the PSU.

There should be a **switch** (S1) between the mains plug and the transformer. This should be the double-pole single-throw type so that, when it is off, the power supply circuit is completely disconnected from the mains. The case should be made of metal or stout plastic able to withstand rough treatment. The Earth line of the mains is connected to a point on the chassis of the PSU, and to the case, if the case is made of metal.

The **diode** is a rectifier diode, designed to pass a relatively large current without damage. The photo shows types of rectifier diode suitable for use in a PSU.

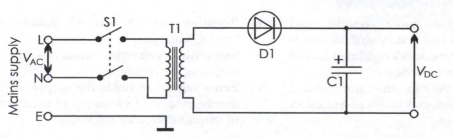

The simplest PSU has a half-wave rectifier, based on a single diode.

Rectification

The output from the transformer is a sinusoid. The diode is able to conduct only during the positive half-cycle of the sinusoid.

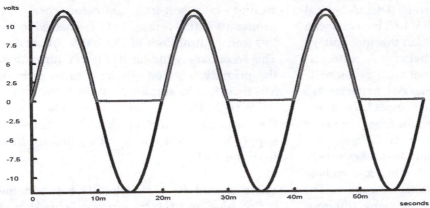

The input to the rectifier circuit is low-voltage AC at 50 Hz, plotted here with a black line. The output (grey line) is pulsed DC because the diode conducts only on the positive half-cycles.

The waveform consists of a series of positive pulses occuring 50 times per second. This waveform is known as **pulsed DC**. Because of the forward voltage drop across the diode, the amplitude of the pulses is 0.7 V less than that of the transformed AC. The circuit produces current only during the positive half-cycles, so it is called a **half-wave rectifier**. It is inefficient because the diode is conducting for only half of the time.

Smoothing

For applications such as driving a small DC motor, the pulsed DC can be used directly from the rectifier.

If only a small current is required, the waveform can be smoothed by connecting a capacitor (C1) across the output terminals. An electrolytic capacitor of large value (for example, 1000 μF) is needed.

Output

To protect the transformer and diode against overloading or short-circuiting, include a fuse on the output side. This is not shown in the diagram on p. 147.

With no load, the output is several volts higher than expected from the turns ratio of the transformer.

Voltage drops to the expected value when a small load is connected.
This PSU is **unregulated**, so the output voltage decreases if the load draws increasing amounts of current.

Rectifiers

The Companion website has animated diagrams that show how rectifiers operate.

Full-wave rectified and Zener regulated supply

This PSU is slightly more complex and expensive than the one just described, but is more efficient and produces a regulated output voltage. It comprises four stages:

- **Transformer:** As in the unregulated supply, this transforms mains power into a lower AC voltage.

- **Rectifier:** converts low AC voltage into low DC voltage.
- **Smoothing capacitor:** smooths the DC output of the rectifier.
- **Zener regulator:** holds the output voltage steady in spite of variations in the amount of current drawn by the load.

The two new improvements in this PSU are the rectifier bridge of four diodes and the Zener regulator circuit. The system diagram now has four stages:

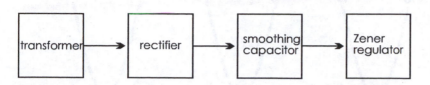

Full-wave rectifier

The four diodes D1 – D4 are connected to form a rectifying **bridge**. This receives AC from the transformer.

During the positive half-cycle, terminal A of T1 is positive of terminal B. Current flows from A, through D1, and through R1 to the load. It passes through the load and returns through D2 to B.

During the negative half-cycle, terminal B is positive of terminal A. Current flows from B, through D3, and through R1 to the load. It passes through the load and returns through D4 to A.

Summarising, D1 and D2 conduct on the positive half-cycle, while D3 and D4 conduct on the negative half cycle.

During both half-cycles the current passes through two diodes, so there are two diode drops, totalling 1.4 V

There is conduction during *both* half-cycles, which is why this is called a full-wave rectifier. Given a 50 Hz supply, the output from the diode bridge is pulsed DC, frequency 100 Hz.

Because of forward votage drop, the amplitude of the pulsed DC is 1.4 V less than that of the AC produced by the transformer.

Smoothing

The diagram above shows the pulsed DC output from the rectifier stage. The diagram below shows the effect of the smoothing capacitor C1. The capacitor charges at each pulse. The voltage across the capacitor reaches its maximum level, which equals the peak voltage from the rectifier. Then, as the output voltage from the rectifier falls, the capacitor retains all or most of its charge.

If a load is connected, some of the charge is conducted away from the capacitor. The voltage across the capacitor falls.

In a properly designed PSU, the fall in voltage is small. Before the voltage has had time to fall far, the next pulse arrives and the capacitor is recharged to the peak level. The effect of charging and partial discharging is a slight **ripple** on the DC level. The graph opposite shows this ripple.

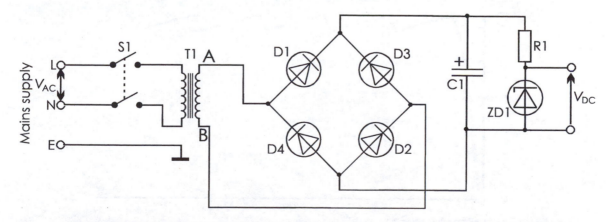

The diode rectifier bridge and the Zener diode give improved performance.

149

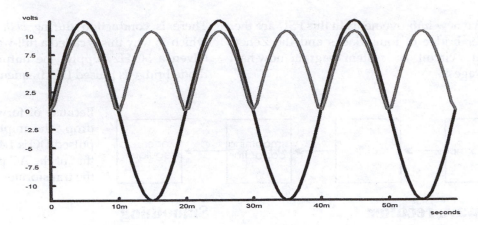

Without the electrolytic capacitor shown in the diagram opposite, the output (grey) waveforms of a full-wave rectifier is pulsed DC with a frequency double that of the input (black). The amplitude is 1.4 V less than the input because of the voltage drop across two diodes.

The depth of the ripple depends on the capacity of C1 and the amount of current drawn by the load. The greater the current, the deeper the ripple.

Ripple may be minimised by using a large-value capacitor (1000 μF or more) for C1, so that the pd across it falls more slowly.

Calculating ripple

It can be seen from the graph below that ripple is a more-or-less triangular waveform superimposed on the steady DC output.

There is a rapid rise as the charge on the capacitor is boosted to the peak voltage, followed by a slow fall as charge is conducted away through the load. To simplify the calculations, we will assume that the rise is almost instantaneous and that the fall lasts for one complete cycle of the waveform. To begin with, we will consider what happens with a half-wave rectifier and no subsequent voltage regulation, as in the circuit on p. 147.

If I is the mean load current and T is the period of the waveform, then the change in charge on the capacitor is:

$$q = IT$$

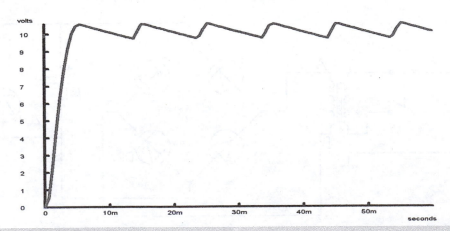

Adding a capacitor to the output side of the rectifier produces DC with a small 100 Hz ripple.

The *fall* in the voltage across the capacitor is:

$$v = IT/C$$

Given that $T = 1/f$, where f is the frequency, this equation can be rewritten:

$$v = I/Cf$$

We now consider what happens in different types of power circuit. In the half-wave rectifier power supply, the output voltage falls from V_p (peak voltage) to $V_p - v$ at an approximately constant rate. This being so, the mean output voltage is:

$$V_{mean} = V_p - \tfrac{1}{2} \times \frac{I}{Cf} = V_p - \frac{I}{2Cf}$$

The peak voltage is reduced by an amount that is proportional to the current being drawn and is inversely proportional to the capacitance and frequency. This explains why the smoothing capacitor should have as large a capacitance as practicable.

With a full-wave rectifier, the ripple frequency is double that of the AC frequency. If f is the AC frequency, the fall in output voltage is:

$$v = I/2Cf$$

This results in a mean output voltage of:

$$V_{mean} = V_p - \frac{I}{4Cf}$$

This shows that the fall in voltage for a given load is halved when a full-wave rectifier is used.

The descriptions above refer to unregulated power supplies. The effect of ripple is much reduced in stabilised supplies, such as the Zener stabilised supply of p. 149 and the regulator stabilised supply of p. 153. The variations in voltage due to ripple are largely eliminated by the stabilising action of the Zener diode or the regulator.

Zener stabilised supply of p. 149 and the regulator stabilised supply of p. 153.

> **Self Test**
>
> The output of a half-wave rectifier has peak voltage 6.5 V. The smoothing capacitor is 2200 μF, and the frequency is 50 Hz. What is the mean output voltage when the current drawn is (a) 25 mA, and (b) 120 mA?

Regulation

The output voltage of this PSU is held steady under varying load by using a regulator circuit. This is based on a **Zener diode**. The Zener diode needs about 5 mA or more passing through it to maintain its Zener action. This is the reason for the resistor R1. The value of this is chosen to allow at least 5 mA to flow through the diode, even when the load is drawing its maximum current.

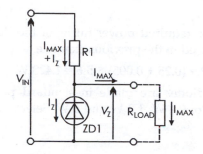

> When the load is drawing the maximum current, there must be at least 5 mA passing through the Zener diode.

In the diagram above, the load is drawing its maximum current, I_{MAX}. The current through the Zener diode is I_Z. The current through R1 is $I_{MAX} + I_Z$.

The pd across the Zener diode is its Zener voltage, V_Z. If V_{IN} is the smoothed voltage across the capacitor of the PSU, the pd across R1 is $V_{IN} - V_Z$. By Ohm's Law, the minimum value of R1 is:

$$R_1 = (V_{IN} - V_Z)/(I_{MAX} + I_Z)$$

Example

A PSU has a smoothed voltage of 10.8 V. The Zener voltage is 5.6 V. The maximum current that the load may take is 250 mA. Calculate the value of R1.

R1 = (10.8 − 5.6)/(0.25 + 0.005) = 5.2/0.255
$$= 20.4\ \Omega$$

Use the next larger E24 value, 22 Ω.

Zener power rating

It is essential for the Zener diode to be able to carry the full current $I_{MAX} + I_Z$ when there is no load connected to the PSU.

Under this condition, the pd across the Zener diode is V_Z and the current through it is $I_{MAX} + I_Z$. The power dissipated in the diode is then:

$$P = (I_{MAX} + I_Z) \times V_Z$$

Example

Find the required power rating of the Zener diode used in the previous example.

$$P = (0.25 + 0.005) \times 5.6 = 1.428 \text{ W}$$

Zener diodes are made in standard power ratings of 0.4 W, 1.3 W and 5 W. Select a 5 W diode.

Resistor power rating

The current through the resistor is $(I_{MAX} + I_Z)$ and the pd across it is $(V_{IN} - V_Z)$. Therefore, the power dissipated in the resistor is:

$$P = (I_{MAX} + I_Z) \times (V_{IN} - V_Z)$$

Example

Calculate the required rating for the resistor in the previous example.

$$P = (0.25 + 0.005) \times (10.8 - 5.6) = 1.326 \text{ W}$$

A 2 W resistor should be used.

Load regulation

Load regulation is expressed as the percentage change in output voltage as the current drawn is varied from zero up to the maximum for which the supply is designed.

Example

Under no load, the mean output voltage of a supply is 12 V. When the load is drawing the maximum current of 2.5 A, the output voltage falls to 11.1 V. What is the load regulation?

Change in output $= 12 - 11.1 = 0.9$ V.

Load regulation is $0.9/12 \times 100 = 7.5\%$.

High current

In the basic Zener circuit, the diode must be rated to carry the maximum output current, plus 5 mA. This means that an expensive power Zener may be required. Further, when the maximum current is not being drawn by the load, a large current is passed through the Zener, which is wasteful.

The diagram below shows an alternative Zener regulator circuit, in which Q1 is connected as an emitter follower (p. 77).

Self test

A Zener-regulated mains PSU using a full-wave rectifier is to provide a current of up to 400 mA at 7.5 V. The transformer is rated to produce an output of 12.5 V AC at 1 A.

a What is the current rating of the rectifier diodes?

b What is the voltage of the output from the rectifier?

c Specify what type of Zener diode you would select for this PSU.

d What is a suitable resistance and power rating for the dropper resistor?

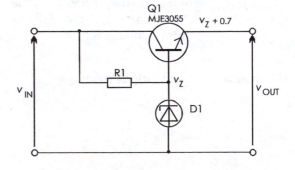

With the MJE3055, this circuit can deliver up to 10 A, regulated.

Input v_{IN} comes from a rectifier or a battery. The Zener voltage V_Z is 0.7 V more than the required output voltage. A low-power (400 mA) Zener is suitable. R1 is chosen so that the current through D1 is more than 5 mA, say 10 mA. The Zener holds the base of Q1 steady at V_Z, and the base-emitter voltage drop produces $(V_Z - 0.7)$ V at the output.

Full-wave rectified and IC regulated supply

This circuit improves on the previous two power supply circuits by using:

* an encapsulated diode bridge rectifier, and
* an integrated circuit voltage regulator.

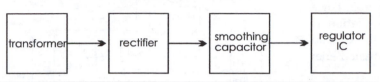

In this circuit the Zener diode regulator is replaced by an IC voltage regulator

Diode bridge

In an encapsulated diode bridge the four power diodes are connected to form a rectifier bridge, as on p. 149.

They are sealed in a capsule with four terminal wires. Two of the wires ('AC' symbols) receive the transformed AC direct from the transformer and the rectified DC output is obtained from the other two wires (+ and –). The bridge in the photo is rated at 6 A.

A bridge rectifier makes circuit construction simpler.

The diode bridge works the same way as a bridge built from four individual diodes but an encapsulated bridge is easier to mount and simpler to connect.

Regulator IC

Many types of regulator IC are available, but one of the most popular is the three-terminal regulator. The 7805 IC is an example of this type.

The 7805 is a member of the '78XX' family, which includes regulators of different power ratings, producing various fixed output voltages. The output is indicated by the final two digits of the type number. The 7805, for example, produces a fixed output voltage of 5 V. Other ICs are available for variable voltage regulation (*see* overleaf).

The precision of a typical voltage regulator is ±4%, which is held in spite of variations in the output current and the supply voltage (*see* next section). This is better performance than can be obtained from a Zener regulator under similarly wide-ranging conditions. The supply voltage should be about 2.5 V higher than the output voltage. This allows good regulation without undue dissipation of power within the regulator IC. The maximum allowable supply voltage is 30 to 40 V.

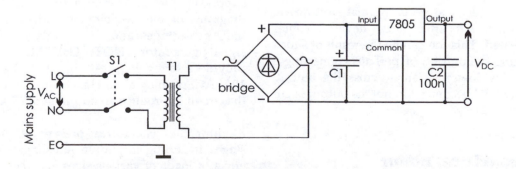

The classic voltage regulator IC is a three-terminal device.

The 7805 is rated to produce a current of 1 A. Similar regulators are available for currents of 100 mA and 500 mA. A heat sink may be needed with the higher current ratings. The IC has a metal tag with mounting hole for attaching this.

Load and line regulation

Load regulation was defined on p. 152. For a typical IC regulator, the load regulation is about 0.1%. This compares with a load regulation of several percent for a basic Zener diode regulator.

Line regulation is a measure of the change in output voltage resulting from a change in input voltage. It is defined as:

$$\frac{\text{change in output voltage}}{\text{change in input voltage}} \times 100\%$$

Again, IC regulators have superior performance. A typical IC regulator has a line regulation of 0.01%, compared with several percent for a basic Zener regulator.

Safety features

All IC regulators have a **current limiting** feature. If excess current is drawn from the regulator, perhaps because its output is short-circuited, the output voltage is automatically limited. This limits the size of the current that can be drawn from the device and so protects it and the load from damage.

Another safety feature is **thermal shut-down**. This comes into effect if the device becomes overheated. This too can be the result of short-circuiting the output, or perhaps connecting a unsuitably low resistance across the output. When the IC becomes too hot, its output is cut to zero.

Adjustable regulator

The LM317 is typical of the 3-terminal adjustable regulators.

The regulator can accept a maximum voltage input of 40 V. The output voltage is adjustable from 1.2 V to 32 V. The input voltage must be at least 3 V greater than the intended output voltage. The output voltage is set by the value of the resistor R in the diagram below.

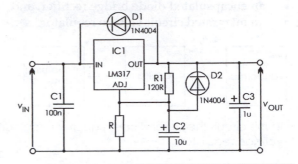

The diodes protect the regulator against short-circuiting either the input or output terminal to the 0 V rail.

The value of R is calculated from :

$$R = 96 \times v_{\text{OUT}} - 120$$

Example

In an LM317 regulator circuit, what value of R is required to produce an output voltage of 10.5 V?

$$R = 96 \times 10.5 - 120 = 888 \ \Omega$$

The nearest E24 value is 910 Ω. If this is not close enough, use 820 Ω and 68 Ω in series, preferably with 1% tolerance.

Activities — Power supplies

Investigate the action of two of the power supply circuits and write a report including diagrams of the two circuits and comparing their advantages and disadvantages. Use a signal generator (**NOT THE MAINS**) to provide a low-voltage AC input. A signal of 10 V AC, running at 50 Hz, is suitable. Design the circuit to produce an output of 5 V DC.

Monitor the waveforms present at various stages, including amplitude and shape. Try the effect of loads of various sizes and their effect on smoothing and ripple. Estimate the load and line regulation of the circuits.

Zener diodes   Box 20

A typical diode, such as a rectifier diode, can withstand a reverse voltage of several hundred volts before breaking down. Breakdown results in irreversible damage to the diode.

A Zener diode breaks down at a precisely known reverse voltage that is often only a few volts. At this **Zener voltage**, V_Z, current flows freely through the diode without damaging it. The Zener voltage is determined during manufacture. Diodes are available with specfied Zener voltages ranging from 2.0 V up to 100 V or more.

As the power supply circuit on p. 149 illustrates, a Zener diode makes a simple voltage regulator. However, the Zener voltage is not perfectly constant. It rises slightly as the current increases. It also depends on temperature. For these reasons, a Zener regulator has relatively low precision.

Questions on power supplies

1   Describe how a half-wave rectifier works and sketch the waveform that it produces. Compare its voltage amplitude with that of the AC it rectifies.

2   Why is a full-wave rectifier more efficient that a half-wave rectifier? Sketch the waveform of   the rectified current. Compare its voltage amplitude with that of the AC it rectifies.

3   Describe the action of the smoothing capacitor in a full-wave power supply.

4   A power supply unit has a smoothed voltage of 8 V. The Zener voltage is 3.9 V. The PSU is rated to provide up to 400 mA. Calculate the required resistance and power rating of the dropping resistor.

5   In the previous question, what should be the power rating of the Zener diode?

6   What are the advantages of an IC voltage regulator when compared with regulation by a Zener diode.

7   List and discuss the safety features and precautions that should be considered when building a mains-powered PSU.

8   Design a regulated power supply to provide 1 A at 12 V DC with 4% precision.

9   Design a mains PSU for powering a 6 V DC electric motor, at a maximum of 4 A.

10 Design a power supply for powering a logic circuit that requires 5 V ± 0.5 V, at 3 A. (This is typical of the requirements for a TTL circuit).

11 Explain the cause of ripple on the output of a full-wave rectified and smoothed power supply circuit. Given: peak voltage = 15 V, the current = 2 A, smoothing capacitor = 4700 µF, and frequency of the rectified AC = 100 Hz, what is the mean output voltage? What can be done to increase this voltage without altering the load current?

12 A simple Zener regulator is loaded with different resistors. The table shows the corresponding Zener voltage. If the circuit is rated to supply up to 240 mA, what is its load regulation?

Load (Ω)	V_Z
15	4.05
20	4.76
25	5.30
30	5.70
35	5.70
No load	5.70

13 The table shows the output voltage of a simple Zener regulator as the supply voltage in increased from 8 V to 18 V. What is the line regulation of this circuit?

V_S	V_{OUT}	V_S	V_{OUT}
8	4.62	14	5.79
10	5.66	16	5.84
12	5.73	18	5.90

18 Logical circuits

Logical circuits perform **logical operations**. There are three basic logical operations:

- NOT, sometimes called INVERT.
- AND
- OR

Logical operations are performed by circuit units called **gates**. Because a typical logical circuit needs several, or maybe many, gates they are made as integrated circuits. The drawing below shows the symbol for an AND gate and a typical logic IC containing four AND gates. The gates are separate but they share the power supply pins (0 V and +5 V).

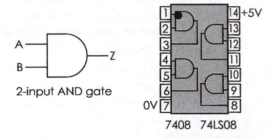

2-input AND gate

7408   74LS08

The 7408 IC contains four 2-input AND gates in a 14-pin package. A circular dimple in the top surface indicates which is pin 1. Or there may be a notch in the end that is nearer to pins 1 and 14.

AND gate operation

The drawing at top right shows a circuit for demonstrating the action of a 2-input AND gate. The IC has three other AND gates, but these are not being used, so their inputs are connected to the 0 V rail or the positive rail.

In the drawing we see a commonly used way of providing input to a logic circuit. The inputs of the gate are held at 0 V by the pull-down resistors R1 and R2. We say that the inputs are 'low'. When a button is pressed, the corresponding input goes to +6 V. We say that the input is 'high'. These buttons are **interfaces** between the gate and the human user.

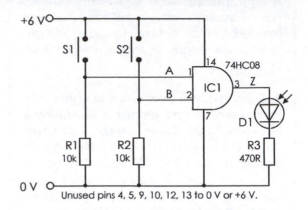

Unused pins 4, 5, 9, 10, 12, 13 to 0 V or +6 V.

A circuit for demonstrating the action of a 2-input AND gate. Manual input is by the two push-buttons with pull-down resistors. The state of the output is indicated by an LED.

The other interface between gate and the user is an LED. This has a series resistor to limit the current and protect the gate. The LED is not lit when the output of the gate is at 0 V (= 'low'). The LED comes on when the output of the gate goes to +6 V (= 'high').

Gate

A logic gate has nothing to do with the gate of an FET or a thyristor.

When connected as above, the gate acts as a **source** of current that turns the LED on when the output is high. A gate can also act as a **sink**, as shown in the drawing below.

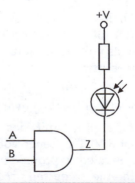

If the gate is wired as a current sink, the LED comes on whenever the output of the gate is low. At other times the LED is off.

The behaviour of the gate is investigated by pressing various combinations of the two buttons S1 and S2. Because there are two inputs to this gate, there are four possible combinations. We record the output of the gate for each combination. This table summarises the results:

Inputs		Output
B	**A**	**Z**
0	0	0
0	1	0
1	0	0
1	1	1

A '0' in the input columns indicates that the button is not pressed, so the input is low. A '1' indicates that the button is pressed, so the input is high.

A '0' in the output column indicates that the gate output is low, and the LED is off (the gate is wired as a source). A '1' indicates that the output is high and the LED is on.

This table of results is identical with the truth table for the AND operator. In words, the output is high only if inputs A AND B are high. If an AND gate has three or more inputs, the output is high only if *all* inputs are high.

Another way of stating this result uses **Boolean algebra**:

$$Z = A \bullet B$$

This statement uses the AND operator, '•'. It tells us that output Z is the result of ANDing input A with B.

A similar statement covers the result from a four-input AND gate:

$$Z = A \bullet B \bullet C \bullet D$$

The output is high only if all four inputs are high.

In Boolean algebra, we may omit the '•' operator and write instead: $Z = AB$, or $Z = ABCD$.

OR gates

To investigate an OR gate, we may use the same circuit as on p. 156, but with a 74HC32 IC instead. This contains four 2-input OR gates. The symbol for a 2-input OR gate is drawn below:

The behaviour of the OR gate is summarised in this table:

Inputs		Output
B	**A**	**Z**
0	0	0
0	1	1
1	0	1
1	1	1

This is the truth table for the OR operation. Compare this with the AND truth table in the left column. The inputs are the same as before, but the outputs are different. In words, the output is high when either A OR B OR both are high.

The Boolean expression for this result is:

$$Z = A + B$$

This statement uses the OR operator, '+', which is the same as the 'plus' operator used in arithmetic. This operator must not be omitted from Boolean equations.

OR gates are available with three or more inputs. The result from a four-input OR gate is:

$$Z = A + B + C + D$$

NOT gate

The NOT gate can have only one input terminal and its output is always the inverse (or opposite) of its input. The 74HC04 IC contains six NOT gates. These can be investigated as described above.

This is the symbol for a NOT gate:

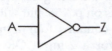

The small circle at the output indicates that the output is inverted. The truth table of the gate is simple:

Input A	Output Z
0	1
1	0

In symbols, the operation of the gate is:

$$Z = \overline{A}$$

In Boolean algebra, a bar over a variable is used to show that it is inverted.

NAND and NOR gates

The three basic gates that we have already described can be connected together in a wide variety of ways to perform more complex logical operations. We shall look at some of these in following chapters. For the moment, we will describe one of the most useful combinations of gates, an AND gate that feeds its output to a NOT gate. This NOT-AND, or NAND gate is illustrated below:

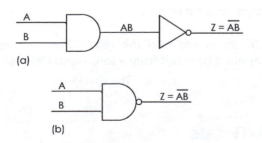

(a)

(b)

A NAND gate is the equivalent of an AND gate followed by a NOT gate. NAND gates are also made with three or more inputs.

The truth table for the two-input NAND gate may be obtained by writing the table for AND and then inverting the output:

Inputs		A•B	Z = $\overline{A \cdot B}$
B	A		
0	0	0	1
0	1	0	1
1	0	0	1
1	1	1	0

The third column of the table holds A AND B. This is inverted in the fourth column, as indicated by the line drawn over 'A•B'.

The NOR gate is equivalent to an OR gate followed by a NOT gate. In this table we look at a 3-input NOR gate:

Inputs			A+B+C	Z = $\overline{A+B+C}$
C	B	A		
0	0	0	0	1
0	0	1	1	0
0	1	0	1	0
0	1	1	1	0
1	0	0	1	0
1	0	1	1	0
1	1	0	1	0
1	1	1	1	0

It is essential to include all possible combination of inputs in the first three columns. Enter the inputs as the digits of the eight binary 3-digit numbers, 000 to 111. This rule applies to 2-input gates, as can be seen opposite. It also applies to larger numbers of inputs. For instance, with a 4-input gate the truth table has 16 rows, running from 0000 to 1111.

The drawing below illustrates the 3-input NOR gate.

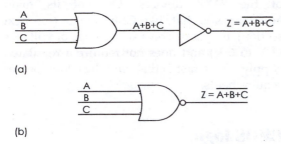

(a)

(b)

A 3-input NOR gate (b) is equivalent to a 3-input OR gate followed by a NOT gate (a).

Exclusive-OR gate

An exclusive-OR, or EX-OR gate has two inputs A and B. Its output is '1' when A or B *but not both*, are '1'. This is the truth table:

Inputs		Z =
B	**A**	$\mathbf{A \oplus B}$
0	0	0
0	1	1
1	0	1
1	1	0

Note the symbol $\oplus$ for the exclusive-OR operation. Except for the last line, the table is identical to the table for OR. The symbol for an EX-OR gate is:

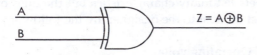

The EX-OR gate is sometimes known as the 'same or different' gate. Its output is '0' when its inputs are the same and '1' when they are different. This makes the gate useful for comparing two logical quantities.

The exclusive-NOR gate is equivalent to an EX-OR gate followed by a NOT gate. It has only two inputs. Its output is the inverse of EX-OR.

Logic families

There are three main logic families:

- Transistor-transistor logic (TTL).
- Complementary MOSFET logic (CMOS).
- Emitter coupled logic (ECL)

Each of these families has its own versions of the standard logic gates such as NAND and NOR. Each family includes a range of ICs with more complex functions such as flip-flops, adders, counters, and display drivers. The families differ in the ways the basic gates are built, and this gives rise to family differences in operating conditions and performance. We will look at each family in turn.

Transistor-transistor logic

This was the first family to become widely used and set the standards for logic circuits for many years.

The devices in the original TTL family were all given type numbers beginning with '74'. This was followed by two or three digits to distinguish the types. For example 7400 is a quadruple 2-input NAND gate, and 7493 is a 4-bit counter/divider. To refer to the family itself we substitute 'XX' for the last two or three digits.

The original 74XX series is now virtually obsolete, being used mostly for repairing old equipment, and is becoming much more expensive than it used to be.

There are over 20 more recent versions of the series differing in various features, such as power, speed, immunity from noise, EMI reduction, and supply voltage. Some of the newer versions operate with low power on a 3.3 V supply, which makes them suitable for battery-powered portable equipment, as well as being more economical of power generally.

One of the most popular series is the low power Schottky series, with device numbers 74LSXX. This has a faster operating time than standard TTL. Both standard and 74LSXX ICs require a regulated power supply of 5 V ± 0.25 V.

The output stage of TTL gates has two transistors operating in opposition. One switches the output to high, the other switches it to low. In operation, there is a changeover stage during which *both* of the transistors are switched on. This causes a large current to flow as the gate changes state. This may overload the supply line, causing a fall in voltage which may interfere with the action of nearby ICs. For this reason the supply to TTL ICs must be decoupled by wiring capacitors between the supply lines. Decoupling usually requires a 100 nF disc ceramic capacitor for every 5 ICs.

Certain TTL logic devices are made with an **open collector output**. This has a transistor with its base and emitter connected to the gate circuit but its collector left unconnected. The output must have a pull-up resistor connected to it (1 kΩ is often suitable).

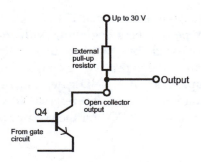

The output of TTL gates that have an open collector output can be pulled up to voltages higher than the standard 5 V operating voltage.

The resistor is connected to a voltage higher than the standard 5 V operating voltage. Depending on the type of gate, the higher voltage can be as much as 15 V or 30 V. Provided that the gate can sink sufficient current, this allows high-voltage low-current devices to be switched under logical control.

The 74HCXX series is another popular branch of the TTL family. It comprises CMOS versions of the 74XX devices. One of its main advantages is that, like CMOS logic (see next section), it operates on a wide range of voltages (2 V to 6 V) and does not require a regulated supply. It is fast-acting and has low power requirements.

CMOS logic

This is based on complementary MOSFET transistors. By complementary, we mean that the transistors operate in pairs, one of the pair being n-channel and the other p-channel. Both are enhancement MOSFETs.

CMOS is available in two main series, in which the type numbers range from 4000 and 4500 upwards. The most commonly used series, the 4000B series, has buffered outputs. The 4000UB series has unbuffered outputs. Although the inclusion of buffers adds to the time that the gate takes to operate, the buffers make the output more symmetrical, with equal and faster rise and fall times. This is important when CMOS gates are used in pulse circuits and timing circuits.

TTL and CMOS compared

For all except the fastest logic, the choice of family usually lies between TTL and CMOS. Here we compare the characteristics of the most popular TTL series, the low power Schottky series, and the CMOS 4000B series. As already mentioned, the TTL 74HCXX series combines some of the advantages of both. Data sheets list many characteristics but the ones of most interest to the designer are the following:

* Operating voltage.
* Power consumed, usually specified as the power per gate.
* Input voltage levels, the level above which the gate recognises the input as logic high, and the level below which it is taken as logic low.

- Output voltage levels, the minimum high level, the typical high level, the maximum low level and the typical low level produced by the gate.

- Input currents, the maximum when a gate is receiving a high input and the maximum when it is receiving a low input.

- Output currents, the minimum when a gate has a high output and the minimum when the gate has a low output.

- Fanout, the number of inputs that can be driven from a single output (can be calculated from input and output currents).

- Propagation delay, the time, usually in nanoseconds, between the change of input voltages and the corresponding change in output voltage.

- Speed, as indicated by the maximum clock rate it is possible to generate.

The 74LSXX series, the 74HCXX series and the CMOS 4000B series have the characteristics listed in the table below. Their input and output voltages are plotted at top right.

The 74LS series operates on 5 V DC, which must be held to within 5%. This is the voltage commonly used in computers and similar equipment.

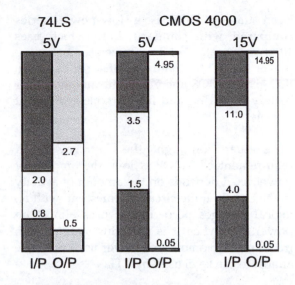

The shaded areas indicate the voltage ranges that are defined as logic low and logic high for TTL and for CMOS with a 5 V or a 15 V supply voltage. The output voltage levels of gates are set to be well within the limits accepted at the inputs. Levels for 74HC00 at 5 V are the same as for CMOS.

By contrast, CMOS is able to work on a wide voltage range, which makes it much more suitable for battery-powered equipment. It is also easier to interface with logic circuits and with other devices.

Characteristic	74LS00	74HC00	CMOS
Operating voltage	5 V ± 0.25 V	2 V to 6 V	3 V to 15 V (max 18 V)
Power per gate	2 mW		0.6 μW
Input current, high (max)	20 μA	0.1 μA	0.3 μA
Input current, low (max)	–0.4 mA	–0.1 μA	–0.3 μA
Output current, high (min)	–0.4 mA	–4 mA	–0.16 mA (–1.2 mA)
Output current, low (min)	8 mA	4 mA	0.44 mA (3 mA)
Fanout	20	50 (nominal)	50 (nominal)
Propagation delay	9 ns	10 ns	125 ns (40 ns)
Fastest clock rate	40 MHz	40 MHz	5 MHz

Even though 74LS is a low-power series compared with standard TTL (which uses 10 mW per gate), 74LS series needs much more power than CMOS, and the 74HC series. Again this gives CMOS and 74HC the advantage for battery-powering and for portable equipment in general.

Note that although the *average* power requirement of CMOS is low, the amount of power used depends on the amount of activity in the circuit. If the circuit is quiescent, with no logical changes occurring, it uses almost no power. But when it is changing state at high frequency the amount of power used may rise almost to the level used by TTL.

CMOS has insulated input gates (using 'gate' in the other sense) like any other FET device, so it is not surprising to find that input currents are very low in CMOS and 74HC. The output currents are lower than those of 74LS, the figures being quoted for a 5 V supply and (in brackets for CMOS) for a 15 V supply.

When calculating fanout we have to consider what happens under both high and low logic conditions.

Example

Consider a pair of 74LS gates, gate A which is feeding its output to an input of a second 74LS gate, gate B. If A is high, then, according to the table, it can supply at least 0.4 mA, or 400 μA. Only 20 μA are needed to supply the input of B, so A can supply 400/20 = 20 gates.

Now suppose that the output of A is low. It can draw at least 8 mA from any outputs to which it is connected. But gate B needs only 0.4 mA drawn from it to qualify as a low input. So the number of gates that A can bring down to a low level is 8/0.4 = 20. At both high and low levels A can drive the inputs of 20 gates. Its fanout is 20.

This is far more inputs than a designer would normally need to connect to a single output so, for practical purposes, the fanout is unlimited.

Similar calculations can be done for CMOS but here we have the extra complication that output currents depend on operating voltage. The high input resistance of CMOS gates means that the fanout is large. A fanout of 50 is taken as a rule-of-thumb number, and is one that will very rarely be needed.

Self Test

What is the fanout when:
(a) a 74LS gate drives a 4000B series gate?

(b) a 4000B gate drives a 74LS gate?

Looking at the diagram opposite, we note a major difference between 74LS and CMOS. On the left we see that a 74LS gate takes any input voltage (I/P) greater than 2.0 V to be logic high. Any input lower than 0.8 V is low. Inputs between 0.8 V and 2.0 V are not acceptable and give unpredictable results. Acceptable inputs are provided by 74LS outputs as shown in the O/P column.

A 74LS gate is guaranteed to produce an output of 2.7 V or more when it goes high. Thus any high output will register as a high input, with a margin of 0.7 V to spare. This allows for spikes and other noise in the system and gives a certain degree of **noise immunity**. Provided that a noise spike takes a high output level down by no more than 0.7 V, it will still be taken as a high level. In this case the noise immunity is 0.7 V. Noise immunity is very important in logic systems. A '0' that accidentally becomes interpreted as a '1' can have devastating effects on the logic.

At low level the maximum output voltage is 0.5 V, though typically only 0.25 V. This is 0.3 V (typically 0.55 V) below the input level that is registered as logic low, so again there is a useful amount of noise immunity.

In the case of CMOS, the high and low input levels vary with operating voltage. But the output levels are always within 0.05 V of the positive or 0 V supply lines. This is because the family is built from MOSFETs. When turned on, they act as a very low resistance between the positive supply line and the output. When turned off, they act as a very low resistance between the negative line and the output. This gives CMOS very high noise immunity.

The table shows that 74LS is much faster than CMOS and this allows clocks and other time-dependent circuits to run at much higher rates with 74LS. Propagation delay is reduced in CMOS if the supply voltage is made higher. The values quoted are for the more frequently used 4000B series. Part of the delay is due to the output buffers. If speed is important, the 4000UB series has a propagation delay of only 90 ns when used at 5 V.

Emitter coupled logic

The prime advantage of this family is its high speed, which is due to the fact that the transistors are never driven into saturation. This makes it the fastest of all logic families with a propagation delay of 1 ns per gate and a maximum clock rate of 500 MHz.

As well as avoiding transistor saturation, the circuits have low input impedances to avoid the speed-reducing effects of capacitance. Because of these measures, power consumption is high, as much as 30 mW per gate. Also, the logic output levels are close together at –0.8 V and –1.6 V. This leads to low noise immunity. Consequently, an ECL circuit needs an ample power supply carefully protected against voltage spikes. Another factor which makes ECL difficult to work with is that the fast edges on the signals mean that connecting wires must be treated as transmission lines, with precise routing on the circuit board.

Because of the difficulties associated with ECL, its use is restricted to large computer systems in which its high speed is an indispensable benefit.

Activity — Logic levels

Investigate the input and output logic levels of logic gates. Use CMOS 4011, and 74LS00 dual-input NAND gates. You need:
- A breadboard.
- A supply of short lengths of single-core wire with insulation stripped from both ends.

- A power supply, 5 V DC for 74LS series IC, or any DC voltage between 3 V and 15 V for CMOS.
- Integrated circuits to provide the gates.
- Data sheets to show the pin connections of the ics.
- Output indicator, either a testmeter or a logic probe. You can also use an LED in series with a resistor, as on p. 156.

If NAND or NOR gates are used with inputs wired together they act as INVERT gates. Connect the two inputs of a gate together and supply them with input from the wiper of a variable resistor connected across the supply lines as a potential divider. The CMOS IC can be operated on any supply voltage in the range +3 V to +15 V.

Unused CMOS inputs must be connected to the supply or 0 V before testing. Unused TTL inputs should be connected to the supply through a 1 kΩ resistor; several inputs can share a single resistor.

Increase the voltage slowly from 0 V up to the supply voltage. Measure the input voltage at various stages and note whether the output is low or high. Record the level at which the output becomes firmly low. Decrease the voltage slowly from supply voltage to 0 V. Record the level at which the output becomes firmly high.

For the CMOS gate, repeat this investigation at two or three different supply voltages. Record your results in a table to compare the two families.

Activity — Truth tables

Investigate the behaviour of the basic logic gates, using CMOS, 74HC or 74LS gates. For each gate, draw up a blank truth table and mark in all possible combinations of inputs, using 1 and 0 or H and L to indicate the two possible logic states. A suitable test circuit is illustrated on p. 156. Then test the combinations of gates shown opposite and identify the function that the combination performs.

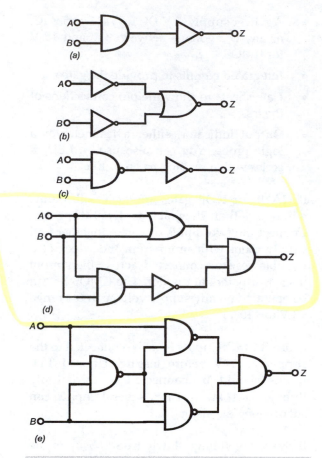

(a)

(b)

(c)

(d)

(e)

Test these combinations of gates as described in the Activity opposite.

On the Web

The TTL and CMOS families of logic circuits have grown with the development of new manufacturing technologies. There are over 25 subfamilies of CMOS ics, for example. Some are designed to operate on low voltage, to make them suitable for hand-held equipment. Others operate at high speed, and some are specialised for processing digital video data.

Search the Web for 'Logic families', and note the features of a few of the newest additions to the range. For what applications are these new logic devices being used?

Questions on logical circuits

1 Write out the 2-input truth tables for AND, NAND, OR and NOR.

2 What is the difference between the CMOS 4000B series and the 4000UB series?

3 Compare the 74LS series with the 4000B series from the viewpoint of power supplies.

4 Compare the 74LS series with the 4000B series from the viewpoint of operating speed and noise immunity.

5 List 10 applications of logic ICs, and name those which would be most suitable for TTL logic and those most suitable for CMOS logic.

6 Explain how ECL achieves its very high speed and what problems this raises.

Multiple choice questions

1 The logic series with the fastest operating time is:

    A 74XX
    B ECL
    C CMOS 4000B
    D 74LSXX.

2 The fanout of 74LS devices is:

    A 10
    B 50
    C 16
    D 20.

3 The power used by a CMOS gate is:

    A less than 1 μW
    B 2 mW
    C more than 5 mW
    D 30 mW.

4 The fanout of a CMOS gate is very high because of:

    A the high output current from the gate
    B the high noise immunity
    C the high gain of complementary MOSFETs
    D the high input resistance of MOSFET gates.

19 Logical operations

There are four basic logical operators, TRUE, NOT (or INVERT), AND and OR. The last three of these were described in Chapter 18. The first operator, TRUE is illustrated by its truth table:

Input A	Output Z
0	0
1	1

The output equals the input. This is the symbol for a TRUE gate:

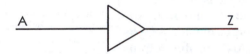

There may seem little point in having a TRUE gate, but a TRUE gate that is also a buffer is useful at times. The input may require only a small current, but the output may be able to source or sink a larger current than a standard logic gate. A buffer can be used to drive an LED, a buzzer, or a similar output device.

The actions of logical operators are shown in truth tables, in which A and B are inputs and Z is the output. Logic 'true' is represented by 1, and 0 represents 'false'. In practical circuits logical operations are performed by gates.

In **positive logic** circuits, we represent 'true' by a high voltage (fairly close to the positive supply voltage) and 'false' by a low voltage (close to zero volts). All the logic circuit descriptions in this book assume positive logic, which is the system most often used.

The truth tables for AND, OR and some other commonly-used gates are shown below side by side for comparison. Reading from left to right, the outputs are A AND B, A OR B, A EX-OR B, A NAND B, A NOR B and A EX-NOR B. The outputs for the last three gates are simply the inverse of the outputs of the first three. The truth tables are shown with two inputs, but gates can have more than two inputs. The truth tables for these can be worked out from the table above.

Example

In the AND column there is only one way to get a 1 output. This is by having *all inputs at 1*. This rule applies to any AND gate. For example an 8-input AND gate must have *all inputs at 1* to get a 1 output. Any other combination of inputs gives a 0 output.

Boolean symbols

The table also serves as a summary of the symbols used in Boolean algebra. Remember that the symbol for AND ($\bullet$) may be omitted, so that AB means the same thing as $A \bullet B$.

The symbol for NOT or INVERT is a bar over the variable, so $\overline{A}$ means NOT-A.

It makes a difference whether two variables *share* a bar or have *separate* bars. For example: $\overline{A \bullet B}$ means A NAND B but $\overline{A} \bullet \overline{B}$ means NOT-A AND NOT-B. Write out the two truth tables to see the difference.

Inputs		Output Z					
B	A	$A \bullet B$	$A+B$	$A \oplus B$	$\overline{A \bullet B}$	$\overline{A+B}$	$\overline{A \oplus B}$
0	0	0	0	0	1	1	1
0	1	0	1	1	1	0	0
1	0	0	1	1	1	0	0
1	1	1	1	0	0	0	1

Predicting output

Truth tables are an effective way of predicting the output from a circuit made up of gates. Usually we need to try all possible combinations of inputs and work out the output for each combination. If there are two inputs there are four combinations, as listed in the table opposite. If there are 3 inputs there are 8 combinations. If there are 4 inputs there are 16 combinations. With even more inputs the number of combinations becomes too large to handle easily and it is simpler to use logic simulator software on a computer. Sometimes there are combinations of inputs that we know in advance can not occur. These can be ignored when working out the outputs.

Example

Predict the outputs of this circuit for each possible combination of inputs:

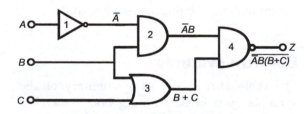

This logic circuit is analysed by several different techniques in this chapter.

First work from left to right, entering the logical expressions at the output of each gate, as has already been done in the diagram. The output of gate 1 (NOT) is the invert of A. The inputs of gate 2 (an AND gate) are $\overline{A}$ and B, so the output is $\overline{A} \cdot B$. The output of gate 3 (an OR gate) is $B+C$. The inputs to gate 4 are $\overline{A} \cdot B$ and $B+C$ so its output is their NAND.

Now follow this through in a truth table (below), for all eight possible combinations of A, B, and C. You can get all the combinations simply by writing out the eight 3-digit binary numbers from 0 to 7 in the first 3 columns.

The next four columns show what each gate does for each combination of inputs. The output of Gate 1 is the inverse of the A column. The output of Gate 2 is obtained by ANDing the output of Gate 1 with the B column. The output of Gate 3 is the OR of B and C. Finally, the output of Gate 4 is the NAND of the outputs of Gates 2 and 3.

Summing up, the last column shows the output of the circuit for all possible combinations of inputs.

Self Test

In the circuit shown on the left, find the outputs of Gate 4 for all possible combinations of inputs if:

(a) Gates 2 and 3 are both NOR gates.

(b) Gates 2 and 4 are both OR gates.

Inputs			Outputs			
C	B	A	Gate 1 $\overline{A}$	Gate 2 $\overline{A} \cdot B$	Gate 3 $B+C$	Gate 4 Z
0	0	0	1	0	0	1
0	0	1	0	0	0	1
0	1	0	1	1	1	0
0	1	1	0	0	1	1
1	0	0	1	0	1	1
1	0	1	0	0	1	1
1	1	0	1	1	1	0
1	1	1	0	0	1	1

A second way of solving this problem is to use the logical identities (p. 368). Most of these are obvious but three in particular are important in circuit design.

- The redundancy theorem:
$$A + AB = A$$

- The race hazard theorem:
$$CA + \overline{C}B = CA + \overline{C}B + AB$$

- De Morgan's Theorem, which has two forms:
$$\overline{A+B} = \overline{A} \bullet \overline{B}$$
$$\overline{AB} = \overline{A} + \overline{B}$$

These may be proved by writing out their truth tables.

Example

Prove the redundancy theorem.

Inputs		AB	A+AB
B	A		
0	0	0	0
0	1	0	1
1	0	0	0
1	1	1	1

Fill in columns B and A in the usual way. In the third column write A AND B. Then, in the fourth column, OR the third column with column A. It can then be seen that the second and fourth columns are identical. Therefore:

$$A + AB = A$$

Example

Prove the first identity of de Morgan's Theorem.

Inputs		A+B	$\overline{A+B}$	$\overline{A}$	$\overline{B}$	$\overline{A} \bullet \overline{B}$
B	A					
0	0	0	1	1	1	1
0	1	1	0	0	1	0
1	0	1	0	1	0	0
1	1	1	0	0	0	0

The truth table lists all four input combinations and the various logical operations. The contents of the fourth and seventh columns are identical, so proving the theorem.

Self Test

Use the truth table technique to prove the Race Hazard Theorem and the second identity of De Morgan's Theorem.

Identities are used to find the outputs of logic circuits.

Example

Find the output of the circuit on p. 166 by labelling the output of each gate with the logical result of its action. Work across the diagram from the inputs to the output. This gives the logical expression for the output shown on the extreme right of the figure. Next, simplify this, using De Morgan's Theorem and other logical identities:

$$\overline{\overline{A \bullet B} \bullet (B+C)} = \overline{\overline{A \bullet B}} + \overline{B+C}$$
$$= \overline{\overline{A} + \overline{B}} + \overline{B} \bullet \overline{C}$$
$$= A + \overline{B} + \overline{B} \bullet \overline{C}$$
$$= A + (\overline{B} + \overline{B} \bullet \overline{C})$$
$$= A + \overline{B}$$

Line 1 recognises the expression as two parts that are NANDed together and converts them (using De Morgan) into two 2-term expressions inverted and ORed. Line 2 uses De Morgan again to split the first expression (NANDed) into two inverted ORed variables, and to split the second expression (NORed) into inverted ANDed variables.

Line 3 simplifies NOT-NOT-A to A.

Line 4 put brackets around an expression in B and C, which can then be simplified to $\overline{B}$ in line 5.

Self Test

Check this simplification, using a truth table.

The result of this simplification shows that the output of the circuit does not depend on the value of C. This means that this 4-gate circuit can be replaced by two gates, as in the diagram opposite.

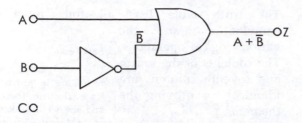

The circuit on p. 166 can be replaced by this 2-gate version.

4-input circuit

This circuit has four inputs, so there are 16 possible combinations of input states:

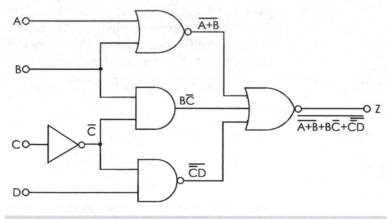

We use identities to determine what combination or combinations of inputs makes Z go high.

The expression for Z is:

$$\overline{\overline{A + B} + B \bullet \overline{C} + \overline{\overline{C} \bullet D}} = (A + B) \bullet B\overline{C} \bullet \overline{\overline{C}D}$$

$$= (A+B) \bullet (\overline{B}+C) \bullet \overline{C}D$$

$$= (A+B) \bullet (\overline{B}\overline{C}D) + (C\overline{C}D)$$

$$= (A+B) \bullet \overline{B}\overline{C}D$$

$$= A\overline{B}\overline{C}D + B\overline{B}\overline{C}D$$

$$= A\overline{B}\overline{C}D$$

There is only one combination for which the output is high. This is when A AND D are high, AND B AND C are low.

The identities are used as follows. On line 1 we use De Morgan to replace a NOR expression with the inverts of its terms, ANDed together. The middle term, which is a NAND expression is again replaced by an OR expression in line 2. On line 3 we replace $(\overline{B} + C) \bullet \overline{C}D$ with $(\overline{B}\overline{C}D + C\overline{C}D)$. Because C ANDed with its invert is 0, the second of these terms is eliminated on Line 4. Line 5 expands what is left into two terms, and this time we have B and its invert ANDed, so the second term is eliminated. This leaves only one term.

This result can be confirmed by writing the truth table of 16 lines. Only one line contains '1' in the output column.

In this example it is seen that all four inputs need to be at the correct level to produce the required output. This suggests that there may be no way in which the number of gates may be reduced. However, it is not easy to be sure of this.

Karnaugh maps

A Karnaugh map has the advantage that it tells us if it is possible to simplify the logic circuit and, if so, how to do it.

Example

Use a Karnaugh map to simplify the logic circuit on p. 166.

Taking the output column of the truth table on p. 166, we enter these results into a three-variable Karnaugh map. This has two variables (AB) on one edge and variable C on the other edge (see over). This gives 8 cells in the map, each corresponding to one of the rows in the truth table. Note that the columns and rows are headed in such a way that only one bit in the expression changes as we go from one column to the next.

The next stage is to encircle groups of 2, 4 or 8 '1's. In this example, it is possible to encircle two overlapping groups of four 1's as shown in the map below.

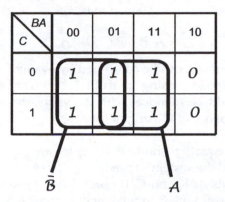

The Karnaugh map of the logic of the circuit on p. 149 shows us how to perform the same logic with fewer gates. There are two groups of four '1's, representing A = 1 and B = 0 (or $\overline{B}$ = 1). The output is 1 when A = 1 OR when $\overline{B}$ = 1. In symbols, Z = A + $\overline{B}$.

The groups are A and $\overline{B}$, so the circuit simplifies to A OR $\overline{B}$. The Karnaugh shows that the circuit has the same action as the 2-gate circuit of p. 168. It does not have a C input, as the state of C makes no difference to the output.

Designing circuits

Karnaugh maps may be used for designing logic circuits to a given specification.

Example

A company car park includes two covered bays, one suitable only for a small car. The bays are out of view of the main car park entrance, so it is decided to install an illuminated sign to indicate when there is a vacancy in the covered bays. The system has a sensor in each bay to detect if it is occupied. At the entrace to the park there is a sensor to detect whether a car is large or small. There is also a radio sensor to detect a priority car.

Thus the system has four inputs:

- A is the sensor in Bay 1, the bay for both large and small cars. A = 1 when the bay is occupied.

- B is the sensor in Bay 2, the bay for small cars. B = 1 when the bay is occupied.

- C is the car size sensor: C = 1 for large cars.

- D is the priority car sensor. D = 1 for priority cars.

Sensors A and B are in the bays. Sensors C and D are at the entrance to the park. Also at the entrance is the illuminated sign Z which says 'Vacancy'. This is the sole output of the system. It is assumed that a car offered a vacancy will take it (if this is not assumed we need duplicate sensors C and D in the bays, which needs six sensors and makes the logic far more complicated!)

The rules of the system are that there is a vacancy if both bays are empty. Small cars can park in either bay, but large cars can park only in Bay 1. One of the bays must be kept vacant in case a priority car needs to park there. This means that only one bay may be occupied by a non-priority car.

The first step in solving problems of this kind is to set out a truth table (opposite). The sign is illuminated when Z = 1. This indicates that there are seven combinations of inputs in which a vacancy occurs.

At this stage the way to build the logic circut may be obvious, and there is no need for a Karnaugh map. However, the logic is not obvious in this example. so we proceed to the map (overleaf).

This problem needs a 16-cell map, so there are columns for A and B, and rows for C and D. The seven 1's of the output column of the truth table are carried down to the corresponding cells of the map. We then look for groups of 1's, preferring large groups to small groups as they make the logic simpler. There are no groups of 8, but two groups of 4. Note that groups may overlap and also that groups may wrap round from one side of the map to the opposite side (example, the group $\overline{AD}$) or from top to bottom.

Inputs				Output
D	C	B	A	Z
0	0	0	0	1
0	0	0	1	0
0	0	1	0	0
0	0	1	1	0
0	1	0	0	1
0	1	0	1	0
0	1	1	0	0
0	1	1	1	0
1	0	0	0	1
1	0	0	1	1
1	0	1	0	1
1	0	1	1	0
1	1	0	0	1
1	1	0	1	0
1	1	1	0	1
1	1	1	1	0

The logic of the car park system. Above: the truth table. Below: the Karnaugh map.

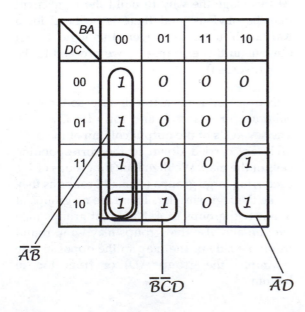

$\overline{A}\overline{B}$

$\overline{B}\overline{C}D$

$\overline{A}D$

There is a third goup consisting of a pair of 1's.

The output Z is high, and the 'Vacancy' sign is switched on whenever the combination of inputs corresponds to one of the cells in the groups. Any one OR other of the groups will do, so the logical equation is:

$$Z = \overline{A}\,\overline{B} + \overline{B}\,\overline{C}\,D + \overline{A}\,D$$

This result can be confirmed by evaluating the expression using a truth table. The result should be identical with the values shown in column Z of the original table.

The equation shows that 7 gates are needed to build the system. It needs 3 NOT gates to invert inputs A, B, and C. It needs 3 AND gates (one of them with 3 inputs), and it needs a 3-input OR gate to produce the value of the whole expression.

It may be possible to reduce the number of gates, as was done for the circuit on p. 166. Sometimes a Karnaugh map shows how to reduce the gate count, but in this case it does not. Instead we can try to simplify the expression by using identities. This simplification is left to the student as an exercise. The gate count can be reduced to 5, possibly to fewer.

Breadboards and simulators

A fourth way to solve a circuit is to connect IC gates together on a breadboard, run through all possible input states, as listed in the truth table, and see what output is produced at each stage. You can do the equivalent thing by setting up the circuit on a simulator.

Example

The screen display overleaf is the result of a simulation of the circuit of p. 166, using 74LS series devices. To provide the input we used a 5 MHz clock and fed its signal to a 3-stage binary counter (more about these in Chapter 22). The clock and counter are seen on the left of the screen.

Times are in nanoseconds, the count incrementing every 200 ns. The vertical dashed lines on the output display represent 200 ns intervals. The inputs (plots A, B and C) begin at the count of 000 and the trace runs while they increase to 111 and return to 000 again. The count then runs through 000 to 100 at the extreme right side of the screen.

The output from the NAND gate is plotted as 'Out' on the display. This is Z and equals 1 *except* when A is low and B is high ($A = 0$ and $B = 1$). The state of C has no effect on output. This result agrees with the truth table on p. 166.

Working with NAND and NOR

Usually we are given a circuit or a truth table and we design a circuit to perform the logic with the minimum of gates. The top circuit on p. 168 is an example of a minimised circuit. Although it requires the minimum of components, takes up the least possible board space and uses the minimum of power, it may not be ideal in practice. When a circuit is built from ICs there is the practical point that an IC usually contains several identical gates. A single IC may contain six NOT gates, or four 2-input gates performing the other basic logical operations.

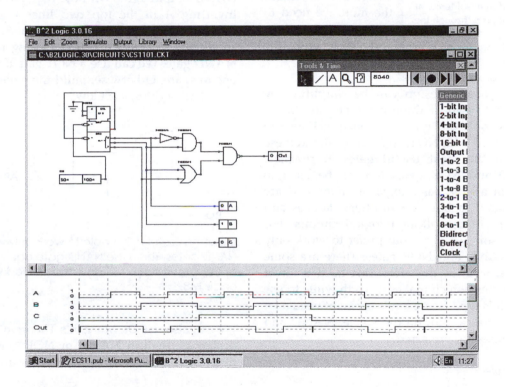

A run on a simulator confirms that the truth table of p. 166 is correct. The display shows logic circuits (left) that produce the eight different combinations of inputs to the four logic gates (centre). The graphs show the sequence of logical input and output levels.

The trace for output Z shows an unexpected effect. There is a **glitch** in the output, shown as a brief low pulse occurring as input changes from 011 to 100. The reason for this is discussed on p. 199. It is not a fault in the logic of the circuit.

It is uneconomical of cost, board space and power to install a 6-gate or 4-gate IC on the circuit board if only one or two of its gates is going to be used. It is better to try to restrict the design to a few different types of gate and so reduce the number of ICs required.

This circuit requires only two gates, NOT and AND but unfortunately they are of different kinds. To set up this circuit on its own requires two ICs containing 10 gates, of which we use only 2. If this circuit is part of a larger logic system, we may find that the system does not need any other AND gates. Or, if it does need a few AND gates, they are situated a long way from the NOT gates. This means having long tracks wandering all over the board. Tracks take up space and, if they are long, may delay the signal sufficiently to cause the logic to fail.

The more long tracks there are, the more difficult it becomes to design the board, and the more we need to resort to wire links and vias to make the connections.

> **Via**
>
> On a printed circuit board a via is a connection (usually a double-headed pin soldered at both ends) that goes through the board, linking a track on one side of the board to another track on the other side.

Circuit board design can be simplified by keeping to two or three types of gate. Better still, use only one type. All logic functions can be obtained with NOT, AND and OR, but these are not the most useful gates in practice. NAND and NOR gates too can be used to perform all the basic logic functions and are also useful in building flip-flops, latches and other more complicated logical circuits. For this reason, we generally prefer to work only with NAND and NOR gates. There are some circuits, such as decoders, that need a lot of AND gates, which makes it worth while to use ICs of this type. Some kinds of circuit need a lot of NOT gates. But, on the whole, NAND and NOR gates are preferred.

Example

The first step in converting the circuit of p. 168 to NAND and NOR is to note that the second half of the truth table repeats the first half, so it is unnecessary to take C into account. This leaves two inputs, A and B, so there are four possible combinations of input state. Now work backwards from the output. The outputs for the four combinations comprise one 0 and three 1's. The truth tables (p. 165) show that this mix of outputs is a feature of NAND gates. Therefore the output gate of the converted circuit must be NAND.

Compare the normal output of NAND with the output of the circuit on p. 168:

| Inputs | | Outputs | |
B	A	NAND	Circuit
0	0	1	1
0	1	1	1
1	0	1	0
1	1	0	1

Inverting A in the bottom two lines makes the NAND output agree with the required output. Inverting A in the top two lines makes no difference as both lines have output 1. So the solution is to invert A before feeding it to the NAND gate. We can use a NOT gate if there is one to spare. Otherwise, build the circuit from two NAND gates, as below:

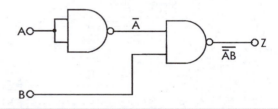

It is often useful to be able to work only with NAND gates. This is the NAND gate equivalent of the circuit on p. 166 and its simplified version on p. 168.

We can also use De Morgan's Theorem to help design a circuit in NAND or NOR gates. The diagrams below show equivalent circuits.

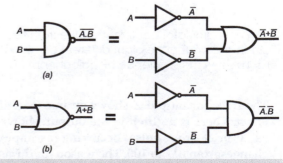

Replacing (a) OR with NAND, and (b) AND with NOR, after inverting the inputs.

Exclusive-OR and exclusive-NOR ICs are available, usually four gates to a package, and sometimes a circuit needs several such ICs (see the parity tree in Chapter 20). But often we need such a gate only once or twice in a circuit and it is better then to build it from other basic gates. The Karnaugh map technique is no help as there are no entries from the truth tables that can be grouped together.

The expression for exclusive-OR from the truth table itself is:

$$A \oplus B = A\overline{B} + \overline{A}B$$

This can be translated directly into gates, using three of the basic types, NOT, AND and OR:

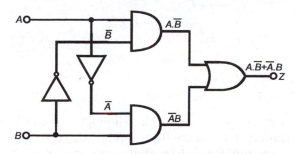

An exclusive-OR gate that is built from NOT, OR and AND gates, requires three different ICs.

In practice we might not find all three types already available as spare gates, and they might be too far apart on the circuit board to make it convenient to connect them.

An alternative exclusive-OR circuit circuit can be built entirely from NAND gates:

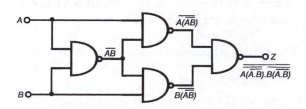

This version of the exclusive-OR gate uses only NAND gates to build it. It is assembled from a single IC

When trying to solve any particular logical problem, it is useful to adopt more than one of the four possible approaches. The results from one approach act as a check on those produced in another way. So, to confirm the all-NAND version of the exclusive-OR gate we will try to arrive at the same result by using logical equations. As usual, De Morgan's identities are important in the proof.

We begin with the equation quoted on the left:

$$A \oplus B = A\overline{B} + \overline{A}B$$

The proof continues by including two expressions that are certain to be false, for they state that A AND NOT-A is true, and also that B AND NOT-B is true. Their value is zero so they do not affect the equality. Next, common factors are taken to the outside of brackets. The final two lines rely on De Morgan:

$$= A\overline{B} + \overline{A}B + A\overline{A} + B\overline{B}$$
$$= A(\overline{A} + \overline{B}) + B(\overline{A} + \overline{B})$$
$$= A(\overline{AB}) + B(\overline{AB})$$
$$= \overline{\overline{A(\overline{AB})}.\overline{B(\overline{AB})}}$$

The array of bars over the final expression is complicated but, if you work carefully from the bottom layer of bars upward, you will see that this expression is made up entirely of NAND operations and is the exact equivalent of the circuit at bottom left.

Activity — Logical operations

Any of the logical circuits in this chapter may be set up on a breadboard for testing. You need:

- A breadboard.

- A supply of short lengths of single-core wire with insulation stripped from both ends.

A power supply, 5 V regulated DC for 74LS series ICs, or any DC voltage between 3 V and 15 V for CMOS. Any DC voltage between 2 V and 6 V for 74HC series.

- Integrated circuits to provide the gates.
- Manual to show the pin connections of the ICs.
- Output indicator, either a testmeter or a logic probe. You can also use an LED in series with a resistor to limit the current to a few milliamperes.

Complete the wiring and check it *before* applying power. With CMOS, all unused inputs *must* be connected to *something* before switching on, usually either to the positive or to the 0 V supply line. Unused outputs are left unconnected.

Take care to avoid pins becoming bent beneath the IC when inserting the ICs into the sockets in the breadboard. The pin may not be making contact but you will not be able to see that there is anything wrong.

CMOS ICs need careful handling to avoid damage by static charges. Keep them with their pins embedded in conductive plastic foam until you are ready to use them. It is good practice to remove charges from your body from time to time by touching a finger against an earthed point, such as a cold water tap. Follow any other static charge precautions adopted in your laboratory.

Apply 1 or 0 inputs by connecting the appropriate pins to the positive or 0V supply lines. Preferably use push-buttons with pull-down resistors, as on p. 156.

Activity — Using a simulator

Simulation software (pp. 170-171) can be used for finding the answers to questions on logical operations. They can also be used to check the answers to questions that you have solved by other methods, such as truth tables, identities and Karnaugh maps.

For example, the question might be:

Investigate the action of the circuit shown in the diagram at top right. Write its truth table and describe the simplest equivalent circuit.

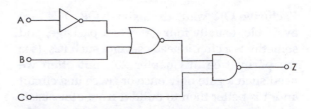

(Above) The circuit to be investigated.

(Below) The same circuit set up on a simulator. The user sets each of the three inputs on the left to '0' or '1'. The simulator displays the output on the right

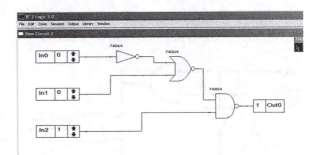

Ater running through all eight possible combinations of the three inputs and recording the output for each combination, the value of Z is found to be '1' for all combinations except for $\overline{A} \bullet B \bullet C$, when it is '0'.

This pattern of outputs (all '1' except for one '0') is typical of a NAND gate (see table, p. 165). Therefore the circuit is equivalent to a 3-input NAND gate with input A inverted.

Draw some circuits based on 3 or 4 logic gates, each with 2 or 3 inputs (exclusive-OR and exclusive-NOR gates can have only two inputs). Use a simulator to investigate their action. Try to simplify the circuit and use a simulator to confirm your answer.

Use a simulator to solve some of the questions opposite.

Questions on logical operations

1 Simplify this expression and write its truth table:

$$\overline{A \bullet B + \overline{A} \bullet B}$$

To what logical operation is it equivalent?

2　Draw a circuit diagram to represent this

$$Z = \overline{A} \bullet B \bullet C + A \bullet \overline{B} \bullet C + A \bullet B \bullet \overline{C} + A \bullet B \bullet C$$

Use a breadboard or simulator to investigate its logic.

3　Write the truth table for the circuit below. To what gate is it equivalent?

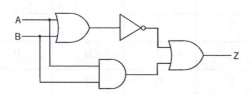

4　Obtain the logical equations from the Karnaugh maps below. Draw the logic circuit for each equation.

B.A / C	00	01	11	10
0	1	1	0	0
1	0	1	1	0

(a)

B.A / C	00	01	11	10
0	1	0	0	1
1	1	0	1	1

(b)

B.A / D.C	00	01	11	10
00	0	0	1	1
01	0	1	0	1
11	0	1	1	0
10	0	0	0	0

(c)

B.A / D.C	00	01	11	10
00	1	0	0	1
01	0	0	0	1
11	0	1	0	0
10	1	0	0	1

(d)

5　Convert the circuits of Question 4 into an all-NAND or all-NOR form.

6　Simplify the following expressions and draw their equivalent circuits:

(a)　$Z = (AB + BC + ABC) \bullet \overline{C}$

(b)　$Z = (A + B) \bullet (A + C)$

(c)　$Z = A \bullet 1 + \overline{A} \bullet 1 + B \bullet 0$

(d)　$Z = \overline{A}\overline{B}\overline{C} + A\overline{B}\overline{C}$

(e)　$Z = AB + A\overline{B} + \overline{A}B$

(f)　$Z = \overline{A}B\overline{C}D + \overline{A}BCD + \overline{A}\overline{B}CD$

Multiple choice questions

1　In this list, the operator that is not a basic logical operator is:

A　AND
B　OR
C　NAND
D　NOT.

2　The operator represented by the ⊕ symbol is:

A　NAND
B　NOT
C　NOR
D　Exclusive-OR.

3　The output of a 3-input NAND gate is low when:

A　all inputs are low
B　one input is high
C　more than one input is low
D　all inputs are high.

4　The number of input combinations which can make the output of a 4-input OR gate go low is:

A　8　　C　15
B　16　　D　1.

5　The output of an Ex-OR gate is '0' when:

A　A is '0'.
B　A is '1' and B is '0'.
C　A and B are the same.
D　A and B are different.

6　The combination of inputs that produces a low output from the system below is:

A　All low.
B　A high, B and C low.
C　A and B high, C low.
D　All high.

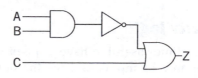

20 Logical combinations

Using the design techniques described in Chapter 19, it is possible to build a wide range of logical functions. In this chapter we look at a number of useful functions that you might need when designing electronic projects. All of these depend on logical combinations, often referred to as **combinatorial logic**. The output of a combinatorial logic circuit depends solely on the present state of its inputs. Its output can be predicted from its truth table.

Combinatorial logic contrasts with **sequential logic** which is fully explained in Chapters 21 and 22.

Many of the functions you may need in project design can be achieved with fewer than ten gates. These you can build from the basic gates in IC packages. You might build them entirely from NAND gates.

Other functions might require dozens, or possibly hundreds, of logic gates. Designing, building and testing such circuits are difficult and demanding tasks.

Fortunately, the more complex logical functions are so frequently used that manufacturers have designed ICs in which complete logical systems are ready-built on a single chip. This is known as **medium scale integration** (MSI), in contrast to the **small scale integration** (SSI) of the NAND gates and the other simple logical devices that we met in Chapter 19. In this chapter we look at some of the more commonly used MSI devices for dealing with logical combinations.

Majority logic

It is sometimes useful to have a circuit with an odd number of inputs and for the output to take the state of the majority of the inputs. For example, with a 5-input majority circuit, the output is high if any 3 or 4 or all 5 inputs are high.

Inputs			Output
C	B	A	Z
0	0	0	0
0	0	1	0
0	1	0	0
0	1	1	1
1	0	0	0
1	0	1	1
1	1	0	1
1	1	1	1

The truth table of a 3-input majority circuit (above) requires three inputs, A, B, and C. The output is high when any two of A, B, and C are high or when all three are high.

Stating this as a logical equation, we obtain:

$$Z = \overline{A}BC + A\overline{B}C + AB\overline{C} + ABC$$

To implement this as written would take three NOT gates, four 3-input AND gates and a four-input OR gate, a total of 8 gates. Here is the Karnaugh map of the circuit:

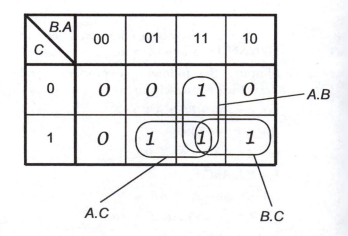

There are three groups of 1's, giving:

$$Z = A{\bullet}B + A{\bullet}C + B{\bullet}C$$

Three 2-input AND gates and one 3-input OR gate, a total of 4 gates, can obtain the same logic. This is only half the number of gates used by the original logic and these have fewer inputs.

The all-NAND version is easy to find because the equation above is converted to the required form by a single step, using De Morgan's theorem:

$$Z = \overline{\overline{A{\bullet}B}{\bullet}\overline{A{\bullet}C}{\bullet}\overline{B{\bullet}C}}$$

NAND the three pairs of inputs, then NAND the three NANDS so obtained. Again we need just 4 gates.

Half adder

Arithmetical operations are often an essential function of logic circuits. This truth table describes the simplest possible operation, adding two 1-bit binary numbers A and B:

Inputs		Outputs	
B	**A**	**Carry (C_o)**	**Sum (S)**
0	0	0	0
0	1	0	1
1	0	0	1
1	1	1	0

The half adder circuit requires two outputs, sum (S) and carry-out (C_o). The logic of each of these is worked out separately. Comparing the outputs in the S column with those in the truth tables on p. 165 shows that the operation for summing is equivalent to exclusive-OR. The operation for producing the carry-out digit is AND. The complete half adder needs only two gates.

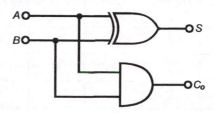

A half adder needs only two gates to find the sum of two 1-bit numbers and the carry out digit.

If required, the half adder can be built entirely from NAND gates by using an exclusive-OR circuit (p. 156) to produce S. The exclusive-OR circuit already produces the NAND of A and B. It is necessary to use only one more NAND gate as inverter to obtain the AND function. This provides C_o.

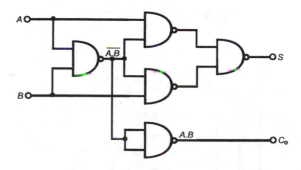

A half adder can also be built entirely from NAND gates.

Full adder

The half adder is so-called because it is incomplete. It is not able to accept a carry-in C_i from a previous stage of addition. The truth table of a full adder is shown on the next page. The first four lines show the addition of A and B when the carry-in is zero. The values of A, B, S and C_o are identical with those in the half adder table. The last four lines show the addition when carry-in is 1.

Inputs			Outputs	
C_i	B	A	C_o	S
0	0	0	0	0
0	0	1	0	1
0	1	0	0	1
0	1	1	1	0
1	0	0	0	1
1	0	1	1	0
1	1	0	1	0
1	1	1	1	1

The first thing to notice is that, for S, the first four lines have an exclusive-OR output, as before. The last four lines have an exclusive-NOR output. In the circuit below we use a second exclusive-OR gate to perform the true/invert function on the output from the exclusive-OR gate of the half adder.

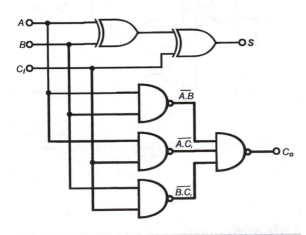

A full adder allows for a carry-in digit from a previous stage of addition.

When C_i is low the second exclusive-OR gate does not invert the output of the first one, so the output is exclusive-OR, and gives S for the first four lines of the table. When C_i is high, the second gate inverts the output of the first one, giving exclusive-NOR, which gives S for the last four lines.

Looking at the column for C_o, this is a majority logic function. C_o is high whenever two or more inputs are high. We use majority logic but base it on four NAND gates. If the two exclusive-OR gates are built from NAND gates, the whole circuit is in NAND gates. A gate is saved because the first exclusive-OR gate already has a NAND gate to provide the NAND of A and B so the output of this is used in producing the carry-out. Altogether, a total of 11 NAND gates is required.

Although it is interesting to build a complex circuit from the basic SSI units, this is the point where we begin to use MSI devices such as the CMOS 4008 and the 74LS283 binary 4-bit full adders.

Parity tree

Parity trees are used for checking if a byte of data is correct. In an 8-bit system, 7 bits are reserved for the data and the eighth bit is the **parity bit**. When the data is prepared, the numbers of 1's in each group of 7 bits is counted. The parity bit is then made 0 or 1, so as to make the total number of 1's odd. This is called **odd parity**. Later, perhaps after having been transmitted and received elsewhere, the data can be checked by feeding it to a parity tree.

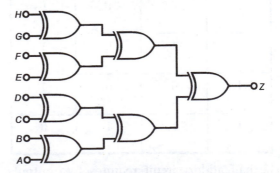

The output of this parity tree goes high if the 8-bit input contains an odd number of 1's.

If the data is correct it still has odd parity and the output of the tree goes high. If one of the bits has changed, the number of 1's is even and the output of the tree goes low, warning that the data is corrupted.

Magnitude comparator

This circuit compares two 4-bit words, and gives a low output if they are identical. The two inputs of each gate receive the corresponding bit from each word.

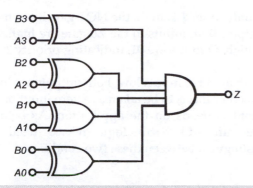

A magnitude comparator is based on an array of exclusive-OR gates.

Self Test

Write truth tables for the comparator (or investigate it on a breadboard or simulator) to find the output when (a) both words are 1011, and (b) when word A is 1100 and word B is 0101.

MSI versions of this circuit have additional logic to detect which of the two words is larger (supposing them to be 4-bit binary numbers) if they are unequal.

Decimal to binary converter

Assuming that only one input is high at any one time, the circuit below produces the 2-bit binary equivalent of the decimal numbers 1 to 3. An extended converter is built with three 3-input OR gates to convert numbers from 1 to 7. A circuit that converts decimal to binary is also called an **encoder**, which is sometimes shortened to **coder**.

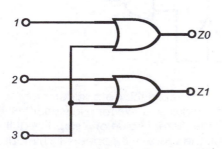

The circuit does not allow for a zero (0) input. This is because, in any given application, a zero input may or may not have a meaning different from 'no input'. In either case the output should be 00. If the range of possible inputs starts at 1, a zero input can not occur and there is no confusion. Then '00' clearly means 'no input'.

Self Test

Write the truth table for the converter. Note that the table has two output columns.

If a zero input is possible, the circuit needs an extra line, known as an 'input present' line. It has an output that is driven to 1 whenever there is an input to the circuit. If the output is 00 and the 'input present' output is 1, this is interpreted as an input of 0. If the output is 00 but the 'input present' output is 0, it means there is no input.

Binary to decimal converter

This performs the reverse function to the previous circuit. Given a binary input, the *one* corresponding decimal output goes high. A circuit such as this, which converts binary to decimal, is also known as a **decoder**.

The truth table is:

Inputs		Outputs			
B	A	3	2	1	0
0	0	0	0	0	1
0	1	0	0	1	0
1	0	0	1	0	0
1	1	1	0	0	0

Looking at the output columns, each is 1 for only one set of inputs. This suggests that NOR gates might be suitable as the output gates. The circuit diagram overleaf shows how each gate is fed with either the TRUE or the NOT form of the two input digits.

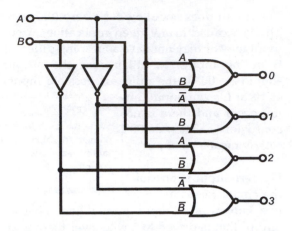

A binary number is fed to terminals A and B of this binary-to-decimal converter, making the corresponding one of its outputs go high.

Priority encoder

A priority encoder is similar to the binary to decimal converter but does not assume that only one input is high at any time. If two or more inputs are high, the outputs show the binary equivalent of the larger-numbered (or priority) input. The truth table is:

Inputs			Outputs		Priority
3	2	1	Z1	Z0	
0	0	0	0	0	0
0	0	1	0	1	1
0	1	0	1	0	2
0	1	1	1	0	2
1	0	0	1	1	3
1	0	1	1	1	3
1	1	0	1	1	3
1	1	1	1	1	3

To follow the operation we start with the highest-numbered input. If input 3 is high, both OR gates receive at least one 1, so both $Z0$ and $Z1$ are high, coding binary 11 and indicating priority 3. Whatever else is on the other lines makes no difference.

If input 2 is high (but 3 is low), $Z1$ is high. The output of the NOT gate is low, so the AND gate output is low. Inputs to the $Z0$ gate are both low, so $Z0$ is low. Outputs are binary 10, indicating priority 2. The state of input 1 makes no difference.

If only input 1 is high, the NOT gate has a high output. Both inputs to the $Z0$ gate are high. $Z0$ is high. Outputs are 01, indicating priority 1.

With all three inputs low, both outputs are low. Whether this is to be taken as a zero input or no input depends on the application. As before, we can add extra logic if we need to distinguish between these two conditions.

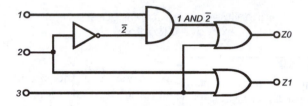

When two or more of the inputs of this priority encoder are made high, the outputs show the number of the higher ranking input.

Data selector

A data selector (or **multiplexer**) has several inputs but only one or two outputs. The data selector circuit below has two inputs, A and B, and one output, Z.

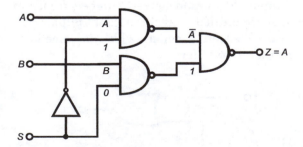

The logic levels in this data selector (or multiplexer) are shown for a low select (S) input. The current value of A appears at the output. The value of B appears if S is made high.

The action of the selector depends on the logic level at the select input, S. If S is low, as in the figure, output Z follows the state of input A. Conversely, Z follows B if S is high. So, by setting S to 0 or 1, we can select data A or data B. The select input is the equivalent of an address, the two possible addresses in this circuit being 0 or 1.

Many data selectors are more complicated than this one, but they all have the same three kinds of input:

- Data inputs, of which there are usually two, four, eight or sixteen.

- Address inputs, used to select one of the data inputs to appear at the output. Depending on the number of data inputs, there are one, two, three, or four address inputs, which are the equivalent of 1-bit 2-bit, 3-bit and 4-bit addresses.

- An enable input, to switch the output on or off. This is also known as the **strobe** input. This is optional and this selector does not have one.

There may be only one output, which shows the selected data input whenever it is enabled. Some data selectors have a second output which shows the inverse of the data.

The selector described here is the simplest possible one, with two data lines, one address line and one output. It does not have an enable input. Its truth table is:

Inputs			Output
B	**A**	**Select**	**Z**
X	0	0	0
X	1	0	1
0	X	1	0
1	X	1	1

The 'X' means 'Don't care'. In other words, the state of that input has no effect on the output.

When S is low, the output shows the state of input A. When it is high, it shows the state of input B. This action is obtained by using a pair of NAND gates. The figure shows the logic levels when the data on A is selected by making S input. Note that the signal is inverted twice as it passes through the circuit and emerges in its TRUE form.

Data distributor

This performs the opposite function to a data selector, and is often called a **demultiplexer**. It has one data input and the data may be routed to any one of several outputs. There are address inputs to select which output is to show the data.

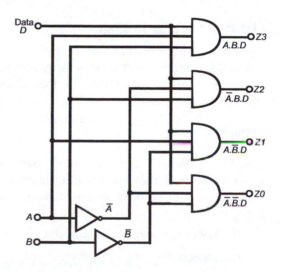

A data distributor (or demultiplexer) routes the data to one of its outputs. The output is selected by supplying its binary address to inputs A and B.

The truth table of a 2-line to 4-line data distributor is shown overleaf. The first line states that, if the data input is 0, all outputs are 0, whatever the state of the select inputs. As before, 'X' means 'Don't care'.

If the data is 1, it appears as a 1 output at the *one* selected output. The unselected outputs remain at 0.

This is the truth table of the data distributor illustrated on p. 181:

Inputs			Outputs			
Data	Select B	A	Z3	Z2	Z1	Z0
0	X	X	0	0	0	0
1	0	0	0	0	0	1
1	0	1	0	0	1	0
1	1	0	0	1	0	0
1	1	1	1	0	0	0

Arbitrary truth tables

Very often a logic circuit has to perform a complicated function that is not simply related to one or more of the standard logical or mathematical functions.

Example

Consider a logic circuit controlling the hot water valve in a domestic washing machine. Whether it is to be open or not depends on many factors:

- the water temperature required,
- the level of the water in the wash-tub,
- the washing cycle selected,
- whether or not a 'half-load' version of the cycle has been selected,
- the stage in the washing cycle, and
- whether or not an operating fault has been detected requiring the water supply to be cut off.

These requirements produce a very complicated truth table, known as an **arbitrary truth table**.

In some applications, a Karnaugh map may solve the problem, and only a few gates are needed. In other cases a large number of gates is unavoidable.

One instance of this is a circuit for driving a 7-segment display. The display (p. 211) represents the decimal digits 0 to 9 by having 7 segments which are turned on or off. This action has an arbitrary truth table, of which the first 4 lines are:

Inputs		Outputs						
B	A	a	b	c	d	e	f	g
0	0	1	1	1	1	1	1	0
0	1	0	1	1	0	0	0	0
1	0	1	1	0	1	1	0	1
1	1	1	1	1	1	0	0	1

We will follow the steps for designing a circuit to decode the first 4 digits, 0 to 3. Column a shows when segment a is to be on. This is the output column of the truth table for:

$$a = \overline{A \bullet \overline{B}}$$

The reasoning is the same as for the example on p. 155, except that A and B are exchanged. The next segment, segment b, is on for all combinations of inputs (though not if we continue the table to include digits 4 to 9). Its equation for the table above is $b = 1$. The equations for the remaining segments are:

$$c = \overline{\overline{A} \bullet B}$$
$$d = a$$
$$e = \overline{A}$$
$$f = \overline{A + B}$$
$$g = B$$

> **Self Test**
>
> Confirm that the equations for digits c to g are correct.

The 0-3 decoder has the circuit drawn opposite. This is fairly complicated but the decoder 0-9 is even more so, because there are two more inputs (C, D) to operate with. We have to add these to the 0-3 decoding because we must make sure that digits 0 to 3 appear only when $C = 0$ and $D = 0$.

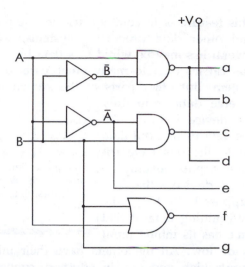

This circuit decodes 2-bit binary inputs 0 -3 and produces outputs for driving a 7-segment display. Segment *b* is switched on for all of digits 0-3, so it is powered directly from the positive supply rail.

In the full 0 to 9 decoder, extra logic is needed for driving segment *b* because it is not lit for digits 5 and 6. Segment *d* needs its own logic because it is not equal to *a* for digit 7. In fact, *all* of the segments need more complicated logic to cover the additional digits 4 to 9.

Hexadecimal

A counting system based on 16 digits, 0 to 9 and A to F, the letters representing decimal values 10 to 15. This covers all the 16 binary values 0000 to 1111.

To make things even more complicated, we could also decode for hexadecimal digits A to F when the input is 1010 to 1111.

Designing a binary-to-seven-segment decoder is an interesting project and helps you to learn about simple gates and how they work. But, in practice, we do not attempt to build decoders from individual gates. Instead, we use a ready-made decoder such as the 74LS47 or the CMOS 4511. These inexpensive ICs contains all the logic for decoding a binary input, and also such features as **leading zero blanking**.

In a display of several digits, leading zero blanking switches off all the segments on the left-most digits when the digit receives an all-zero input. A 4-digit display shows ' 63', for example, instead of '0063'. Another feature is a 'lamp test' input that switches all segments on to check that they are in working order.

When truth tables are so long and complicated, there are other ways in which we can obtain the same function, if it is not already available as one of the standard ICs. One of these ways is to use data selectors. The diagram below shows how we might use a data selector in the control circuit of a car radio. The radio has seven pre-tuned stations, some of which transmit in mono and the others in stereo. In the truth table overleaf, a 0 output means mono and a 1 output means stereo. Stations are numbered 1 to 7 in binary.

In the first line we see that there is no button for selecting Station 0. We can ignore this possibility when working out the logic. In practice, since the '0' input of the IC must be connected somewhere and the adjacent pin is to be connected to the positive supply, the simplest course is to make it positive.

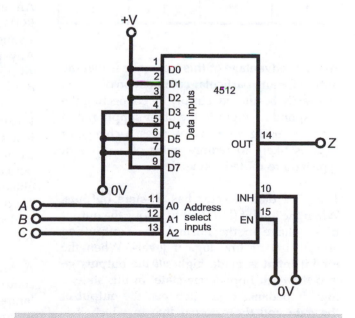

When using a data selector, inputs are wired to the positive supply or to the 0 V line to give the required pattern of highs and lows in the output.

The IC has an enable ($\overline{EN}$) input that must be held low, as shown in the diagram. The bar over the letters 'EN' indicates that this input is **active-low**. This means that the input exerts its effect (that is, to enable data selection) when it is made low. If it is made high, the output goes low independently of the data inputs.

The output of the data selector goes high or low depending on the station number, and switches the radio set into mono or stereo mode, according to the preset data inputs.

	Inputs		Output
C	B	A	Z
0	0	0	Can not happen
0	0	1	1
0	1	0	1
0	1	1	0
1	0	0	1
1	0	1	0
1	1	0	0
1	1	1	1

Another advantage of this technique is that the output for any particular combination of inputs can easily be altered, simply by connecting the corresponding input to the other supply rail. If we were using an array of gates, a change of logic is likely to require gates of a different type, and re-routed connections.

The 4512 data selector has **tri-state outputs**. When the inhibit (INH) input is low, the output pins behave in the normal way, outputting either high or low logical levels. When the inhibit input is made high, *all* the outputs go into the **high impedance** state. In this state, a high impedance exists between the output of the gate and the external line to which it is normally connected. In effect, all outputs become disconnected from the system and no signals come from the device.

This feature is in common use in computers and other data processing systems, where several ICs may be wired to a **data bus.** This bus carries data to many other parts of the system. But other parts of the system must receive data from only one device at a time. All the devices that put data on to the bus have tri-state outputs, but only the one device that is supposed to be transmitting data at that time has its inhibit input made low. All the others have their inhibit inputs high and are, in effect, disconnected from the bus.

> **Data bus**
>
> A set of conductors between any two or more parts of a data-processing system, often each carrying the individual bits of a data group.

Using ROM

Another approach to complicated combinational logic is to use a ROM. A memory chip has address inputs by which any item of data stored in the memory can be read. Chips vary in the number of data items stored and the number of bits in each item.

> **ROM**
>
> A read-only memory device.

An example of the smaller ROMs is the 82S123A, which stores 256 bits arranged in 32 bytes, each consisting of 8 bits. Any particular byte is read by inputting its address (00000 to 11111, equivalent to 0 to 31 in decimal) on five address input pins.

When the 'chip enable' input is made low, the byte stored at this address appears on the eight tri-state output pins. Thus, for each of the 32 possible combinations of input states, we can obtain one of 32 corresponding output bytes. There is no need for complicated logic circuits to decode the inputs. We simply program the ROM with the required output levels.

A device such as the 82S123A can be used to control up to eight different devices such as lamps, displays, solenoids, or motors. It is programmable only once, but many other types of ROM may be reprogrammed if the logic has to be changed.

Programmable array logic

If a logic design is highly complicated it may be best to create it in a **logic array**. An array has several hundreds or even several thousands of gates, all on one chip. They may be connected together by programming the device, using a personal computer. The chip is usually contained in a PLCC package, which is square in shape with a large number (as many as 84) pins arranged in a double row around the edge of the device. Depending on the type of PAL, some of the pins are designated as inputs, some as outputs, and some may be programmable to be normal inputs or outputs or to be tri-state outputs. Many PALs also contain an assortment of ready-made flip-flops and other useful sequential logic units.

Programming PALs is an intricate matter. It is done on a development board or programmer, which provides the programming signals. This is under the control of a personal computer, running special software to produce the required logical functions. Most types of PAL are erasable and reprogrammable.

Logic in software

Mention of software reminds us that there is an increasing tendency to use microprocessors and microcontrollers (such as the PIC and the AVR controllers) to replace complicated logic circuits. We use *software* logic instead of *hardware* logic. This allows even more complicated logical operations to be performed, yet requires no greater expense, board space or power.

As explained in detail in Chapter 36, the microcontroller receives inputs from sensor circuits and control switches. The logic is performed by software running on the microcontroller. The truth table is built into the logic of the software. This determines what the outputs will be. Then the appropriate signals are sent to the microcontroller's output terminals. If the logic needs to be changed or updated, it is relatively simpler and quicker to re-write software than it is to redesign and reconstruct hardware.

Activity — MSI devices

Investigate some of the MSI devices that perform the combinational functions described in this chapter. Work on a breadboard or use a simulator.

Devices to investigate include:

Adder (4-bit, full): 4008, 74LS283.

Magnitude comparator: 4063 (4-bit), 74LS85 (4-bit), 4512 (8-bit).

Data selector/multiplexer: 4019 (2 line), 74LS157 (2-line), 74LS153 (4-line), 74LS151 (8-line), 4512 (8-line).

Data distributor/demultiplexer/decoder: 74LS138 (3-to-8 line), 74LS139 (2-to-4 line), 4555 (2-to-4 line).

Decoder: 4028 (1 of 10), 4511 (7-segment), 74LS47 (7-segment).

Priority encoder: 4532 (8-to-3 line), 74LS148 (8-to-3 line), 74LS147 (10-to-4 line).

Verify that the circuit operates according to the truth table (function tables) in the manufacturer's data sheet.

Questions on logical combinations

1 Design a 3-bit binary to decimal converter, using only NOR gates.

2 Explain the action of a 2-input data selector.

3 Design a 3-input data selector, with an enable input and TRUE and NOT outputs.

4 Design a 4-input data selector, using only NAND gates.

5 Design and verify (on a breadboard or simulator) a circuit for driving a 7-segment display for (a) the digits 0 to 7, and (b) the even digits 0 to 8.

6 Explain how a binary-to-decimal decoder can be used as a data distributor.

7   In a shop, a siren is to be sounded automatically if:

*   the emergency button is pressed, or

*   smoke is detected, or

*   a person is detected inside the shop between dusk and dawn, or

*   if someone enters the storeroom without first closing the hidden 'storeroom access' switch, or

*   the shopkeeper leaves the premises without first locking the storeroom door.

    Design the logic for this system. State what outputs each sensor and switch gives to indicate the 'alarm' condition.

8   In a greenhouse, the irrigation system is to be turned on automatically when:

*   the temperature rises above 25°C, or

*   humidity falls below 70%, or

*   between 1430 hrs and 1530 hrs every day.

    The system must never be turned on when:

*   it is between dusk and dawn, or

*   temperature falls below 7°C,

*   there is someone in the greenhouse, or

*   the manual override switch has been closed.

    Design the logic for this system. State what outputs each sensor and switch gives to indicate the 'irrigate enable' condition.

9   Outline the techniques available for building circuits for arbitrary truth tables when the logic is complicated.

Multiple choice questions

1   The output of a majority logic circuit goes high when:

   A more than half the inputs are 1.
   B all inputs are 0.
   C some inputs are 1 and some are 0.
   D all inputs are 1.

2   The inputs to a half adder are $A = 1$ and $B = 1$. Its outputs are:

   A $S = 1$, $C_o = 1$.
   B $S = 0$, $C_o = 1$.
   C $S = 1$, $C_o = 0$.
   D $S = 0$, $C_o = 0$.

3   The outputs of a full adder are $S = 0$, $C_o = 1$. Its inputs might be:

   A $A = 1$, $B = 0$, $C_i = 0$.
   B $A = 1$, $B = 1$, $C_i = 1$.
   C $A = 1$, $B = 0$, $C_i = 1$.
   D $A = 0$, $B = 0$, $C_i = 0$.

4   An 8-input parity tree has odd parity. Its output goes high when the input is:

   A 11111111.
   B 10011000.
   C 10000001.
   D 00000000.

5   'Demultiplexer' is another name for:

   A a parity tree.
   B a data selector.
   C a decoder.
   D a data distributor.

6   A data distributor:

   A has one data input.
   B has two outputs.
   C is often called a multiplexer.
   D has two address inputs.

7   In the circuit drawn below, if inputs 1 and 2 are made high and 3 is low, the outputs are:

   A Z0 = 0 and Z1 = 1.
   B Both are 1.
   C Z0 = 1 and Z1 = 0.
   D Both are 0.

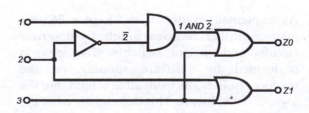

21 Logical sequences

If a logic circuit has *feedback* in it, then its response to a given input may not be the same the second time an input is applied. The circuit is in one state to begin with and then goes into a different state, in which its response may be different. It goes through a *sequence* of states. Logic of this type of circuit is known as a **sequential logic**.

Flip-flops

An example of sequential logic is the **set-reset flip-flop**, or SR flip-flop shown below:

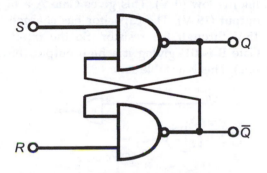

> A set-reset flip-flop has two stable states. It is made to change from one state to the other by applying a low pulse to its normally high inputs.

Its name suggests that it can exist in two states, *Set* and *Reset*, and that it can be made to flip into one state and then flop back into the original state. The essential point is that the circuit is **stable** in either state. It does not change state unless it receives the appropriate input. Because it has *two* stable states it is also known as a **bistable** circuit.

The diagram on the right shows how the flip-flop works:

(a) The flip-flop is in its *Set* state, with Q = 1 and $\bar{Q}$ = 0. The inputs must be held high by the circuit that provides input.

In the Set state, both gates have inputs and outputs that conform to the NAND truth table, so the circuit is *stable*.

(b) The Reset input has been made low. The input to Gate B is now 01. Its output changes from 0 to 1, in accordance with the NAND truth table. Gate A then has a 11 input, and its output is about to change to 0. The flip-flop can not stay in this state; it is unstable.

> **NAND**
> Output is low when all inputs are high.

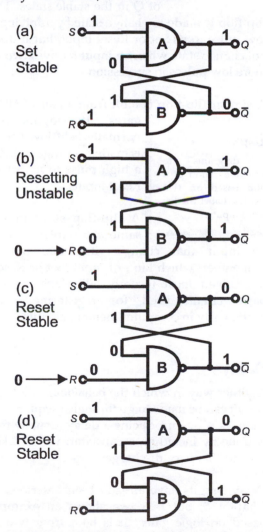

(a) Set Stable

(b) Resetting Unstable

(c) Reset Stable

(d) Reset Stable

> The sequence of stages of resetting an S-R flip-flop.

187

(c) The output of Gate A has gone low, so that the input to Gate B is now 00. This makes no difference to its output, which stays high. The circuit is now in the *Reset* state, with $Q = 0$ and $\overline{Q} = \overline{1}$. It is *stable* in this state because both gates have inputs and outputs that conform to the NAND truth table.

(d) There is no further change of output when the Reset input goes high again. This is because the output of B remains high when it has 01 input. The flip-flop remains stable.

Summing up, $\overline{Q}$ is the inverse of Q in the stable states. The flip-flop is made to change state by applying a low pulse to its Set or Reset input, but it does not change state when an input receives two or more low pulses in succession.

A similar flip-flop is built from a pair of NOR gates. Its inputs are normally held low and it is made to change state by a high pulse to its Set or Reset inputs.

A flip-flop stays in the same state until it receives an input that changes its state. It can 'remember' which kind of input (Set or Reset) it received most recently. Flip-flops are the basic circuit used for registering data, particularly in computer memory circuits.

Delays

Another way in which the behaviour of a logic circuit can be made to go through a sequence of changes is by introducing a **delay**. Usually this is done by including a capacitor, which takes time to charge or discharge.

Delay circuits have already been described in Chapter 5, but here we shows an example based on logic gates. It is built from two 2-input NAND gates. Gate A has its inputs connected together so that it behaves like a NOT gate.

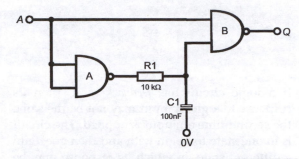

This edge-triggered circuit produces a low output pulse when the input rises from low to high. Its output remains high when input falls from high to low.

The way the delay works is shown in the diagrams below and in the timing diagram opposite. The sequence of stages is as follows:

(a) The normal input to the circuit (dark grey line) is low (0 V). This gives Gate A a high output (15 V). The capacitor has charged to 15 V through the resistor. So the input to Gate B is 01, giving it a high output (black line). This is a stable state.

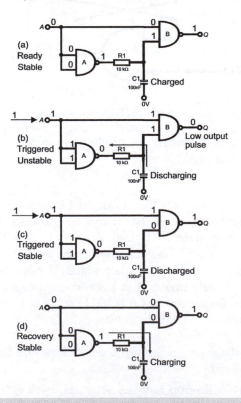

The sequence of stages in the generation of a pulse by the delay circuit. The letters (a) – (d) indicate the stages in the timing diagram.

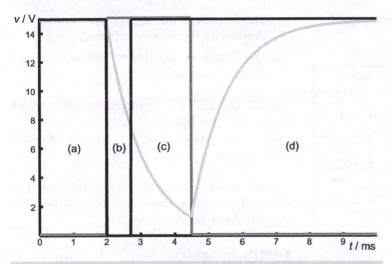

Changes of voltage in the delay circuit show a rise of input at 2 ms (light grey), which immediately produces a low output pulse lasting 0.7 ms (black). Output remains high when input goes low at 4.5 ms. Changes of voltage across the capacitor are plotted in dark grey.

The chief disadvantages of this simple monostable are:

- It must be given time to completely recharge the capacitor before being triggered again.

- The input must go high for longer than the length of the output pulse.

- Pulse length depends on supply voltage.

The delay circuit is stable in its quiescent or 'ready' state, so it is a monostable (p. 41). Once triggered, it produces a pulse and then returns to the same stable state. Monostables with higher precision are available in the 74LS and CMOS series.

(b) At 2 ms the input to the circuit (dark grey line) goes high. This immediately changes the output of Gate A from 1 to 0. The capacitor begins to discharge through R1, current flowing *into* the output of Gate A. The exponential fall in voltage across the capacitor can be seen in the timing diagram (light grey line). Nothing happens to Gate B (black line) at this stage.

(c) At 0.27 ms the falling voltage across the capacitor reaches 7.5 V, so the input to Gate B is now equivalent to 01. Its output goes high, ending the low output pulse.

(d) When the input pulse ends (here at 4.5 ms, but it could be at *any* time after the beginning of stage (c)) the state of Gate A is reversed and the capacitor begins to charge again. It makes no difference to the output of Gate B when its input changes from 00 to 01. So there is no output pulse when the input pulse ends.

Self Test

Describe the action of a pulse generator based on a pair of NOR gates.

This circuit is often used for generating pulses in logic circuits. The length of the pulse depends on the values of the resistor and capacitor.

Monostables

The 74LS221 (below) is a typical monostable pulse generator. The pulse length is determined by the values of the resistor and capacitor and is approximately $0.7RC$ seconds. This is the time taken to charge the capacitor. Using a capacitor of maximum value 1000 µF and a resistor of maximum value 70 kΩ, the greatest pulse length obtainable is 70 s. Output Q is normally low and $\overline{Q}$ is normally high. The monostable is triggered by a falling edge on input A or a rising edge on input B, as indicated by the truth table overleaf.

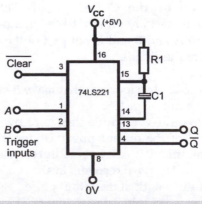

The 74LS221 comprises two identical monostables. The pin numbers in this figure refer to only one of these.

This is the truth table for the operation of the 74LS221 monostable IC:

Inputs $\overline{CLR}$	A	B	Outputs Q	$\overline{Q}$
0	X	X	0	1
X	1	X	0	1
X	X	0	0	1
1	0	↑	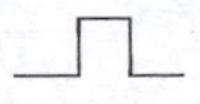	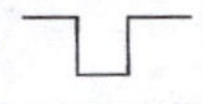
1	↓	1	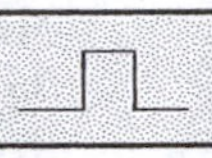	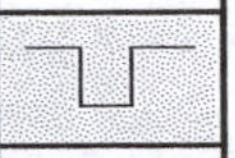
↑	0	1	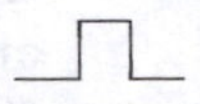	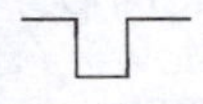

The CLEAR input is active low. When it is made low, the values of A and B have no effect; output Q is low and output $\overline{Q}$ is high. The monostable can also be cleared by making A high or B low, whatever the state of the CLEAR input.

To generate a pulse the CLEAR input must first be made high. Then a falling edge on A (a downward arrow in the table) or a rising edge on B (a rising arrow) triggers the pulse to begin. Once the monostable is triggered, the length of the pulse depends only on the values of R and C. In other words, the monostable is not retriggerable. However, a low level on CLEAR can end the pulse at any time. The last line of the table shows that a pulse can also be triggered by the CLEAR input. This only occurs if certain inputs have already been made to A and B and is not part of the normal action of the circuit.

The 74LS221 has features that make it superior to the circuit on p. 188:

- It requires very little time to recover at the end of the output pulse. Under the right conditions it can have a duty cycle up to 90%. That is, it recovers in about one-ninth of the length of its output pulse.
- B is a Schmitt trigger input. This means that the monostable is reliably triggered on a slowly changing input voltage level.
- The length of the timing period does not depend on the supply voltage.
- The length of the timing period does not depend on temperature.

The CMOS 4528 is a monostable with similar properties, except that it is re-triggerable, and its recovery time is usually several milliseconds.

The 555 timer IC is generally the most popular choice for monostables and for astables, as described in Chapter 5.

CMOS astables

Astables can be built from TTL and CMOS gates. The diagram below shows one built from a pair of CMOS NOT gates. The third gate is not part of the astable but acts as a buffer between the astable and an external circuit such as an LED. R2 is not an essential part of the astable, but serves to give the output signal a squarer shape. It has a value about 10 times that of R1.

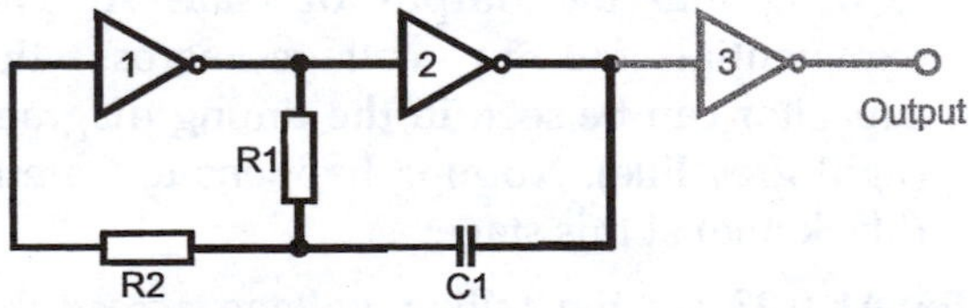

A pair of NOT gates (or their equivalent made from NAND or NOR gates) make an astable multivibrator. The third gate (grey) acts as a buffer.

The action of the astable is as follows, referring to the diagram opposite. (To avoid having to repeat the words 'the supply voltage' several times, assume that it is 15 V):

(a) The input of Gate 1 is 0 V, its output is 15 V, and therefore the output of Gate 2 is 0 V. The capacitor is at a low voltage (actually negative, but see later). Current flows from Gate 1 through R1 and gradually charges the capacitor. The rate of charging depends on the values of R1 and C1.

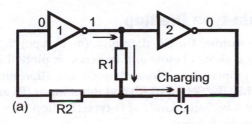

(a)

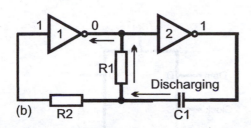

(b)

(b) When the voltage across the capacitor reaches about 7.5 V, the input to Gate 1 (which is at almost the same voltage) is taken to be at logic high. Instantly, the levels change to those shown in (b). The sudden rise in the output of Gate 2 pulls *both* plates of the capacitor up by 15 V. This means that the plate connected to R1 rises to 22.5 V. Now current flows in the reverse direction through R1 and *into* the output of Gate 1. The voltage at this plate falls until it has reached 7.5 V again. The input of gate 1 is taken to be logic low and all the logic levels change back to state (a) again. The sudden fall in the output of Gate 2 pulls both plates of the capacitor down by 15 V. The plate connected to R1 falls to −7.5 V. The astable has now returned to its original state, stage (a).

The frequency of this astable is given by:

$$f = 1/2.2R_1C_1$$

Bistable latch

A latch is a circuit is which the output is normally equal to the data input but which can be held (latched) at any time.

In the diagram below the gates on the right form an S-R flip-flop exactly as on p. 187.

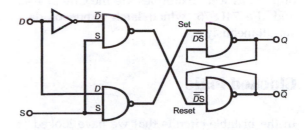

With a data latch, the Q output follows the data input (D) when the STORE input (S) is high, or is latched when S is low.

In that circuit the inputs are normally held high and a low pulse to the Set input sets the flip-flop. In this circuit the Set input of the flip-flop receives $\overline{D \bullet S}$ while the Reset input receives $\overline{\overline{D} \bullet S}$. D is the Data input which at any instant is 0 or 1. S is the Store input. The truth table for D and S is:

| Inputs | | Outputs | | Effect |
S	D	$\overline{D \bullet S}$	$\overline{\overline{D} \bullet S}$	on Q
0	0	1	1	Latched
0	1	1	1	Latched
1	0	1	0	Equals D
1	1	0	1	Equals D

When $S = 0$ the outputs of the NAND gates on the left are 1. They hold the flip-flop in the latched state, since neither input can go low. When $S = 1$, and D goes low, the Reset input goes low, resetting the flip-flop so that Q goes low. When $S = 1$ and D goes high, the Set input goes low, setting the flip-flop so that Q goes high. Whatever the value of D, output Q is equal to D. The other output, Q, always is the inverse of Q and therefore of D.

There are ICs that contain several data latches equivalent to the above. They have individual data inputs, Q outputs and sometimes $\overline{Q}$ outputs as well. Examples are the CMOS 4042 and the 74LS75, which latch 4 bits of data simultaneously.

Clocked logic

In the bistable circuits that we have looked at so far in this chapter, the circuit changes state immediately there is a suitable change of input. This can cause difficulties in large logic circuits.

Logic gates always take time to operate and different amounts of delay may arise in different parts of a circuit owing to the different numbers of gates through which the signals have to pass. It becomes very difficult to make sure that all the bistables change state in the right order and at the right times to respond to the data.

The solution to this is a **system clock**. This controls the exact instant at which the bistables change state. It is like the conductor of an orchestra, keeping all the players in time with each other by beating 'time'. We say that the players (or the bistables) are **synchronised**.

Clocked logic devices usually operate on the **master-slave** principle. They are available as MSI integrated circuits. Each flip-flop actually consists of two flip-flops, the **master** and the **slave.** The master flip-flop receives data and is set or reset accordingly. The state of the master flip-flop can not be read on the output terminals of the IC at this stage. Then, on the next positive-going edge of the clock input, the data on the master is transferred to the slave and appears at the output terminals of the IC.

After this, new data may change the state of the master but this has no effect on the slave or on the data appearing at the terminals. The present output state is held unchanged until the next positive-going clock edge. Then the data currently on the master is transferred to the slave.

Data-type flip-flop

The symbol for the data-type (or D-type) flip-flop is shown below and its action is plotted in the graphs. For a **clocked mode** operation such as this, the Set input (S) and Reset input (R) are held low. Some kinds of D-type flip-flop do not have the R input.

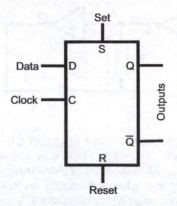

A D-type flip-flop has a data input (D), a clock input (C), and a Q output. It may also have set and reset inputs and an inverted Q output. Data is transferred to the Q output on the rising edge of the clock.

The graphs below are obtained from a simulator, modelling a 74LS74 D-type flip-flop. The clock is running at 10 MHz and the time scale is marked in nanoseconds. The lowest graph shows output Q. The simulator is stepped manually so that the data input can be made high or low at any desired time.

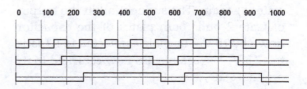

In this plot of the action of a D-type flip-flop, the top trace is the clock, with a period of 100 ns. The data (middle trace) can be changed at any time but the Q output (bottom trace) changes only on the rising edge of the clock. The change is slightly delayed because of the propagation delay of the flip-flop (22 ns in this case).

The run begins with both *D* and *Q* low. *D* is made high at 170 ms. At this time the clock is high. Nothing happens to the output when the clock goes low. But *Q* goes high at the next *rising edge* of the clock. This occurs at 250 ns. Looking more closely at the timing, we note that the rise in *Q* is slightly later than the rise in the clock. This is due to the propagation delay of the flip-flop, here set to a typical value of 22 ns. So *Q* eventually goes high at 272 ns. Subsequent changes in *D*, at 540 ns, 640 ns and 880 ns, always appear at output *Q* at the *next* rising edge of the clock (plus the propagation delay).

D-type flip-flops are used for sampling data at regular intervals of time. They then hold the data unchanged while it is processed. An example is the use of these flip-flops at the input of digital-to-analogue converters (see Chapter 24).

The Set and Reset inputs are used in the **direct mode** of operation of the flip-flop. They are also known as the Preset and Clear inputs. When the set input is made high, *Q* goes high instantly, without waiting for the rising edge of the clock. Similarly, *Q* goes low immediately the Reset input is made high.

Toggle flip-flop

The output of a toggle (or T-type) flip-flop changes every time the clock goes high. Toggle flip-flops are not manufactured as such. It is easy to make one from a D-type flip-flop, by connecting the $\overline{Q}$ output to the D input.

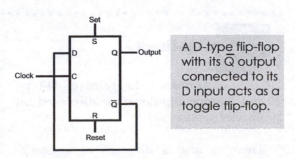

A D-type flip-flop with its $\overline{Q}$ output connected to its D input acts as a toggle flip-flop.

Assuming the D input at the most recent clocking was 0, then *Q* is 0 and $\overline{Q}$ is 1. So the D input is now 1. On the next clocking *Q* follows *D* and becomes high, so $\overline{Q}$ goes low. Now the D input is 0 and *Q* goes to 0 at the next clocking. The state of *D* and therefore of *Q* changes every time the flip-flop is clocked. Below is a graph of the toggle action.

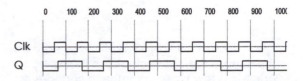

The clock (top trace) is running at 10 MHz. The Q output of the toggle flip-flop changes state slightly after every rising edge of the clock. For this reason, the frequency of the output from the flip-flop is 5 MHz.

The toggle flip-flop demonstrates one of the advantages of clocked logic. After the flip-flop has been clocked, there is plenty of time before the next clocking for the circuit to *settle* into its stable state, and for its output to be *fed back* to the data input. This makes the operation of the circuit completely reliable. Unless the clock is very fast, there is no risk of its operation being upset by propagation time and other delays.

J-K flip-flop

The J-K flip-flop (overleaf) has its next output state decided by the state of its J and K inputs:

- If J = K = 0, the outputs do not change.
- If J = 1 and K = 0, the Q output becomes 1.
- If J = 0 and K = 1, the Q output becomes 0.
- If J = K = 1, the outputs change to the opposite state.

The $\overline{Q}$ output is always the inverse of the Q output. With most J-K flip-flops, the changes occur on the next rising edge of the clock. In the fourth case above, in which J = K = 1, we say that the outputs *toggle* on each rising edge.

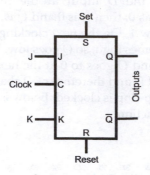

The output of a J-K flip-flop changes state on the rising edge of the clock (on the falling edge in a few types). Whether it changes or not depends on the inputs to the J and K terminals.

J-K flip-flops also have Set inputs and (usually) Reset inputs. These operate in the same way as those of the D-type flip-flop, producing an instant change of output, without waiting for the clock. Several types of J-K flip-flops are clocked by a negative-going edge instead of the more usual positive-going edge.

Activities — Flip-flops

Investigate each of the circuits listed below, either on a breadboard or by using a simulator. Use the suggested ICs and consult data sheets for connection details.

Build a flip-flop using NAND gates as on p. 187 (CMOS 4011 or 74LS00).

Design and build a flip-flop using NOR gates (CMOS 4001 or 74LS02).

Build a pulse generator using NAND gates as on p. 188.

Design and build a pulse generator built as on p. 188 but with NOR gates.

Design and build a pulse generator which produces a low pulse on a *falling* edge, using two NAND gates.

Investigate a pulse generator built from the 74LS221, with pulse length 1.5 ms. Check the pulse length with an oscilloscope.

Design and build a CMOS astable as on p. 190 using either NAND (4011), NOR (4001) or NOT (4069) gates. Select components to give a frequency of 100 Hz and check the frequency by using an oscilloscope.

Build a CMOS astable using exclusive-OR gates (4070). Construct the bistable latch shown on p. 191 and study its action by placing probes at various points in the circuit.

Investigate an MSI data latch (CMOS 4042, 74LS75).

Activities — Clocked logic

Investigate each of the circuits listed below, either on a breadboard or by using a simulator, using the suggested ICs. Consult data sheets for connection details. Most of the ICs specified below contain two or more identical flip-flops, of which you will use only one. With CMOS, it is essential for *all* unused inputs to be connected to the positive or 0 V rail.

Check out a D-type flip-flop (CMOS 4013, 74HC74, 74LS174, or 74LS175). Use a slow clock rate, say 0.2 Hz. Connect the data input to the positive rails through a 1 kΩ resistor to pull it high. Use a push-button connected between the data input and the 0 V rail to produce a low input. The 74LS174 <u>has</u> only Q outputs but the others have Q and Q outputs. Investigate both clocked and direct modes.

Flip-flops

Animated diagrams of D-type and JK-type flip-flops are displayed on the Companion Site.

There is also a diagram of a CMOS astable.

Test a J-K flip-flop (CMOS 4027, 74LS73A, 74LS112A, 74LS379). Connect J and K inputs with pull-up resistors and push-buttons.

Make a toggle flip-flop from a D-type flip-flop or a J-K flip-flop, and investigate its action.

Questions on logical sequences

1 Explain the working of a flip-flop built from two NAND gates.

2 Explain how to make a pulse generator based on two NAND gates.

3 Describe how to make an astable using two CMOS NOT gates, a capacitor and two resistors.

4 Show how to build a data latch using logic gates and use a truth table to explain how it works.

5 What are the advantages of using clocked master-slave logic devices?

6 Draw a timing diagram of a toggle flip-flop, showing the input from the clock and the output from the flip-flop.

7 Calculate the value of the timing resistor required in a CMOS astable, given that it has a 4.7 nF capacitor and the frequency is 5 kHz. What is the nearest E24 value for the resistor? What is the E24 value of the other resistor (R2) in the astable?

8 The diagram on p. 190 shows NOT gates used in the astable. What other types of gate could be used, and how would they be wired?

Multiple choice questions

1 A logical circuit with two stable states is NOT called a:

    A monostable.
    B flip-flop.
    C multivibrator.
    D bistable.

2 If the supply voltage is 12 V, the voltage across the capacitor of a 555 monostable as the pulse ends is:

    A 12 V.   C 3 V.
    B 4 V.   D 8 V.

3 If the supply voltage is reduced to 7.5 V in the monostable described in question 2, the length of the pulse is:

    A unchanged.
    B doubled.
    C halved.
    D divided by 3.

4 If the symbol for an input terminal has a bar drawn over it, it means that:

    A the input must always be low.
    B the input does nothing when it is high.
    C a negative input produces a negative output.
    D the input is made low to bring about the named response.

5 Which of these statements about a toggle flip-flop is NOT true?

    A The $\overline{Q}$ output is the inverse of the Q output.
    B The output follows the D input.
    C Clocking reverses the state of both outputs.
    D The $\overline{Q}$ output is connected to the D input.

6 A J-K flip-flop has its J and K inputs connected to the 0 V line. Its Q output is high. It is triggered by positive-going edges. When will its Q output go low?

    A On the next positive-going clock edge.
    B When the Set input is made high.
    C On the next negative-going clock edge.
    D Never.

7 A circuit in which the output follows the input, but can be held when the store input goes low, is called:

    A a bistable latch.
    B an astable.
    C a toggle flip-flop.
    D a data flip-flop.

8 The $\overline{Q}$ output of a D-type flip-flop is made high by:

    A a low pulse on the D input.
    B a low-going edge on the clock input.
    C a high pulse on the reset input.
    D a low pulse on the preset input.

22 Counters and registers

In Chapter 21 we showed how a data-type (or D-type) flip-flop could be connected to make it into a toggle-type (or T-type) flip-flop. This is simply done by connecting its $\overline{Q}$ output to its D input. As a T-type flip-flop, its output changes state whenever its clock (or C) input changes from low to high — that is, it is triggered on a **rising edge**.

It is possible to connect two or more T-type flip-flops, feeding the Q output of one into the C input of the next. The drawing below shows three D-type flip-flops, operating as toggle flip-flops, and connected in a chain. We are basing the circuit on CMOS 4013 ICs, which contains two D-type flip-flops each, so this circuit needs two ICs, with one flip-flop to spare.

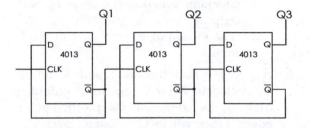

Three D-type flip-flops wired as toggle flip-flops act as a 3-bit up counter divider.

Assume that all the flip-flops are reset to begin with. This could be done by making their direct Reset inputs (not shown in the diagram) high for an instant. The flip-flops begin with their Q outputs low and their $\overline{Q}$ outputs high.

Clocking pulses, perhaps from an astable circuit are fed to the clock input of the first flip-flop. Nothing happens until the clock changes from low to high. Then the flip-flop changes state. Its Q output goes high and its $\overline{Q}$ output goes low.

The change in its Q output from high to low has no effect on the second flip-flop, because it too is rising edge triggered.

There is no further change in the system until the next rising edge of the clock. Then the Q output of the first flip-flop changes back to low and $\overline{Q}$ output changes to high. This change from low to high triggers the second flip-flop to change state. The process continues, with the result shown in these graphs:

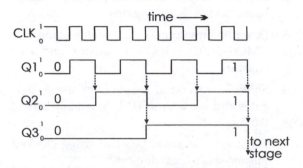

The graphs show the Q output from the flip-flops. The $\overline{Q}$ outputs are the inverse of these.

To sum up the action, we can say that:

- The outputs change state on the rising edge of the clock.

- Each Q output changes state on the falling edge of the previous stage (that is, on the falling edge of its $\overline{Q}$ output).

Dividing and counting

The action of the circuit can be looked at in two different ways:

- **Dividing:** each output changes state at half the rate of the clock or previous stage. The circuit acts as a **frequency divider**. For example, if the clock is running at 100 Hz, the frequencies of Q1, Q2 and Q3 are 50 Hz, 25 Hz and 12.5 Hz respectively.

- **Counting:** The Q outputs run through the sequence 000, 001, 010, 011, 100, 101, 110 and 111, repeating. These are the binary numbers from 0 to 7.

> **Note**
> There is a table on p. 369 that give equivalents in different number systems.

Summing up, a chain of n flip-flops can be used either to divide a frequency by 2, 4, 8, ..., 2^n, or as a counter from 0 to (2^n-1).

Frequency dividers are used for generating frequency-related audio tones of different pitch in alarms and in some kinds of electronic musical instrument. Another common use is in electronic clocks and watches. These depend on high-precision crystal oscillators, which operate only at high frequencies. Typically a 32.768 kHz oscillator is used, and a 15-stage divider divides this by 2^{15}, producing a precise 1 Hz signal. This 'one per second' frequency is used as the basis of the timing action of the clock.

Although it is sometimes convenient to wire up one or two 4013 ICs to divide by 2, 4, 8, or 16, there are MSI devices that have longer chains. An example is the 4020 which has 14 flip-flops connected into a divider/counter chain. This divides by up to 2^{14}, or 16 384.

Almost all the counter circuits and devices can also be used for frequency dividing. This is a fairly straightforward application. So, from now on when describing a counter, we shall ignore its dividing function and refer only to its use as a counter

J-K counter

A J-K flip-flop operates as a toggle flip-flop if its J and K inputs are held high (p. 193). A chain of J-K flip-flops so connected has the same action as a chain of D-type toggling flip-flops.

Up counters and down counters

The counters that we have described so far are known as **up counters**. Each flip-flop is triggered by the Q ouput of the previous stage so, in a 3-bit counter, we run through an up-counting sequence, from 0 to 7. The table shows the sequence:

	Digit		Decimal
C	B	A	
0	0	0	0
0	0	1	1
0	**1**	← 0	2
0	1	1	3
1	← **0**	← **0**	4
1	0	1	5
1	**1**	← 0	6
1	1	1	7
0	← **0**	← **0**	

Remembering that the table shows the Q outputs and that each stage is triggered by a rising Q output, we see that digit B changes from '0' to '1' or from '1' to '0' every time A goes low. Similarly, C changes state every time B goes low. In the table the arrows show when triggering occurs and the digit to the left changes.

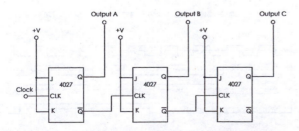

Three J-K flip-flops connected to make a 3-bit binary up counter. This has the same action as the D-type counter on p 196.

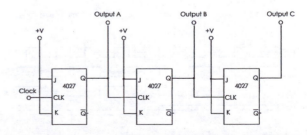

To make a down counter, simply feed the following stage with data from the Q output of the previous stage.

If we change the connections so that each flip-flop is triggered by the Q output of the previous stage, we obtain a **down counter** (p. 197).

	Digits		Decimal
C	B	A	
1	1	1	7
1	1	0	6
1	**0**	← 1	5
1	0	0	4
0	← **1**	← 1	3
0	1	0	2
0	**0**	← 1	1
0	0	0	0
1	← **1**	← 1	7

The arrows show every time a digit changes from 0 to 1, the digit to its left is changed. This produces a repeating downward count from 7 to 0.

MSI counters are made in up-counting and down-counting form. There are also up-down counters, such as the 4029, which count in either direction. They have an up-down input which, makes the counter count up when it is high, and down when it is low.

Ripple counters

The clock is running at 1 MHz in the simulation of a 3-bit up-counter pictured on the left below. From top to bottom, the traces show the clock, and output A, B, and C. At clock rates of 1 MHz or less, propagation delays of a few nanoseconds may be ignored. If we increase the clock rate by 10 times and also plot the outputs on a 10 times scale, we obtain the result shown on the right below.

Propagation delays are relatively more important and we can see them in the plots. The typical delay of the 74LS73 J-K flip-flop used in the simulation is 11 ns. The plot shows that each stage changes state 11 ms after the output from the previous stage has gone low.

This shows up best after 800 ns, where all outputs are 1, then they all change to 0, one after the other. First the clock goes low, then flip-flop A, then flip-flop B and finally flip-flop C. The change from 1 to 0 *ripples* along the chain. For this reason such a counter is known as a **ripple counter**. The rippling action is also illustrated by the arrows in lines 5 and 9 of the table on p. 197 and on the right. The rippling action is even more noticeable in counters with four or more stages.

The table opposite shows the sequence of outputs in a ripple up-counter as the count changes from 111 to 000. The digits change from 1 to 0 in order, beginning with the LSD on the right.

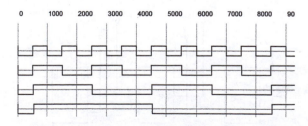

When the clock is running at 1 MHz the output of the 3-bit counter apparently runs correctly through all stages from 000 to 111.

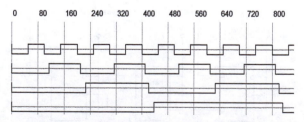

When the clock speed is increased to 10 MHz, propagation delay becomes more noticeable and it can be seen that changes of output from 1 to 0 'ripple' through the three flip-flops. This produces glitches in the counter output.

Digits	Decimal equivalent	
111	7	correct
110	6	error
100	4	error
000	0	correct

While changing from 7 to 0 the counter output goes very briefly through 6 and 4. This may not matter. For example, if the clock is running at 1 Hz, and the counter is driving an LED display that shows the number of seconds elapsed, we will not be able to see the incorrect '6' and '4'. The display will apparently change straight from '7' to '0'. But we might have a logic circuit connected to the output, intended to detect, say, a '6' output. This would easily be able to respond to the brief but incorrect output. This might cause it to trigger some other activity in the circuit at a time when it should not take place.

An instance of an error of this kind was shown on p. 154. In the simulation a ripple counter was used as a convenient way to provide inputs 000 to 111 to the logic network. However, when the counter input changed from 011 to 100 (at 800 ns), it went briefly through a stage 010. As can be seen from the plots at about 500 ns, this is an input which gives a low output from the network. Consequently, there is a low glitch on the output at 800 ns. The logic network has been fast enough to respond to the incorrect output.

Synchronous counters

Clocking all the flip-flops at the same time eliminates the ripple effect. Instead of letting each stage be clocked by the output from the previous stage, they are all clocked simultaneously by the clock input signal.

The diagram at top right shows such a counter made up from three J-K flip-flops. It is known as a **synchronous counter**.

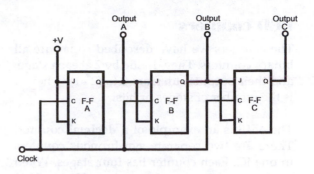

A synchronous counter, in which all the flip-flops are clocked simultaneously, avoids the glitches that occur in ripple counters.

The J and K inputs of the first flip-flop are held high, as in the ripple counter, but the J and K inputs of following flip-flops are connected to the Q output of the previous stage. The plots below show the output of the synchronous counter, plotted with the same clock speed and time scale as the diagram on the right, opposite.

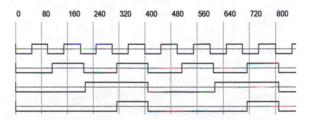

This is the output from the synchronous counter when the clock is running at 10 MHz. Compare this with the output of the ripple counter (p. 198, the diagram on the right).

There is still an 11 ns delay between the clock going low and the changes in the outputs of the flip-flops. But the flip-flops change state at *the same time* (synchronously). This is clearly shown at 800 ns where all three change at once. The count goes directly from 111 to 000 with no intermediate stages.

Counters can also be built from D-type flip-flops, operating as toggle flip-flops.

BCD counters

The counters we have described so far are all binary counters. They divide by 2 at each stage, and their outputs run from 0 to (2^n-1), where n is the number of stages or bits.

The 4518 is an example of a **decimal counter**. There are two separate synchronous counters in one IC. Each counter has four stages. When counting up, its outputs run from 0000 to 1001 (0 to 9) and then repeat. It can also count down. Another useful feature of this counter is that it has four inputs, one to each stage, so that it can be loaded with a value before counting begins.

Two 4518 counters can be cascaded to count from 0 to 99 in **binary coded decimal** (p. 367).

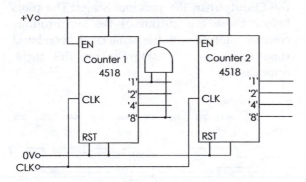

An AND gates links two decimal counters to form a 2-digit BCD counter.

The circuit in the diagram is set up to count clock pulses. However, there is no need in this counter or others to count regularly occurring pulses. It is just as suitable to count pulses that are of varying length and occurring at irregular times. The circuit might be used, for example, to accept input from a photodiode detector and count supporters passing through the turnstile of a football ground.

Note that the input goes to the clock input of *both* counters. The 4518 has an ENABLE input which enables counting when it is made high. The EN input of counter 1 is wired permanently to the positive supply line, so this counter counts all the time.

The outputs of Counter 1 are decoded by a 2-input AND gate. If the count registered on the Counter 1 is between 0000 and 1000 (0 and 8), the output of the gate is low. This means that the Counter 2 can not register any counts — it is disabled.

When Counter 1 reaches 1001 (9), both inputs of the AND gate are high for the first time. The output goes high and enables Counter 2. At the next rising edge of the clock, the Counter 1 changes to '0' and Counter 2 changes to '1'. The output of the AND gate goes low, and the Counter 2 is disabled until Counter 1 again reaches '9'. The graphs below shows the sequence as the counters count from 0 to 15.

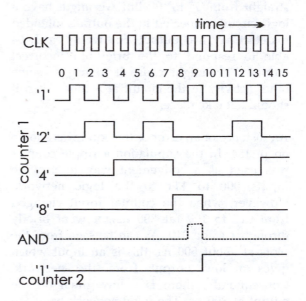

The BCD outputs of the counters and the AND gate as the count runs from 0 to 15.

Some other counters (such as the 4029) have a special OUT output that goes from low to high as the counter changes from 9 to 0. This output is fed to the clock input of the next counter and increments its count. Like the 4518, the 4029 is an up/down counter and is also a binary/decimal counter, making this one of the most versatile counters available.

Counting to *n*

Sometimes we need to count to some number other than 10 or 16. To achieve this we decode the output and reset the counter. The diagram below shows a counter decoded so as to count from 0 to 6 repeatedly.

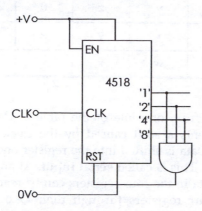

An external AND gate resets the counter as its output becomes 0111.

As the counter runs from 0 to 6, there is no stage at which the three inputs of the AND gate are all high. Its output is low. At the count of 7, all inputs are high, the output of the AND gate goes high, resetting the counter instantly to zero.

Decoding outputs

The circuits just described, in which certain output states are detected by external gates, are examples of output decoding. The external logic is designed to respond to one particular state of the outputs.

This decoding technique can be used to produce a repeated sequence of actions. The traffic light circuit is an example of this. A traffic light runs through a cycle of four stages, so we need a 2-stage counter to generate the sequence. This could be a pair of D-type flip-flops connected as an up counter (p. 196).

The sequence of the counting and the standard sequence of the three coloured lamps is shown in the truth table at top right. We need three decoding circuits, one for each lamp.

Inputs		Outputs		
B	A	R	Y	G
0	0	1	0	0
0	1	1	1	0
1	0	0	0	1
1	1	0	1	0

Taking the output columns separately, the truth table shows that the logic for each lamp is:

- Red: $R = \overline{B}$.
- Yellow: $Y = A$.
- Green: $G = \overline{A}B$.

The decoding circuit is:

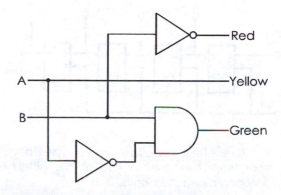

Only three gates are needed for switching traffic lights in sequence.

This example illustrates the technique for generating a repeated sequence of actions:
- Decide how many stages the sequence has:
- Design a counter with that number of stages, using decoding to reset it earlier if there are fewer than 10 or 16 stages.
- Write the truth table.
- Design the logic needed for each output.

Some counter ICs have decoding logic built into them. One is the 4511 7-segment decoder/driver for generating the sequence of outputs to produce numerals 0 to 9 on an LED display. This is described in Chapter 23.

Data registers

Another important use for flip-flops is as data registers or stores. A single D-type flip-flop can be used to store a single bit of data, as already described. An array of four or eight such registers, all connected to the same clock, can be used to store 4 bits or 8 bits of data. Data registers are available as MSI devices, such as the CMOS 4076 and the 74LS175 which store 4 bits.

Data at serial input	Data in registers			
	A	B	C	D
z	0	0	0	0
y	z	0	0	0
x	y	z	0	0
w	x	y	z	0
0	w	x	y	z

Shift registers

Shift registers are often used to manipulate data. The diagram shows a shift register built from four D-type flip-flops. The same clock triggers them all.

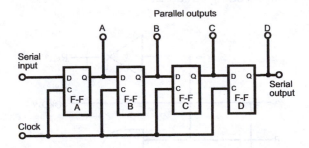

A shift register can be built from D-type flip-flops. When they are clocked, data is shifted one step along the chain.

Going from one line of the table to the next represents a shift caused by the clock going high. Data is shifted into the register one bit at a time. This is called **serial input**. At any time the data in the four registers can be read from the four registers through their Q outputs. These are **parallel outputs.**

The register also has a **serial output**. If we read the data at this output for four more clock cycles the bits are shifted out one at a time, as shown in the table at top right. At each stage, the data held in register D appears at the serial output.

Shift registers are made as integrated circuits with varying features. The one described above operates in **serial-in-parallel-out** mode (SIPO) and also in **serial-in-serial-out** mode (SISO).

At each rising edge of the clock the data present at the input is loaded into flip-flop A and appears at its Q output. At the same time, the data that was previously in A is shifted into flip-flop B. Similarly there is a shift from B to C. Data that was in D is replaced by the data that was in C, so the data that was in D is lost.

If 4 bits of data are presented in turn at the input and the clock goes through 4 cycles, the registers hold all 4 bits of data. This table illustrates what happens when four bits of data (*wxyz*) are shifted serially into a register that is holding four 0's. We begin with the least significant bit, z:

Data at serial input	Data in registers				Data at serial output
	A	B	C	D	
0	w	x	y	z	z
0	0	w	x	y	y
0	0	0	w	x	x
0	0	0	0	w	w
0	0	0	0	0	Empty register

There are also registers that operate in **parallel-in-parallel-out** (PIPO) and **parallel-in-serial-out** (PISO) modes. Registers also vary in length from 4 bits and 8 bits (the most common) up to 64 bits. Most are **right-shift** registers, like the example above, but some can shift the data in either direction.

Shift registers are used for data storage, especially where data is handled in serial form. The bits are shifted in, one at a time, held in the latches, and shifted out later when needed. PISO registers are used when data has to be converted from parallel to serial form. An example is found in a modem connecting a computer to the telephone network. The computer sends a byte of data along its data bus to the modem. Here it is parallel-loaded into a PISO register. Then the data is fed out, one bit at a time, and transmitted serially along the telephone line. The reverse happens at the receiving end. The arriving data is fed serially into a SIPO shift register. When the register holds all the bits of one byte, the data is transferred in parallel to the receiving computer's data bus.

Walking ring counters

Like the counters already described, walking ring counters are based on a chain of flip-flops. These may be D-type or J-K-type. Each stage is triggered by the clock, so the counter is synchronous.

A walking ring counter is like a shift register in which the input t the first stage is fed with the inverted output of the last stage. As the clock runs, data is circulated round the ring of flip-flops, but is inverted as it is fed back to the beginning of the chain.

The output of a ring counter based on three flip-flops is illustrated at top right. Any one output goes through a repeating cycle of three clock periods in the low state, followed by three periods in the high state, repeating every six periods.

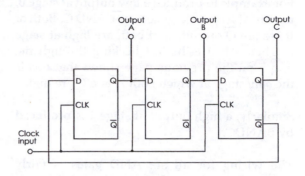

A walking ring counter is based on a chain of flip-flops, with the inverted output of the last stage fed back to the input of the first stage.

The graph below shows the outputs of the three flip-flops.

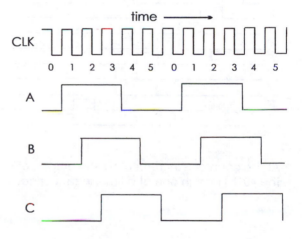

Each output has a period of six clock periods. Each lags one period behind the flip-flop before it.

A walking ring is a useful way of generating a number of out-of-phase digital signals. Another application is the 1-of-n counter. This has a number of outputs, which go high one at a time, in order. In the example above, there will be six outputs, one for each stage of the cycle.

A decoding circuit is used to produce the 1-of-n outputs. The Q and $\overline{\text{Q}}$ outputs of the walking ring are anded together in pairs

For example to produce a low output at stage 0, use an AND gate to generate $\overline{A}$ AND $\overline{C}$. Both of these are $\overline{Q}$ outputs and both are high at stage 0. Their AND is high. Checking through the graph on the previous page shows that this is the *only* stage at which both $\overline{A}$ and $\overline{C}$ are high.

Similarly, a high output at stage 1 is produced by $\overline{B}$ AND C.

The wiring for all six AND gates is fairly complicated but, fortunately, a 10-stage walking ring complete with decoding is available as an IC, the 4017. The graph below shows its output sequence

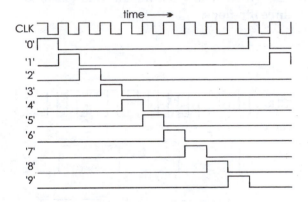

As the counter is incremented, the outputs of the 4017 go high one at a time, in sequence.

This type of counter is useful for triggering a sequence of events. As an illustration of the way in which it might be used, consider a fruit-canning machine. The process is divided into ten stages, repeating for each can processed. The stages might be those listed in the table at top right.

The clock drives the whole system. At each stage, a high pulse going to one of ten logic circuits triggers off the appropriate stage. At some stages the pulse may simply trigger a monostable to run a motor for a set length of time. At other stages it may trigger a more complex logical operation.

Stage	Operation
0	Put empty can on platform.
1	Fill with cooked cherries from a hopper.
2	Shake can to settle cherries.
3	Top up with more cherries.
4	Weigh can; reject if underweight.
5	Fill can with syrup.
6	Place lid on can.
7	Seal lid on to can.
8	Fix label on can.
9	Send can to packer.

This is an imaginary and simplified example, but it demonstrates the principles involved in using this counter. It is assumed that the frequency of the clock allows time for each operation to be completed. However, it would be possible to disable the clock automatically at any stage until that stage had been completed.

This IC has a 10-stage circuit but it can easily be made to operate on a shorter cycle. It has a reset input. This input is normally held low but, when it is made high, all flip-flops reset, so taking the device to stage 0. It continues counting from 0 when the reset input is made low again.

If an output from the counter is wired to the reset input, the counter resets each time that that stage is reached. For example, if output 8 is connected to the reset, the counter resets immediately output 8 goes high. Outputs 0 to 7 go high in turn, repeating. It acts as a modulo-7 counter.

The output graph on the left assumes that all Q outputs will be low when the circuit is first switched on. This does not necessarily happen. In the 6-stage walking ring illstrated on p. 203 the circuit goes through six diffent states. These are 000, 100, 110, 111, 011, and 001.

With three flip-flops there are eight possible states. States 010 and 101 are not normally part of the sequence. These are the **disallowed states.**

On the first clocking, state 010 changes to 101 (shift first two digits to the right and invert the third to become the first). Conversely, state 101 changes to 010. If the counter is in one of the disallowed states at switch-on, it will alternate indefinitely between one and the other. The solution it to use two more AND gates to decode the disallowed states. Their outputs go to the reset input to clear the ring to state 000.

A walking ring can also be used to generate other output sequences, including arbitrary sequences and those with two or more outputs going high.

A Gray code generator is an example of a counter with arbitrary output. With a 4-bit binary counter the sequence is 0000, 0001, 0010, ... 1101, 1111. Sometimes two digits change at the same time. For example, when 0001 (count 1) changes to 0010 (count 2). Two bits change when 0011 (count 3) changes to 0100 (change 4). If the bits do not change at exactly the same time we will get intermediate stages, for example:

$$0011 \rightarrow 0001 \rightarrow 0101 \rightarrow 0100$$

Expressed as counts:

$$3 \rightarrow 1 \rightarrow 5 \rightarrow 4$$

The two transitory stages may last long enough to be decoded and so introduce glitches into the system.

In a Gray code, the outputs always change one at a time, which avoids the possibility of glitches. An example of a 3-bit Gray code sequence is:

$$000 \rightarrow 001 \rightarrow 011 \rightarrow 111 \rightarrow 101 \rightarrow 100$$

repeating 000 ...

Decoding the output from the ring is more complicated than for a 1-of-n counter, and is likely to need more gates. Decoding must also detect and correct for disallowed states.

Static RAM

For storing a large amount of data we use an array of many flip-flops on a single chip. A large array of this kind is usually called a **memory**. Each flip-flop stores a single bit of data. Each flip-flop can have data stored into it individually.

The flip-flop is either set or reset, depending on whether the bit is to be a 0 or a 1. This is called **writing**, because the process of storing data has the same purpose as writing information on a piece of paper.

Each flip-flop can also be **read** to find out whether the data stored there is 0 or 1. A RAM chip may have several hundreds or even thousands of flip-flops on it. To write to or read from any particular one of these, we need to know its **address**. This is a binary number describing its location within the array on the chip. This is discussed in more detail later.

Because at any time we can write to or read from *any flip-flop we choose*, we say this is **random access**. A memory based on this principle is a **random access memory, or RAM.** This is different from a SISO shift register, from which the bits of data must be read in the same order in which they were written.

Once data is stored in a flip-flop it remains there until either it is changed by writing new data in the same flip-flop, or the power supply is turned off and all data is lost. This type of RAM is known as **static RAM, or SRAM.**

Dynamic RAM

There is another type of RAM known as **dynamic RAM, or DRAM.** In this, the bits are not stored in flip-flops but as charges on the gate of MOSFET transistors. If a gate is charged, the transistor is on; if the gate has no charge, the transistor is off. Writing consists of charging or discharging the gate. Reading consists of finding out if the transistor is on or off by registering the output level (0 or 1) that it produces on the data line.

The storage unit of a DRAM takes up much less space than a flip-flop, so more data can be stored on a chip. Also it takes less time to write or read to a DRAM, which is important in fast computers.

The main problem with DRAMs is that the charge on the gate leaks away. The data stored there must be **refreshed** at regular intervals.

Under the control of a clock running at about 10 kHz, the charge is passed from one transistor to another, and the amount of charge is topped up to the proper level at each transfer. The data is kept 'on the move', giving this type of memory the description 'dynamic'. The need to refresh the memory regularly means that a certain amount of the computer's operating time must be set aside for this, and the RAM can not be used for reading and writing data during this time. The operating system is therefore more complicated than that needed for SRAM, which is available for use at any time.

Flash memory

A flash memory chip consists of arrays of MOS transistors each of which has two gates. The transistor that forms one of the gates is completely insulated so that any charge of electrons placed on it remains there indefinitely. The transistor remains on or off indefinitely. Its state is read by a technique that employs the other gate. The two possible states of each transistor are equivalent to the binary logic values '0' and '1', and in this way the chip stores arrays of binary data.

The advantages of flash memory are:

- It needs no power supply to keep the data in memory. This makes it ideal for portable equipment.
- It is removable. An owner of a digital camera can have several flash cards in use. Flash memory used in a computer can be removed and locked away for security.

- It is not damaged by mechanical shock, unlike hard disks made of glass (p. 231).
- Writing and reading are fast, so it is ideal for memories holding large amounts of data. Flash memories holding hundreds of megabits are easily obtainable; chips storing 8 gigabits have been developed.

The disadvantages are:

- Individual bytes can not be erased; flash memory is erased in blocks.
- They eventually wear out; typically, a memory can be erased 10 000 times.
- They are more expensive per stored bit than a hard disk.

The advantages outweigh the disadvantages, so flash memory has become very popular.

Flash memory is used for storing data in a digital camera. A typical flash card stores 256 Mb, which is enough for hundreds of high-resolution digital colour photos. Later, after the data has been transferred to a computer, or has been printed out on to paper as a photo, the data can be erased and a new set of photographs recorded.

A memory stick is a small device generally with a built-in plug that fits into a socket on a CD or DVD player or on a computer. It may contain as much as 1 GB of flash memory. It is a convenient way of transporting data between computers as well as filing away text and images. The text and illustrations of this book need about 200 Mb for storage, so such a memory stick can store five books the size of this one.

Flash memory is also widely used in MP3 players, Personal Digital Assistants and mobile phones.

Memory control

Different types of RAM store different numbers of bits arranged in different ways. We take the 2114 RAM as an example. This is not *typical* since it is one of the smallest memory ICs commonly available. But it illustrates the principles of storing data.

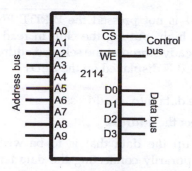

The ten address inputs of the 2114 SRAM allow for storage of 2^{10} (= 1024) words of data, each of 4 bits.

The 2114 stores 4096 bits (usually referred to as 4 kbits), which are writable and readable in groups of four. We say that the chip stores 1024 4-bit **words**. The chip has four data terminals (D0 to D3) through which a word may be written or read. Since there are 1024 words, each needs its own address. 1024 is 2^{10} so we need 10 address lines, A0 to A9. Then we can input every address from 0 (binary 0 000 000 000, all inputs low) to 1023 (binary 1 111 111 111, all inputs high).

As well as the connections to the data and address busses, there are two pins connected to the control bus. One of these is CHIP SELECT ($\overline{\text{CS}}$). The bar over its name and symbol indicate that this must be made low to have the desired effect. When it is high the chip is unselected; its data terminals are in a high-impedance state so the chip is effectively disconnected from the data bus. Before we can write to or read from the RAM the $\overline{\text{CS}}$ input must be made low.

The other control input to the RAM is the WRITE ENABLE ($\overline{\text{WE}}$) input. The data terminals act either as inputs or outputs, depending on whether the $\overline{\text{WE}}$ pin is made high or low. With $\overline{\text{CS}}$ low and $\overline{\text{WE}}$ high, the chip is in **read mode**. Next we put a 10-bit address on the address bus. This selects one of the words and the 4-bits of data that it comprises are put on to the data bus.

With $\overline{\text{CS}}$ low and $\overline{\text{WE}}$ low, the chip is in **write mode**. Data present on the bus is stored in the addressed word in memory.

Often in computing circuits we handle data in bytes, consisting of 8-bits. Some RAMs are organised to store data in this format. For example, the 6264 RAM stores 64 kbits, organised as 8192 bytes, each of 8 bits. This needs 8 data input/outputs and 13 address lines.

In most SRAMs the bits are addressed individually, not as words or bytes. The 4164 SRAM for example stores 64 kbits as single bits. The chip needs only one data input/output, but 16 address lines.

On the whole SRAMs have shorter access times than DRAMs. The 2114 is relatively slow, taking 250 ns to write or read data. Most DRAMs are faster than this, averaging around 100 ns. By contrast, most SRAMs have access times of 60 ns.

Activities — Counters and registers

Investigate each of the circuits listed below, either on a breadboard or by using a simulator, using the suggested ICs. Consult data sheets for connection details.

Build a ripple counter from three or four D-type flip-flops, as on p. 196 using 4013 or 74LS112A ICs (negative-edge triggered).

Investigate the action of a 4-stage MSI ripple counter such as the 74LS93 and the 7-stage CMOS 4024 counter.

Run the 4-stage MSI ripple counter at, say 10 kHz, then use an audio amplifier to 'listen' to the output from each stage.

Design and build a circuit to check the output of a 4-stage ripple counter for glitches. An example is the output '6' which occurs during the transition from '7' to '8'. Connect a set-reset flip-flop to the output of the detector circuit to register when the glitch has occurred.

By using logic gates to detect an output of 1100 (decimal 10), and reset the counter, convert the 74LS93 to a BCD (binary-coded decimal) counter with outputs running from 0000 to 1001.

Design and build a circuit to count in seven stages, from 000 to 110.

Investigate the action of a 4-stage MSI synchronous counter, such as the CMOS 4520 and the 74LS193A (up/down).

Use the circuit you devised in the previous investigation to check for output glitches in a synchronous counter.

Build a shift register from D-type flip-flops as on p. 202.

Demonstate the action of shift registers such as CMOS 4014 (PISO, SISO), CMOS 4015 (SIPO, SISO), 74LS96 (SIPO, SISO, PIPO, PISO), 74LS165A (PISO).

Investigate the action of a SRAM such as the 2114 or 6116. A circuit which can be set up on a breadboard or simulator is shown below. The CHIP SELECT input is held low permanently, and so are all but one of the address inputs. Addressing is controlled by the flip-flop. When this is reset the address is 0 and when it is set the address is 1. The flip-flop is set or reset by briefly shorting one of its inputs to 0 V.

When S1 is not pressed the WRITE ENABLE input is high, putting the chip in read mode. Data already stored at the selected address as a 4-bit word is displayed on the LEDs.

To write data into an address:

- Select the address.
- Set up the data that is to be written by temporarily connecting the data terminals to +5V (1) or 0V (0). On a breadboard, use flying leads.
- Press and release S1. This writes the data into the RAM.
- Remove the connections (flying leads) from the data terminals.
- Read the newly stored data on the LEDs by pressing S1 to put the chip into read mode.

If you select the other address, then return to the same address, the stored data will still be there. Experiment with storing different values at the two addresses.

Build another flip-flop using the other two NAND gates of IC1. Connect its output to A1. It is then possible to address four words (00, 01, 10 and 11). In these addresses store the data for a traffic-light sequence, assuming D0 is red, D1 is yellow and D2 is green (D3 is not used). Or store the codes for the first four digits of a 7-segment display.

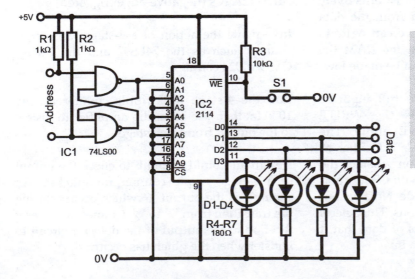

A breadboarded circuit for investigating the action of the 2114 SRAM. Similar circuitry may be used for investigating the action of other small SRAMs, such as the 6116.

By running through the addresses in order, you can simulate a changing traffic light or make a 7-segment display run from 0 to 3. Devise a circuit to operate the traffic lights or the 7-segment display automatically.

Try this activity with other small SRAMs with 4- or 8-bit words.

Questions on Counters and registers

1 Explain what a D-type or J-K flip-flop does and how three such flip-flops may be connected to make a ripple counter.

2 What is the main drawback of a ripple counter? Explain how a synchronous counter overcomes this problem.

3 Describe two logic circuits in which the output of a counter is decoded to trigger a special action.

4 Describe the action of a decade counter with 1-of-10 outputs, and give an example to show how it can be used.

5 Draw a circuit diagram for a serial-in parallel-out shift register, using 4 D-type flip-flops. Explain its action by means of a table.

6 Explain the difference between static and dynamic RAM. What are the advantages and disadvantages of each type?

7 Design and build (or simulate) a 3-bit up-down counter.

8 Explain the principle of a walking ring counter. Name some walking ring ICs and some of their applications.

9 Design and build (or simulate) the decoding circuit for the 6-bit walking ring circuit illustrated on p. 203.

10 Design and build (or simulate) a circuit for operating traffic lights, based on a walking ring.

11 What is flash memory? How does it work? What are its advantages and disadvantages? List three types of electronic equipment that use flash memory.

Multiple choice questions

1 A RAM has 11 address inputs and 4 data inputs/outputs. The number of bits that it stores is:
   A 8192.
   B 2048.
   C 44.
   D 4096.

2 To make a decade counter count from 0 to 6 repeatedly, we decode its output to reset the counter at stage:
   A 6.
   B 5.
   C 7.
   D 10.

3 A 4-bit counter has outputs A, B, C and D. To make it flash an LED at one quarter the clock rate, we connect the LED to output:
   A A.
   B B.
   C C.
   D D.

4 A SISO 4-bit shift register has A = 0, B = 1, C = 0 and D = 0. How many clock pulses are needed to make its serial output go high?
   A 2.
   B none.
   C 3.
   D 1.

On the Web

Find out more about memory. Select three or four of the newest memory devices and compare them for storage capacity, access time, permanent or temporary storage, maximum number of write operations, power requirements, cost per megabyte, physical size, and any special features.

23 Display devices

Light-emitting diodes

Light-emitting diodes, or LEDs, are widely used as indicators and in displays. Although filament lamps and neon lamps are still used, they have been replaced by LEDs in very many applications. LEDs outlast filament lamps, they are cheaper and they require very little current. They are available in a wide range of shapes, sizes and colours.

An LED has the electrical properties of an ordinary rectifying diode. The most important practical difference (apart from the production of light) is that they are able to withstand only a small reverse voltage.

 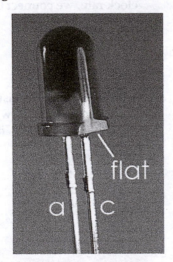

A typical LED has two terminal wires, anode and cathode. The cathode is often identified by a 'flat' on the rim of the LED body.

For most LEDs the maximum reverse voltage is 5 V, so care must be taken to mount them the right way round.

LEDs are usually operated with a forward current of about 20 mA, with a maximum up to 70 mA, depending on the type. But, if power is limited, for instance in battery-powered equipment, most types of LED give reasonable light output with as little as 5 mA.

The current through an LED must generally be limited by wiring a resistor in series with it (p. 15).

The brightness of an LED is specified in candela (cd) or millicandela (mcd). Typical LEDs are in the range 6-10 mcd. Super-bright LEDs have brightnesses of tens or even hundreds of millicandela, while the brightest types are rated at up to 30 candela. Viewing angle is also important, for some types achieve high brightness by concentrating the light into a narrow beam. This makes them less conspicuous in a display if the viewer is not directly in front of the panel. Standard LEDs have a viewing angle of 120°, but the angle may be as small as 8° in some of the brighter types.

The original LEDs were available only in red, green and yellow, but continuing development has produced a wide range of colours. There are several different tints of red (including ruby red and sunset red) as well as blue and white. Infrared-emitting LEDs are also available.

Bicolour LEDs are useful as indicators. A single bicolour LED can indicate the mode in which an electronic device is operating. For example, on a digital camera a bicolour LED may glow red when it is in recording mode and green when it is in play mode.

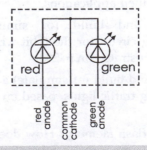

One type has three terminal wires. The LEDs have a common cathode. A positive voltage applied to either of the other two wires is used to light the corresponding LED.

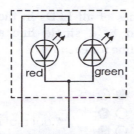

In the two-terminal type ,the LEDs are connected anode to cathode. Which LED lights depends on which terminal is made positive.

Seven-segment displays

When an appliance, such as a clock or a frequency meter, needs to display numerical values, it may use an LED display in the seven-segment format.

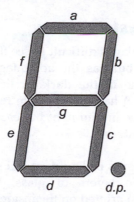

A seven-segment LED array is used for displaying numerals and letters.

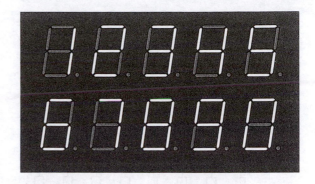

This is how a 7-segment display shows the numerals 0 to 9. The 6 and 9 may sometimes have no tails (segments a and d respectively).

Seven-segment displays are made in a wide range of types. Most are red but some are green or yellow. The height of the display ranges from 7 mm to 100 mm. Most have a single digit but there are some with 2 or 4 digits. The decimal point is an eighth LED, usually round in shape.

All displays fall into one of two categories, **common anode** or **common cathode**. In the common anode display (below), the anodes of each LED are connected internally to a common terminal. The segments are lit by drawing current through their cathodes to current *sinks*. Each segment needs a series resistor.

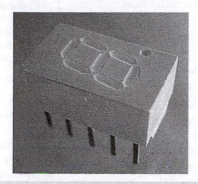

Seven bar-shaped LEDs make up a 7-segment display for numerals 0 to 9. There is also a decimal point, usually on the right, though some displays have a second one on the left.

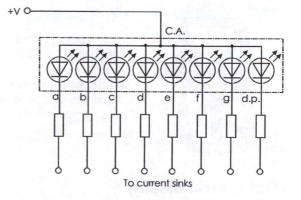

With a common-anode display, all anodes are connected to the positive supply and current is drawn through the individual cathodes through series resistors.

The common-cathode display (below) has the cathodes of all segments connected internally to a common terminal. Current *sources* are connected to each anode, through a series resistor.

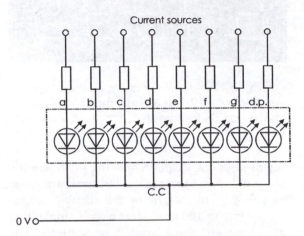

Current sources

With a common-cathode display, all cathodes are connected to the 0 V rail and current is supplied to the individual anodes through the series resistors.

Starburst display

The starburst LED display consists of 14 segments arranged as in the diagram below. It is able to produce any numeral or alphabetic character and many other symbols.

The coding circuit for a starburst logic display is more complicated than that needed for a 7-segment display.

Dot matrix display

This often has 35 circular LEDs arranged in 5 rows and 7 columns. The cathodes of all LEDs in the same column are connected to a single line and there is a separate line for each column.

The anodes of all LEDs in the same row are connected to a single line and there is a separate line for each row.

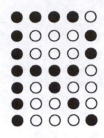

A dot matrix display is capable of generating a wide range of characters and symbols.

The LEDs forming the character are lit one at a time but in rapid succession. This action is repeated so quickly that each illuminated LED appears to glow steadily and all of them seem to be on at the same time.

Dot matrix displays require complicated logic circuits to drive them. There are special LSI or VLSI ICs for this purpose, or the task may be performed by a microcontroller. The dot matrix system is ideal for travelling-message displays and other effects in which the characters appear to move sideways or up or down, to flash, or to change size.

Self Test

Draw diagrams to show how the letter of your name can be formed on **(a)** a starburst display, and **(b)** a dot matrix display.

LSI and VLSI

Large scale and very large scale integration.

Liquid crystal display

An LED display is difficult to see if exposed to bright light but has the advantage that it is highly visible in the dark. A liquid crystal display (LCD) has the opposite features. It is difficult to see in dim light but easy to read in direct sunlight.

An LCD consists of liquid crystal sandwiched between two layers of glass. Conducting electrodes are printed on the inside of the front glass, in the format of a 7-segment display. These are almost invisible. Very narrow printed conductive strips connect each segment to its terminal at the edge of the display.

There is a continuous but transparent coating of conductive material on the inside of the back glass, forming the **back plane**.

Normally the liquid crystal is transparent but, if an electric field exists between the back plane and one of the electrodes, the material becomes (in effect) opaque in that region. The connecting strips are too narrow to produce an image. Thus the characters appear black on a light grey ground. The display can be read by reflected light. In darkness, it may have a back light, but the display is not so easy to read under these conditions as an LED display.

The action of the LCD depends on an alternating electric field between the back plane and the electrodes. The segment electrodes must be made alternately positive and negative of the back plane. If this is not done, the electrodes become plated and the device eventually stops working.

The driving circuit requires an oscillator running at about 30 Hz to 200 Hz. This can be built from two NOT gates (p. 172). It is the *field* (or potential difference) that produces the effect, but the amount of *current* passing through the material is very small. Because of this, the power required for driving LCDs is very much less than that used for an LED display. A typical 4-digit LCD needs only 2 or 3 µA compared with at least 10 mA *per segment* for an LED display. LCDs are ideal for portable battery-powered equipment such as watches, clocks, calculators, kitchen scales, and multimeters.

The fact that the electrodes only have to be printed on glass means that it is inexpensive to create a PCB design for a new product. Setting up the manufacture of a new design of LED display is much more difficult and expensive. As a result of this, a varied range of LCDs is available. Displays are made with 2 to 6 digits, usually including a colon in the centre for timing applications. There may be a plus or minus sign for metering applications and short messages or symbols indicating conditions such as 'alarm on', 'snooze', 'flat battery', and 'overflow'.

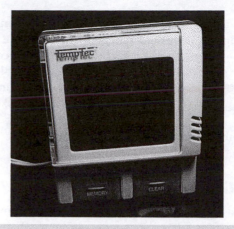

An indoor/outdoor thermometer uses an LCD to display the temperature at two sites. It runs for months, day and night, on a single 1.5 V AAA dry cell.

It is also inexpensive to create special-purpose displays such as that of the 2-site maximum/minimum thermometer pictured above.

LCDs are also made in the dot matrix format. Some of these are used to display messages of four or more lines of 20 or more characters. In calculators and meters the LCD may be used for displaying graphs. Such complicated displays are usually controlled by a microcontroller.

Colour LCDs are used in applications such as computer screens (below), digital cameras and mobile phones.

For instance, the display of the laptop computer overleaf has 600 rows, each of 800 pixels, each able to produce 16 million different colours. LCD computer monitors are becoming widely used with desk-top computers too, replacing the cumbersome cathode-tube monitors.

> **Pixel**
> Picture element.

Activities — LED displays

Investigate the action of various MSI devices used for driving seven-segment displays. Connect a display to the outputs of the IC, remembering to include series resistors of suitable value. Then try the effect of different combinations of inputs.

Devices to work with include:

74LS47: BCD decoder-driver, with open-collector outputs (below, top).

The outputs are active-low, so a common-anode display is required. The input is a 4-bit binary number in the range 0000 to 1001 (decimal 0 to 9). Inputs 1010 to 1111 produce 'nonsense' on the display. This device has ripple-blanking and lamp-test inputs. Try cascading two or more devices and use the ripple-blanking input and output (opposite). If RBI is low, the most significant digit in the display shows blank instead of '0' (that is ' 64' for example, instead of '064'). At the same time the output RBO goes low, so blanking the next digit to the right if it, too, is a zero.

4511: CMOS BCD decoder-latch-driver. (below, bottom). Outputs are active-high, so use a common-cathode display. Series resistors are essential. The display goes blank for inputs 1010 to 1111. It has a STORE input to latch the display while input changes. This gives time for the user to read a rapidly-changing input.

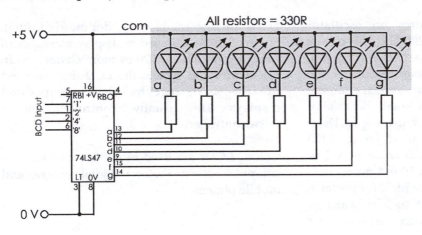

How to connect a common-anode display to a 74LS47 BCD decoder-driver. The ripple-blanking input and output may be left unconnected in a single-digit display.

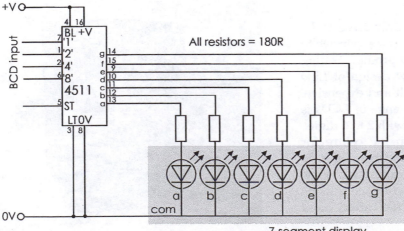

How a common-cathode display is controlled by a 4511 BCD decoder-driver. This IC does not have ripple blanking, but has data latches.

7-segment display

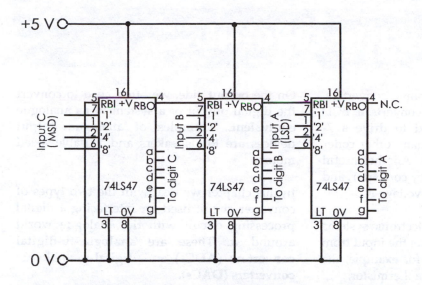

A 3-digit display, with ripple blanking, using 74LS47s. They receive BCD inputs from three cascaded decade counters. Their output goes to three 7-segment LED displays.

Problems on display devices

1 Make a list of the different types of displays and indicators used in appliances, including domestic, business, industrial, leisure and sporting appliances.

2 What is a light-emitting diode? Outline its characteristics.

3 Draw a diagram of a seven-segment display and explain how LEDs arranged in this format are used to display the numerals 0 to 9.

4 Compare seven-segment, starburst, and dot matrix LED displays, giving examples of applications for which each type is best suited.

5 Compare the features of LED displays and liquid crystal displays.

6 Draw the format of a 7 × 5 dot matrix display, showing how it displays a letter *d*.

7 Outline the kinds of electronic display that you would select for use in (a) a bedside clock, (b) an egg-timer, and (c) a pocket calculator. State what type you would use, what format, and any special features. Give reasons for your choices.

Multiple choice questions

1 Outputs *c*, *d*, *e*, *f*, and *g* of a 7-segment decoder IC are at logic high, the rest are at logic low. If a common-cathode display is connected to the IC, the numeral displayed is:

   A 9   C 0

   B 2   D 6.

2 2 Outputs *e*, and *f*, of a 7-segment decoder IC are at logic high, the rest are at logic low. If a common-anode display is connected to the IC, the numeral displayed is:

   A 4   C 1

   B 3   D 0.

3 LCDs are the best kind of display for a digital watch because they:

   A are easy to read in the dark

   B are resistant to damage

   C take very little current

   D can be made very small.

On the Web

Find out how colour LCDs work. Search for data on the latest high-power LEDs that run at 1 watt or more.

What is a plasma screen? How do plasma screens compare with LCD screens for performance and cost?

To answer these questions, use a browser and search with keywords or phrases such as 'light emitting diode', 'colour lcd', and 'plasma screen'. Try to find out more about other types of display device.

24 Converter circuits

We have already looked as some converters, such as the 74LS47, which converts a BCD input into the codes needed to drive a 7-segment display. There are many other coder and decoder ICs and circuits. Another useful circuit is a voltage-to-frequency converter, and its converse, the frequency-to-voltage IC.

An increasing proportion of electronic systems process data in digital form. Yet the input from sensors is often analogue — for example, the signals from a microphone or a thermistor.

On the output side, we often need to convert the digital output of a system to its analogue equivalent. Examples of analogue output devices are loudspeakers and variable-speed motors.

In this chapter we deal with the two types of converter circuit used for interfacing a digital processing circuit with the analogue world around it. These are **analogue-to-digital converters (ADCs)** and **digital-to-analogue converters (DACs)**.

Analogue-to-digital converters

Flash converter

For many applications, such as converting analogue audio signals into digital signals in real time, a high-speed converter is essential. A flash converter consists of a number of comparators connected at their (+) inputs to the input voltage v_{IN}, which is the voltage to be converted (right). Their (−) inputs are connected to a potential divider chain consisting of 8 resistors. One end of this is at 0 V and the other is connected to a reference voltage v_{REF}. The voltage at successive points in the chain increases in steps from 0 V to v_{REF}.

There are three cases:

- If v_{IN} is 0 V or close to 0 V, the (+) inputs of all comparators are at a lower voltage than the (−) inputs. As a result, the outputs of all comparators are low (equivalent to logic low, or 0).
- At the other extreme, if v_{IN} is close to v_{REF}, the (+) inputs of all comparators are at a higher voltage than the (−) inputs. All outputs are high (equivalent to logic high, or 1).
- At intermediate input voltages there are comparators at the lower end of the chain, which have high outputs, and comparators at the top end, which have low outputs.

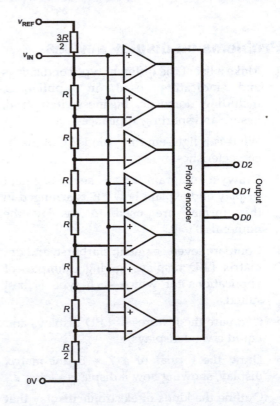

A flash converter is the fastest analogue-to-digital converter, but it needs several hundred comparators to give high precision.

The result is that, as v_{IN} is increased from 0 V to v_{REF} the outputs change like this:

00000000 when v_{IN} = 0 V
00000001
00000011
00000111
00001111
00011111
00111111
01111111
11111111 when v_{IN} = v_{REF}

The outputs are passed to a priority encoder. This has the same action as the circuit on p. 162, but has 8 inputs and 3 outputs. The outputs indicate the highest-ranking high input. As v_{IN} is increased from 0 V to v_{REF} the binary output increases from 000 to 111 (decimal 0 to 7). The output from the encoder indicates the voltage range that includes v_{IN}.

In the diagram on the right, the reference voltage is 8 V to make calculating the ranges easier to understand, but it could be any other reasonable voltage. The value of R (within reason) does not affect the working of the circuit. The essential point is that all resistors must be precisely R, or R/2, or 3R/2. This is usually easy to arrange if all the resistors are made on the same chip.

The resistors in the chain all equal R, except for those at the top and the bottom. Using the formula for potential dividers and numbering the resistors from 0 to 7, starting at the 0 V end, the voltage at a point between the nth resistor and the (n+1)th resistor is:

$$v_n = v_{REF} \times \frac{nR + R/2}{6R + R/2 + 3R/2} = v_{REF} \times \frac{2n+1}{16}$$

This formula gives the voltages marked beside the (–) inputs of the comparators. Because the lowest resistor is R/2, the voltage steps up from 0.5 V in 1 V steps. Thus the ranges run from 0 V to 0.5 V (output 0), 0.5 V to 1.5 V (output 1), 1.5 V to 2.5 V (output 2) and so on up the chain. At all stages, except the last, the output equals the average voltage over the range. Or, in other words, it indicates the voltage to the nearest volt.

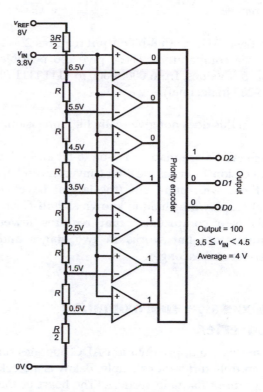

The comparators at the lower end of the chain (least significant bits) have high output, and those at the upper end (most significant bits) have low output. Here, with v_{IN} = 3.8 V, the lower four have high outputs and the upper three have low outputs. The output of the encoder is 100 (decimal 4), which is equal to the input voltage, to the nearest volt.

A flash converter takes only the time required for the comparators to settle, plus the propagation delay in the gates of the encoder. Typically, a flash converter produces its output in 10 ns to 2 μs, depending mainly on the number of bits. It can convert signals in the megahertz ranges and does not need the signal to be held in a sample-and-hold circuit while it does the conversion. The main disadvantage of the 3-bit circuit that we have used for illustration is that it has only 8 possible output values, 0 to 7. Yet the analogue input varies smoothly over a range of several volts, with an almost infinite range of values. The ADC has 7 converters to produce a 3-bit output. The rule is that it takes $2^n - 1$ comparators to produce an n-bit output.

Example

A flash ADC with 8-bit output requires $2^8 - 1$ = 255 comparators. These give 256 possible output values, from 0000 0000 to 1111 1111 (0 to 255 in decimal).

Even this does not give really high precision.

Flash ADCs are made for 4-bit, 6-bit and 8-bit conversion. Some of the 8-bit converters and all of those with more bits (the largest have 12 bits) employ a technique known as **half-flash**. This is a compromise that requires fewer comparators but works in two stages and therefore takes longer.

Successive approximation converter

A successive approximation ADC operates on an entirely different principle. Below is a block diagram of its main sections. The heart of the DAC is the successive approximation register which holds a binary value, usually with 8 or 16 bits. The control logic is driven by the clock and sets the register to a series of values in a systematic way. It gradually 'homes' on the value that is the digital equivalent of v_{IN}.

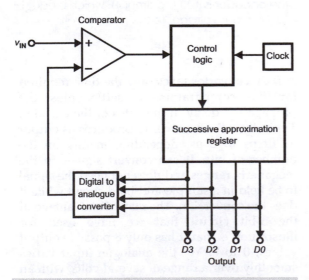

A successive approximation ADC systematically adjusts the content of the register until the output of the DAC is as close as possible to the input voltage v_{IN}.

When conversion begins, the register is set to its 'half-way' value. With 4 bits, this is 1000. If the maximum input voltage is 2 V (approximately, see later) this is equivalent to 1 V. As well as going to the terminal pins of the IC, the output from the register goes to a digital-to-analogue converter (DAC), which is also present on the chip. This converts the digital value to analogue and produces an output of 1 V. This goes to a comparator, where it is compared with v_{IN}. The result of this comparison tells the logic whether to increase or decrease the value in the register, so as make it closer to v_{IN}. This is what is meant by successive approximation.

Example 1

Given that v_{IN} = 4.8 V, the successive approximations are as shown in the table opposite.

The control logic takes each bit of the register, one bit per clock cycle. It starts with the most significant bit. To increase the register, the next bit to the right is set to '1'. To decrease the register, the current bit is reset to '0' and the next bit to the right is set to '1'. In this example, the output at the end of 4 clock cycles is 1010, equivalent to 5 V.

The result is to the nearest bit. In this example 1000 is equivalent to 4 V, so the least significant bit is equivalent to 0.5 V. The output is read as 5 V, which means that the input may actually have any value between 4.5 V and 5.5 V.

Example 2

Given that v_{IN} = 3.8 V, the table opposite shows the stages of the conversion.

The first approximation takes the DAC output down too far, but it gradually returns to 4 V, which is the closest to the input of 3.8 V.

As with the flash converter, precision is low with a small number of bits. But increasing the bit number is only a matter of lengthening the register. ADCs are made with up to 18 bits and a conversion time of 20 µs.

Clock cycle	Register set to	DAC output (V)	v_{IN} lower or higher?	Next step?	Next register setting
1	1000	4	higher	increase register	1100
2	1100	6	lower	decrease register	1010
3	1010	5	lower	decrease register	1001
4	1001	4.5	higher	increase register	1010

Example: The operation of the converter when v_{IN} = 4.8 V.

Clock cycle	Register set to	DAC output (V)	v_{IN} higher or lower?	Next step?	Next register setting
1	1000	4	lower	decrease register	0100
2	0100	2	higher	increase register	0110
3	0110	3	higher	increase register	0111
4	0111	3.5	higher	increase register	1000

Example 2: The operation of the converter when v_{IN} = 3.8 V.

Some ADCs of this type are not as fast as this, even those with fewer bits, and may take up to 100 µs to convert.

The searching routine of the successive approximation method is unsuited to a rapidly changing input. For example, if the input is 3.8 V as in Example 2, the first clock cycle decreases the DAC output from 4 V to 2 V. Suppose that v_{IN} rises to 6 V before the second clock cycle. This could happen because the signal is changing rapidly or perhaps because of an interference 'spike' on the signal. The first digit is now set at '0' so the DAC output can never rise higher than 4 V, even with carry-over. The output error is 2 V.

It is not possible for the searching routine to home on a moving target. For this reason, it is usual to employ a sample-and-hold circuit to hold v_{IN} while it is being converted.

Summing up: The SAC is slower than a flash converter (20 µs, compared with 10 ns to 2µs), but more precise (18 bits compared with up to 12 bits). It needs a sample and hold circuit, which a flash converter does not need.

Dual slope converter

This technique employs an integrator (p. 105). In the first stage of conversion, the control logic switches the integrator to v_{IN} (see diagram, right). Assuming that v_{IN} is negative, the output of the integrator ramps up at a rate dependent on v_{IN}. The control logic allows the voltage to ramp up for a fixed period of time t_1 (see plot, right). The voltage V_R reached in that time is proportional to v_{IN}. It is also proportional to R and C.

In the second stage, the counter is first set to zero, then the input of the integrator is switched to V_{REF}, which is a fixed positive voltage. This makes the output ramp down but, since V_{REF} is constant, the slope of the downward ramp is fixed. As long as the integrator output is positive, it allows clock pulses pass through the AND gate, to be counted by the counter.

When the ramp reaches zero at t_2, pulses no longer pass through the AND gate and counting stops. The length of time that has been taken to ramp down at a fixed rate from V_R to zero is proportional to V_R, which is proportional to v_{IN}. Thus the digital count is proportional to v_{IN}.

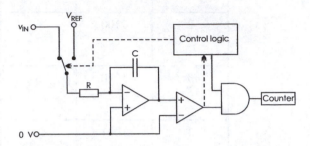

A dual slope converter is based on an op amp integrator.

Both rates of ramping depend not only on voltages but of the values of R and C. However, the values of R and C affect both upward and downward ramps, so their exact values do not matter and neither do changes in their values due to ageing.

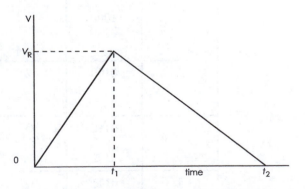

The two slopes by which the value of v_{IN} is estimated.

In addition, the precision and stability of the clock is unimportant since, if , for example, the clock runs slow, the upward ramp lasts longer. This makes V_R greater, so the downward ramp is longer too. Being timed by a slow-running clock, the effect of the slow clock cancels out.

Another advantage of this converter is that an integrator has the properties of a lowpass filter. Noise and spikes on the input line are filtered out. As long as V_{REF} is precise, results of high accuracy are guaranteed.

The main disadvantage is that this technique is slow compared with flash converters. A conversion typically takes 10 µs. It is often used in the measuring circuits of high-precision multimeters, or in other measuring applications where only a low refresh rate is needed.

Sigma delta converter

Rapid analogue-to-digital conversion is required at the input of digital audio circuits. The analogue signal must be sampled and converted many thousands of times per second. The sigma delta converter converts samples at a much higher rate (**oversampling**) but converts the samples very quickly and, by using simple circuitry, into a single bit. The output of the converter at any instant is either logical high or logical low. It produces a rapid succession of highs and lows, known as a **bit stream**.

It is the varying proportion of highs to lows that corresponds with the varying value of v_{IN}.

Imagine the converter (below) without the feedback loop. The output of the adder takes the present sampled value of v_{IN}.

At the integrator this value is added to previous samples. If v_{IN} is positive, the output of the integrator (V_I) ramps steadily up. The output of the one-bit quantiser (V_Q) is high (=1) if its input is zero or positive. It is low if its input is negative. If V_I is positive and ramping upward, V_Q is a series of high bits. If v_{IN} then becomes negative, the integrator ramps down, but V_Q stays high for some time, until the V_I has ramped down below zero. The time taken for this depends on for how long v_{IN} was previously high.

To remove this effect of previous levels of v_{IN}, we introduce the feedback loop. The adder continuously compensates by subtracting the present output from v_{IN} before it is integrated. V_Q is high if v_{IN} is positive and rising, a continuous series of lows if v_{IN} is negative and falling. It is a mixture of highs and lows if v_{IN} is changing from rising to falling or from falling to rising. In other words, V_Q represents the *differences* between successive samples, not their absolute values.

V_Q would also be a mixture of highs and lows if v_{IN} was rapidly alternating between rising and falling, but the sampling frequency is well above the highest frequency of interest so this is not of importance.

The next stage of processing is usually a **digital decimation filter**. This converts the high-frequency 1-bit signal into a multibit signal at a lower frequency. It averages the bits by taking them in groups. For example, suppose the output from the quantiser is:

00110010001110110101

Group the bits in fives:

00110 01000 11101 10101

Select the majority bit (p. 158):

0   0   1   1

The output from the decimation filter is a 4-bit value:

0011

Twenty successive bits at high frequency have been converted into a single 4-digit value at 1/20 of the frequency. This is now suitable for further processing or storage

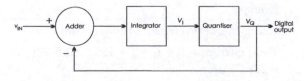

Block diagram of a sigma delta ADC. A sequence of sampled values are fed in at a fast rate on the left, and a bit stream of 0s and 1s emerges on the right.

Digital-to-analogue converters

Op amp adder

The operational amplifier adder (p. 101) is a simple way of converting digital signals to their analogue equivalents. In this application (below) the data inputs are logical inputs, so their voltage is either 0 V (logical 0) or a fixed high voltage v_H (logical 1). The resistors are **weighted** on a binary scale so that:

$$R_3 = 2R_F \qquad R_1 = 2R_2 = 8R_F$$
$$R_2 = 2R_3 = 4R_F \qquad R_0 = 2R_1 = 16R_F$$

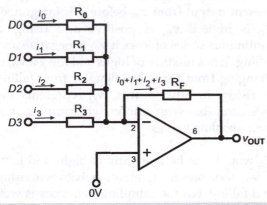

An operational amplifier adder with binary weighted inputs makes a digital-to-analogue converter.

Example

If $R_F = 10$ kΩ, then $R_3 = 20$ kΩ, $R_2 = 40$ kΩ, $R_1 = 80$ kΩ, and $R_0 = 160$ kΩ.

For any given data input, the current flowing through the input resistor is v_H divided by the resistance. If resistances are doubled at each stage, as above, currents are halved. For a high data input, the current through R_2 is half that through R_3, the current through R_1 is half that through R_2 and so on.

Example

If $R_F = 10$ kΩ and $v_H = 5$ V, then:

$$i_3 = v_H/R_3 = v_H/2R_F = 5/20000 = 250 \text{ μA}$$

Similarly, because of the doubling of resistances, the currents corresponding to other high-level data inputs are:

$$i_2 = 125 \text{ μA} \qquad i_1 = 62.5 \text{ μA} \qquad i_0 = 31.25 \text{ μA}$$

When a data input is low, the current through the resistor is zero. When more than one data input is high, the current through R_F is the total of the input currents. The output voltage of the converter is therefore proportional to the sum of the inputs, the inputs being weighted on the binary scale.

Example 1

Given a data input 1010, calculate the output, if other conditions are as set out above.

The total current is the sum of i_3 and i_1. Total current = 250 + 62.5 = 312.5 μA. $R_F = 10$ kΩ. The circuit is inverting, so output is -312.5×10 [μA × kΩ] = -3.125 V.

The output represents the binary input on a scale in which v_H is equivalent to binary 10000 (decimal 16). We therefore have to multiply the output voltage by $-16/v_H$.

Example 1, continued

In the example, the multiplier is $-16/5 = -3.2$. Multiplying the output voltage:

$$-3.125 \times -3.2 = 10 \text{ (exactly)}.$$

This is the same value as binary 1010.

Example 2

Using the same circuit with data input 0111.

Total current = $i_2 + i_1 + i_0$ = 125 + 62.5 + 31.25 = 218.75 μA

Output voltage = 218.75 × 10 [μA × kΩ] = -2.1875 V

Multiplying the output voltage:

$$-2.1875 \times -3.2 = 7 \text{ (exactly)}$$

This is the same value as binary 0111.

This method of conversion is almost instantaneous, taking only the time required for the amplifier to settle.

The main limitation is that the circuit needs an input resistor for each binary digit. With four resistors, as in the figure, the maximum value that can be converted is 1111. Only the integer values from 0 to 7 can be converted. To convert any integer in the range 0 to 255 requires 8 data inputs. When there are many data inputs, the difficulty is obtaining a set of resistors that is sufficiently precise in value. In the 4-bit converter a 1% error in R_3 will make the current through it too large or too small by 25 μA. This is almost as great as the current through R_0, leading to an error of almost ±1 bit.

R-2R ladder converter

Most DACs are based on the R-2R 'ladder'. The ladder is fabricated as part of the IC and consists of resistors of value R and 2R only. The exact value of R does not matter provided that all R resistors have the same value and all 2R resistors are exactly twice that value. This is much easier to do than to provide a set of precisely weighted resistors as required for the adder converter described above. Usually R is in the region of 10 kΩ and 2R is 20 kΩ.

The diagram below illustrates the circuit of a 4-bit converter, though most DACs of this type have 8, 12 or even as many as 18 bits. The switches shown in the circuit are actually CMOS switching devices under logical control.

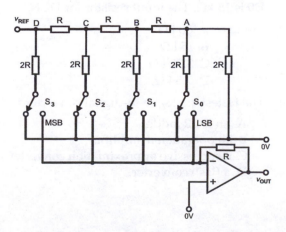

The switches may be switched either to the 0V rail (representing a data input of 0), or to the inverting input of the op amp (representing a data input of 1). Because the inverting input of the op amp is a virtual earth, it makes no difference to the flow of current in the resistor network whether the switches are to 0 V or to the inverting input. Whichever way the switch is set, the current flowing through the switch is the same. In the diagram, the positions of the switches correspond to a data input of 1010.

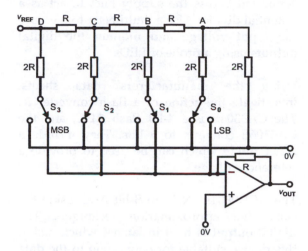

The R-2R 'ladder' is the basis of this DAC.

Because of the resistor values, the current flowing through switch S2 is half that flowing through S3. Similarly, the current flowing through S1 is half that through S2, and so on. We have binary weighted currents (as in the previous DAC) and the op amp sums these. This makes the negative voltage output of the op amp proportional to the data input. Its value does not depend on the value of R. As before, there is a scaling multiplier, which is $v_{REF}/16$ in a 4-bit converter. The output voltage is multiplied by this to obtain the value of the digital input.

Example

If $v_{REF} = 2.5$ V, the voltage output corresponding to a digital input of 1010 (decimal 10) is $10 \times 2.5/16 = -1.5625$ V.

Activity — ADCs

On a breadboard or simulator, set up a 2-bit flash converter, complete with priority encoder. Measure the input voltage ranges needed to produce the output states 00, 01, 10 and 11.

Investigate the action of a 4–bit, 6-bit or 8-bit ADC using the manufacturers' data sheet as a guide. Supply the input from a variable resistor connected across the supply lines to act as a potential divider. Use a multimeter to measure the input voltage and monitor the digital outputs using a probe or LEDs.

Using the manufacterers' data sheets, investigate the action of a flash converter ic. The CA3304E is a 4-bit flash ADC, and the CA3306E converts to 6 bits. They include a zener diode which can be used to provide a reference voltage.

The ACD0804LCN is an 8-bit ADC using the successive approximation technique. The digital output is held in latches which makes this device suitable for connection to the data bus of a microcontroller system. Investigate its action on a breadboard.

Activities — DACs

On a breadboard set up a 4-bit operational amplifier adder as on p. 202. Resistances could be $R_F = 10k\Omega$, $R_3 = 20\ k\Omega$, $R_2 = 40\ k\Omega$ (39 kΩ and 1 kΩ in series), $R_1 = 80\ k\Omega$ (68 kΩ and 12 kΩ in series) and $R_0 = 160\ k\Omega$. Use resistors of 1% tolerance. Connect inputs to the positive supply line or to 0 V to obtain high- and low-level inputs. Measure the output with a multimeter. This circuit could also be modelled on a simulator.

Set up a demonstration R-2R ladder converter on a breadboard or simulator. Measure voltages at various points, including points A to D. The DAC0800LCN is an 8-bit digital-to-analogue converter. Investigate this on a breadboard, referring to the manufacturers' data sheets.

Questions on converter circuits

1 Describe the action of an 8-bit flash converter. What is the main advantage of this type of converter?

2 Explain the principle of the successive approximation technique of analogue-to-digital conversion.

3 Compare the flash and successive approximation techniques of analogue-to-digital conversion.

4 Describe how an operational amplifier may be used as a digital-to-analogue converter. What is the difficulty of using this technique when the number of bits is large?

5 Describe the action of a digital-to-analogue converter based on the R-2R ladder technique.

Multiple choice questions

1 The number of comparators in a 6-bit flash converter is:
- A 6
- B 31
- C 63
- D 64.

2 The number of clock cycles taken by an 8-bit successive approximation converter to produce its output is:
- A 8
- B 4
- C 7
- D None of these.

3 An op amp adder is used as a digital-to-analogue converter. The input resistor for D0 is 15 kΩ. The input resistor for D2 is:
- A 30 kΩ
- B 15 kΩ
- C 3.75 kΩ
- D 7.5 kΩ.

4 The fastest analogue-to-digital converter is:
- A an R-2R ladder
- B an operational amplifier
- C a successive approximation converter
- D a flash converter.

25 Integrated circuits

One of the most important developments in electronic component manufacture is the integrated circuit or IC. Instead of having to build circuits from individual resistors, capacitors and transistors, standard circuits are produced with all its components and their connecting conductors on a silicon chip.

Even for a complicated circuit, the chip may measure only a few millimetres across. Integration means a great saving in circuit board area — an important factor when circuits are becoming more and more complex yet there is an increasing demand to produce small, portable, lightweight equipment such as digital cameras and mobile phones. With non-portable items there is the demand to cram more and more complex circuitry into them without making them any larger.

The chip of this comparator IC is very small but the package must be large enough for the eight terminal pins that connect to the chip inside.

If circuit boards are smaller, the conductors running between the ICs are shorter. The time taken for a signal to travel from one IC to another is reduced. This is very important in circuits running at frequencies in the gigahertz range. The circuit can not run at its fastest if it has to wait for signals to arrive.

The first ICs contained circuits with relatively few components. This was called small-scale integration (SSI). As manufacturing processes improved, larger and more complex circuits were fabricated on a single chip. There was medium-scale, large-scale and very-large-scale integration (VLSI).

Examples of MSI include the op amps and power amplifiers that we have alread studied. One of the latest examples of VLSI is the Intel Teraflop IC. This is a multi-core microprocessor comprising 80 processors working in parallel. The processor in a typical desk-top PC has only 1 and a games console may have 8 processors for high speed operation and detailed graphics. The density of processing power of the Teraflop has been achieved by reducing the dimensions of its 400 million transistors to only 45 nm across.

Some commonly used ICs have already been used in earlier chapters. Here we look at a few more to illustrate the importance of integration in electronic circuit design.

Low-scale integration

It is an easy step from making a single transistor to putting several transistors on the same chip and connecting them into simple circuuit modules. The fact that the transistors are all on the same chip means that their characteristics (such as gain) are identical. This eliminates the need to match transistors when building a circuit from individual devices.

The simpler CMOS logic ICs are typical examples of LSI. The drawing below shows the circuit of an inverter.

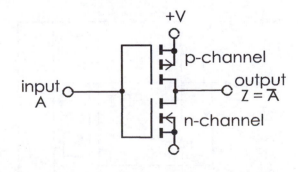

The inverter comprises two MOSFETS, one n-channel and one p-channel. Depending on the logic level of the input A, one or the other of the transistors is switched on. The output is connected to either the positive supply or the 0 V line. The CMOS4016 IC has six such inverters, with individual inputs and outputs, but sharing power line connections.

The inverter is part of another CMOS device, the analogue bilateral switch.

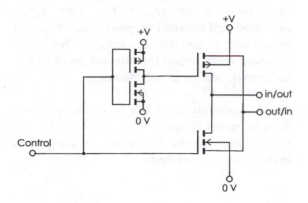

The switch is built from four MOSFETs. When the control input is at logic high, both transistors are switched on. Analogue signals can flow in either direction (which is why this is called 'bilateral').

The 4016 and 4066 ICs contain four switches, each with its own control input. These switches can be joined together in various switching configurations, such as SPDT, DPDT, and so on.

A motor power controller can be built from four discrete power MOSFETs connected as in this diagram:

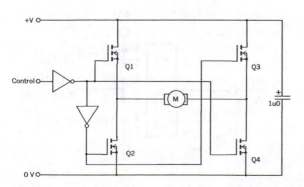

The four power MOSFETs form a bridge. The control input switches on the transistors in pairs. Either Q1 and Q4 are on or Q2 and Q3 are on. This switches the current through the motor in one direction or the other, acting as a reversing switch.

This basic bridge circuit is at the heart of many motor control ICs, but most have additional facilities. Four protective diodes, one across each transistor are a necessary addition. The Semiconductors L6201 Full Bridge Driver has circuits to detect when the IC gets too hot and shuts it down. This could happen if the motor stalls because of mechanical overloading. Logic circuits are added for turning the output on and off. This can be done at high frequency with variable mark-space ratio (pp. 43-44) so the driver controls speed as well as direction. With these extra features the IC is an example of medium scale integration and of the switch mode approach to circuit design.

Switch mode power regulators

A switch mode power supply IC regulates a power supply by switching it on and off at high speed. The output is regulated by varying the mark-space ratio, as described on p. 297. In some data sheets this is referred to as pulse width modulation.

Switch mode has the advantage that the output transistors are either off or fully on. They dissipate no power when off. When on, the resistance of the transistors is very small so little power is dissipated, and the transistors are not liable to overheat.

The heart of a switch mode regulator is the pulse width modulator and the power MOSFET. On the same chip there are also circuits for a number of other functions. These include overload protection, thermal shutdown, and overvoltage protection. Any switching operation can generate electromagnetic interference (EMI) but this can be reduced by adjusting the frequency of modulation.

Switched capacitor filters

The principle of a switched capacitor filter is illustrated by this diagram:

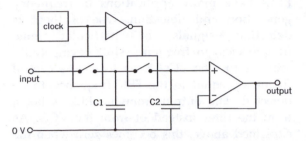

The basic low-pass filter comprises two capacitors, with CMOS switches that are turned on alternately by a built-in clock oscillator. The clock runs at several tens of kilohertz.

C1 samples the input voltage when the first switch is closed and the second switch is open. This voltage may be positive or negative depending on the phase of the input signal at that instant. A moment later the switches change state and the charge on C1 is equalised with the charge on C2, which came from the previous sampling.

The resulting voltage on C2 is read by the op amp, which is connected as a unity gain voltage follower. The op amp has a very high input resistance, so virtually no charge is lost from C2. The output from the op amp is the filtered signal.

The process continues with C1 repeatedly sampling the input, then sharing its charge with C2. The mathematics of the circuit is complex but it can be shown that it behaves like a low-pass filter. At low frequencies, there are only small voltage changes between samplings so the filter is able to respond to these changes. At higher frequencies, it is less able to respond so the signal is reduced or lost.

The important point about the switched capacitor filter is that its frequency response depends on the frequency of its clock. The −3 dB cut-off frequency can be accurately tuned by adjusting the clock rate. Typically the cut-off frequency is $f_{clock}/100$.

Filters of more complicated design but operating on the same principle are manufactured as integrated circuits. A typical example is the MAX7480 8th order filter with a Butterworth response, which has a very flat pass-band and sharp cut-off. It is tunable over the range 1 kHz to 2 kHz. Its action is equivalent to a passive filter built up from four capacitors and four inductors. The saving of space and weight gives the switched capacitor filter a very big advantage.

Phase locked loop

The action of a PLL is to lock on to an input logical signal of given frequency, even when the signal is distorted, has a noisy background, and is mixed with other signals of different frequencies.

The essential parts of a PLL are shown in the drawing overleaf. The first stage is the phase detector. It receives two logic-level signals, one at the input to the PLL and the other fed back from its output.

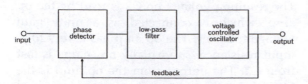

A basic phase locked loop.

The simplest phase detector is an exclusive-OR gate. This is intended for square-wave signals with 50% duty cycle. If the two signals are of the same frequency and in phase with each other, the output of the phase detector is zero.

If the two signals are of different frequencies, the inputs to the ex-OR gate will be unequal and the output of the gate goes high. The greater the difference between the two signals, the more the output of the gate will be high.

The output from the gate is passed through a low-pass filter to average out the overall response. This smoothed ouput goes to a VCO. The central frequency of this is set by resistors and a capacitor to be in the middle of the range that is to be detected, but varies on either side of this, depending on its input voltage — the signal from the phase detector by way of the low-pass filter.

When a signal is first sent to the PLL it is unlikely to be at the same frequency as the VCO. The detector/filter therefore generates an output voltage. This either increases or decreases the frequency of the VCO until it locks on to the input signal. The output of the PLL is therefore of the same frequency as the original input signal. The important difference is that the output from the PLL is a clean square wave and has no noise and no signals of other frequencies.

Often a PLL contains an extra sub-circuit, placed between the VCO and the phase detector. This is a frequency divider (pp. 196-197). If it divides by n, the PLL locks on with the VCO frequency n times that of the incoming signal. This is a way of generating a high-frequency signal from a low frequency but high-precision clock.

Another type of phase detector is used in some PLLs. This can operate with signals of other duty cycles and it allows the PLL to lock on to a wide range of frequencies.

PLLs have many applications in frequency generation and detection. One of these is detecting signals of particular tone (frequencies) in frequency shift keying (FSK, see p. 250). The VCO is set to have a central frequency equal to the FSK frequency to be detected. The output from the PLL is taken from the filter instead of from the VCO. As explained above, this becomes zero when the loop is locked on. A low output pulse indicates that a tone burst is being received.

Another application of PLLs is as frequency modulation (FM, p. 248) detection. The PLL receives the FM signal and locks on to the carrier frequency. As the carrier frequency is modulated the VCO frequency remains locked to it by variations in the output from the filter. The filter output is the demodulated audio or video signal

One of the most popular PLLs is the CMOS 4046 IC, which contains both types of phase detector. PLLs are so useful in circuit design that there are many other ICs of this type on the market.

Activity — Integrated circuit

Study the data sheets of two or three different types of IC. Many are available on the World Wide Web. Write a brief summary of what each IC does and how it works and list its applications.

Set up a ready-built or breadboarded version of one of the ICs and observe it in action. You may also find models of certain ICs on a simulator. Try running one or two of these.

26 Audio and video systems

Systems

The concept of systems was introduced in Chapter 2. Those simple circuits are the simplest possible of systems, usually consisting of no more than a single electronic component at each stage.

Now, in Part 2 of the book, we look at more complicated systems. These are built up from many of the circuits described in Chapters 8 to 25. Yet the basic idea of a three-stage system of **input**, **processing** and **output**, still applies to these more complicated systems.

Audio and video systems comprise a wide range of devices and installations from iPods to Home Theatres.

Many other electronic devices, such as mobile phones and digital cameras, also have audio and/or video capabilities. In this chapter we outline just the main features of this diverse and extensive class of systems.

Input

The input signal may be analogue or digital. Sources of analogue signals include microphones, radio tuners, and the light-sensors of optical fibre systems.

Digital signals are provided by CD/DVD players, magnetic disk drives, flash memory cards and computers.

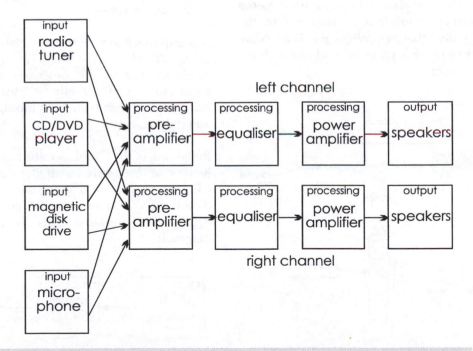

A typical domestic stereo system receives input from four signal sources. Each of these has 2-channel output to the two processing channels. Two microphones are required, one for each channel, for stereo recording. The diagram does not show all the connections that can be made. For example, the output from the tuner can be fed to the disk drive when recording on-air television programmes.

Radio tuner

A radio tuner receives radio-frequency signals that carry the audio signals in a way described on pp. 280-283. It may be monophonic or stereophonic. A stereo tuner has two outputs, for left and right channels and produces analogue signals of a few hundred millivolts in amplitude.

CD/DVD player

A CD/DVD player reads audio or video signals stored in digital form on a **compact disc or digital versatile disc**. The disc consists of two plastic layers sandwiched together with a shiny aluminium layer between them.

The information on the disc is stored on the aluminium layer as a spiral track along which are minute regularly spaced dimples. Each dimple represents logic '1' and a space where there should be a dimple represents logic '0'. The track runs from the centre of the disc to the outside. The dimples and spaces are detected by directing a low-power laser on to the spinning disc (below). When the laser beam strikes a space it is reflected back through the optical system.

The beam is deflected by prismatic beam-splitter and falls on a photodiode. The reflection is detected and recognised as digital '0'. When the beam strikes a dimple, it is scattered sideways and does not reach the photodiode. This condition is recognised as digital '1'.

As the laser gradually moves from the centre of the disc toward the rim, the speed of rotation of the disc is steadily reduced to obtain a constant track speed.

The signal coming from the photodiode is a bit stream (p. 221). This is processed by logic circuits to separate the data for the left and right audio channels and video channel, to check for errors and try to correct them.

Eventually, the logic produces a sequence of multi-bit words that are sent to a digital to analogue converter (p. 222). The analogue output of this then goes to the pre-amplifiers for further processing.

It is important to note that the system diagram on p. 229 shows the CD/DVD player as a single stage, which is described as an input stage. However, it is really built up from many sub-systems, with their own input, processing and output stages.

In the case of the CD player, the first subsystem is the one that reads the digital data from the CD. It then processes it digitally. Finally, the data is converted into an analogue signal that drives the speakers. This is the system diagram:

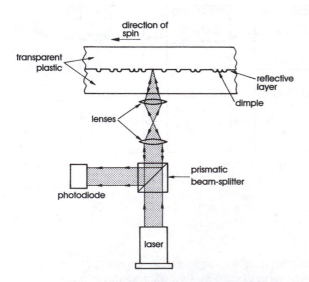

The laser unit scans the underside of the spinning disc, detecting the dimples (=1) and spaces (=0).

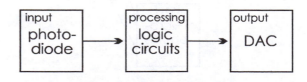

Simplified system diagram of a CD player. There are several stages processed by logic circuits, though these are shown as only one stage here.

A CD/DVD player comprises systems other than the main audio-video chain. One of these acts to keep the laser beam 'on track'. In the 3-beam system, there are two additional lasers in the laser unit, each with its photodiode. These beams are focused slightly ahead of and slightly behind the main beam. If the main beam is centred on the track, the two side beams receive signals of equal intensity. If the main beam starts to wander off the track, the signals from the side beams are unequal, and corrective action is taken.

There is another sub-system that continually monitors the diameter of the laser beam and focuses it as sharply as possible. In some players there are systems for selecting which tracks to play, possibly playing them in random order. In multi-disc players there are systems for selecting which discs to play. Thus the single 'input' stage shown on p. 229 is in reality a complex of interlinked sub-systems. Though we shall not analyse them to such depth, the same complexity is found in most other electronic systems.

Hard disk drive

A hard disk drive has one or more disks (or 'platters') on the same spindle. The disks are made from non-magnetic material, usually aluminium or glass, and are 3.5 inches in diameter in the most popular drives. The disks are coated with a thin film of ferromagnetic (pp. 53-54) material, such a cobalt alloy. The disks may be coated on one or both sides.

Digital data is written on to or read from the disk by heads that can be positioned rapidly to scan any one of a set of concentric circular tracks, as the disk rotates. The disk spins at high speed (up to 15 000 rpm in the fastest drives) and a thin film of air close to the surface of the disk spins with it. The read/write head is shaped so as to 'fly' in this air film. It takes up a position only a few nanometres above the disk surface. Being so close, it is able to write to or read from very small areas of the disk. This allows the disk to store large amounts of data. The high capacity drives used in work stations store several hundred gigabytes.

As in all ferromagnetic materials, the magnetic film becomes organised into microscopic regions called domains in which all the molecules are oriented in the same direction. Each domain is the equivalent of a very small magnet, with its own magnetic field. As the diagram below shows, in a unrecorded track the domains are arranged irregularly with their magnetic fields oriented in random directions. There is no overall magnetic field.

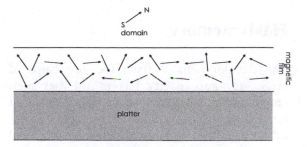

A vertical section through part of an unrecorded magnetic disk, showing the domains. They are drawn much larger than they really are.

When data is being written, the surface film passes close to the writing head as the disk spins. The writing head is esssentially an electromagnet, generating a strong field in one direction to correspond to a '1' and in the other direction for a '0'. The field is strong enough to realign the domains so that a majority are aligned in one direction or the other.

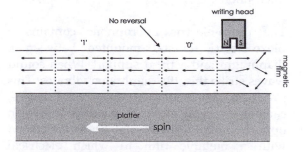

Writing a sequence of binary digits. At regular intervals the polarity is either reversed or not reversed, Reverse/reverse = '1'. No-reverse/reverse = '0'.

Reading data is the reverse of writing but generally uses a special reading head mounted on the same flyer as the writing head. The reading head may be inductive in which case the changing magnetic field induces changing currents in the coil of the head. In modern drives the head is magnetoresistive. The sensor consists of a small block of special semiconductor material. The resistance of the block, and the size of the current flowing through it, varies with the magnetic field.

Flash memory

The development of flash memory chips of large storage capacity has meant that such chips can be used instead of CD, DVDs and hard disk drives. Solid state storage has the advantage that it is unaffected by physical forces that would upset the running of a drive and might even damage its working parts. Glass platters can be shattered by mechanical shock.

Flash memory is discussed fully on p. 206.

Microphone

There are several different types of microphone but the two most commonly used are the piezo-electric microphone (or crystal microphone) and the electret microphone, sometimes referred to as a 'dynamic' microphone.

The piezo-electric microphone contains a piezo-electric crystal mounted between a diaphragm and a fixed support. When sound waves strike this, the diaphragm vibrates. This produces distortions in the crystal lattice, causing a varying emf to be generated between opposite faces of the crystal. These are plated with a metallic film to which electrical connections are made. A crystal microphone does not require a power supply and produces a voltage suitable for feeding direct to a preamplifier.

A crystal microphone 'insert', intended for installing inside a protective case. The crystal is shielded from stray magnetic fields by an aluminium screen. This should be connected to the 0 V line of the amplifier.

Crystal microphones are cheap. However, their response is not linear, so there is distortion in the output signal.

The electret microphone is similar to a capacitor microphone, since both depend on varying the capacitance between a vibrating plate and a fixed plate. The difference is that the electret has a dielectric between the plates. During manufacture, the microphone is heated, and it then cooled with a strong electric field between its plates.

An electret microphone 'insert' usually has a built-in FET amplifier to provide the first stage of amplification.

When cool, an electric field still remains in the dielectric. This means that the microphone does not need an external power supply. However, the signal from the microphone is very small. Most microphones have a built-in FET amplifier, in which case a power supply is provided. In the circuit below, the output from the microphone/amplifier passes through a capacitor to the preamplifier.

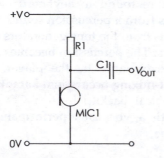

The main advantage of an electret microphone is that its response is linear, so the reproduction is of high quality.

Computer audio/video

As well as computers, such as PCs and laptops, we also include special-purpose appliances such as MP3 players, mobile phones, and digital cameras, under this heading.

These are all essentially data-processing systems which are able to accept or store recorded audio and/or video signals in digital form. Their output is sent to a speaker (see next section), and/or to a display device. This is often a colour LCD.

Sound files are stored in various formats. The MP3 format is popular because its files are more compressed than many other formats, so take less memory to store and take less time to transfer. However there are dozens more formats for audio and video files, and new ones are being devised all the time.

On the Web

Find out about recording formats, their good and bad features, their main applications and their popularity. Keywords to try include: MP3, MIDI, CMF, and format.

Analogue audio systems

Processing

There are two identical processing chains in stereo audio systems.

The preamplifier stage receives a voltage signal of a few millivolts switched through from one of the input stages. Current is low at this stage to avoid generation of noise (p. 242).

Consider a signal source, such as a microphone, connected to the input of a preamplifier.

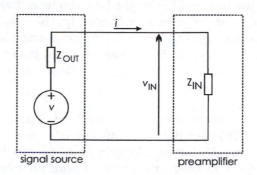

In the diagram above, v is the instantaneous value of the emf produced in the signal source. Z_{OUT} is the output impedance of the source. This might be 600 Ω for an electret microphone or 1 MΩ for a crystal microphone. The output impedance of an op amp is only about 75 Ω.

v_{IN} is the voltage input to the preamplifier (this we want to maximise) and Z_{IN} is the input impedance of the preamplifier.

Consider the current that flows round the circuit at any instant:

$$i = \frac{v - v_{IN}}{Z_{OUT}} = \frac{v_{IN}}{Z_{IN}}$$

From which we obtain:

$$v_{IN} = v \times \frac{Z_{IN}}{Z_{IN} + Z_{OUT}}$$

To obtain maximum voltage transfer, v_{IN} must equal v as closely as possible. This means that the value of $Z_{IN}/(Z_{IN} + Z_{OUT})$ must be close to 1. For this to happen, Z_{IN} must be very much greater than Z_{OUT}. In other words, the input impedance of the amplifier must be much higher than the output impedance of the source. If this is not so, part of the voltage signal will be lost on transfer.

For maximum input impedance, the first stage of the preamplifier might be a common drain amplifier or source follower (p. 67) or a common collector or emitter follower amplifier (p. 77). When working with op amps, the first stage should be a unity gain voltage follower (p. 97).

Preamplifiers are also available as ICs. The SSM2017 is an 8-pin IC with a variable gain set by a single resistor. The gain may be set in the range 1 to 1000. However, as with op amps and other types of amplifier, there is a relationship between bandwidth and gain. As the graph on p. 92 demonstrates, the gain-bandwidth product is constant for any given amplifier. Increasing the gain reduces the bandwidth. There is a trade-off between bandwidth and gain.

For this reason, it is preferable to use a multi-stage amplifier when high gain is required. Two or three stages, each with moderate gain, gives wider bandwith than a single stage with high gain. Ideally, the bandwidth of an audio system should be from 30 Hz (or a little lower) up to 20 kHz.

Mixer

Although this is not found in a domestic stereo system, such as illustrated on p. 229, a mixer is used in some audio systems to combine signals from two or more signal sources either before or after preamplification.

Examples are found in keyboard systems to mix signals from a percussion sound generator with signals from the tone generators activated by the keys. The percussion becomes a backup for the performance by the player. Another example of mixing occurs in a karaoke system where musical backup from a CD player is mixed with a vocalist's performance at the microphone.

A mixer circuit is based on an op amp adder (p. 101). If resistors R1 to R4 are variable, the signals can be mixed in any proportion. Overall volume can be set by a variable resistor in series with R_F.

Equaliser

This stage controls the frequency response of the channel. It consists of one or more filters which can cut or boost the amplitude of frequencies in various ranges. The filters may be passive or active (Chapter 13).

Within the bandwidth of the system, all frequencies should be amplified by the same amount. The frequency response curve (p. 66) should be flat-topped. Then the system will reproduce the original sound with the loudness of all its harmonics in the correct proportions.

> **Harmonic**
>
> When a musical instrument produces a note of given frequency f, it also produces notes of higher frequencies $2f$, $3f$, $4f$ and so on. These have smaller amplitude. They are called the harmonics or overtones.

We recognise different musical instruments by the assortment of harmonics they produce and their relative amplitudes. If different frequencies are amplified by different amounts the mix of harmonics will be changed and the instrument may not be recognisable when the track is played back.

Although is is possible to design an amplifier with a reasonably level response over its full bandwidth, there remains the problem that different signal sources may have an uneven response. A microphone may emphasise some frequencies and reduce others. CDs have different frequency characteristics.

Some systems have built-in filters that are switched in at the same time as the signal source is selected. There may also be filter circuits to be operated by the user (see photo).

A well-known but simple equaliser circuit is shown below. This has a low input impedance so requires a low-impedance source. A pre-amplifier IC typically has a 75 Ω output impedance, which would be suitable. The treble control cuts or boosts the response above 1 kHz to a maximum of about 20 dB at 10 kHz. The bass control cuts or boosts response below 1 kHz to a maximum 20 dB at 50 Hz.

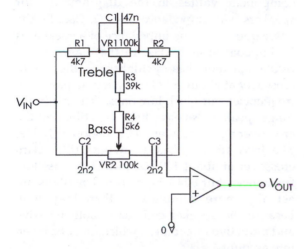

The Baxandall tone control circuit has treble and bass filters in the feedback loop of an op amp.

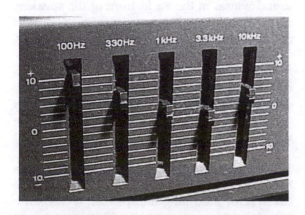

This equaliser allows five frequency bands to be cut or boosted as required by the user.

Power amplifier

The preamplifier stages provide the required amount of voltage amplification. There is often no voltage gain in the power amplifier stage, but a considerable current gain. The result is a gain of power sufficient to drive the loudspeakers. For preference we use Class B or Class AB push-pull amplifiers (Chapter 15) as these are more efficient than Class A.

The gain of the power speakers is usually fixed, the volume control and left-right balance adjustments being made at the preamplifier stage. This means that light-duty components can be used for these circuits.

Speakers

The minimum requirement is a pair of speakers spaced sufficiently far apart so that the left and right channels can clearly be heard more strongly by the corresponding ear.

Most speakers are electromagnetic (see diagram, overleaf). A specially-shaped and powerful permanent magnet provides the field. The poles concentrate the field in the region in which the voice coil moves. The magnet is made from steels specially compounded to provide a high field intensity and are resistant to demagnetisation.

The voice coil is wound on a former made from light plastic or aluminium.

The former is attached to a cone made from fibre-impregnated paper or mylar. There are circular corrugations at the rim to provide a firm but flexible support. When the current from the power amplifier passes through the coil, the magnetic field of the coil interacts with the field of the magnet, causing the cone to vibrate backward and forward. This produces sound waves in the air around the cone.

The frequency at which a speaker produces the loudest sound depends mainly on the diameter of the cone. A single mid-range speaker about 100 mm in diameter is suitable for low-quality reproduction. But no one speaker functions equally well over a wide range of frequencies.

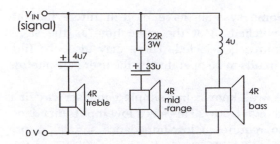

A typical speaker array has three speakers of different physical sizes but with the same coil impedance. The crossover network comprises three passive filters, one for each speaker.

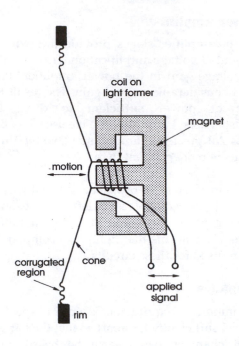

A section through a typical electromagnetic speaker.

Speakers are rated by their impedances and the component values in the diagram are for speakers with impedances of 4 Ω. Usually the value quoted for the impedance of a speaker is its impedance at 1 kHz. In the diagram, the treble speaker has a highpass filter with crossover at about 8 kHz. Capacitors pass high frequencies but block low ones. The medium-range speaker also has a high-pass filter but the crossover point is lower, in the region of 1 kHz. The bass speaker has a lowpass filter with crossover at about 150 Hz. Inductors pass low frequencies but block high ones. The filters are actually more complicated than they look because the speaker coils have both resistive and inductive impedance, which must be taken into account.

For high quality reproduction it is necessary to have an array of two or three speakers for each channel. Between them they cover the range of frequencies perceptible to the human ear. This is the range from 30 Hz to 20 kHz. Below 30 Hz the sensation is one of vibration rather than sound. The smallest speaker is a **tweeter**, for sounds in the highest frequency range, 2 kHz to 20 kHz. The mid-range speaker is larger and covers the range from 50 Hz to to 5 kHz (note that the ranges overlap). The bass speaker, or **woofer**, deals with the low frequency range from 30 Hz to 800 Hz. There may also be a fourth speaker, a **sub-woofer** that handles 20 Hz to 200 Hz.

Each channel has only one power amplifier to drive the array of speakers. To avoid waste of power, signals in the appropriate frequency range must be directed to each speaker. This is done by a **crossover network** (below). This consists of a number of passive capacitor/inductor and inductor-inductor filters. One of the inductors in each case is the speaker coil.

Enclosures

When the speaker cone vibrates it generates sound waves in the air in front of the speaker. At the same time, it generates sound waves in the air behind the speaker. These two sets of waves are 180° out of phase. If waves generated at the rear of the cone are allowed to mix with the waves from the front of the cone, the two sets will largely cancel each other. This causes a very significant reduction in the volume of the sound.

The solution is to mount the speaker on a **baffle**. This is a firm panel or wall with a hole cut in it for mounting the speaker. If the baffle is wide and high enough, it will prevent sound waves from the rear merging with those from the front. Sound is heard at full volume.

More often a speaker is mounted in a box or **enclosure**. This has one or more holes cut in the front for mounting the speakers. The front may be covered with stretched open-mesh fabric or a grille of plastic or metal. These do not obstruct the sound waves appreciably. At the rear, the box is completely self-enclosed so that sound waves from the rear of the speaker can not merge with those from the front.

In some designs of enclosure there is an open vent that lets sound waves escape from the rear but not to interfere with those from the front.

The designing of speaker enclosures must take into account the resonances of the speaker and of the air within the enclosure so as to produce a level response, with no resonant 'booming'.

Enclosure design is a matter of physics, not electronics, and we will leave this discussion at this point.

Questions on audio and video systems

1 Draw a block diagram of a domestic stereo system. Describe the characteristics of the signal that each stage receives and generates.

2 Draw a block diagram of an audio system other than a domestic stereo system. Describe the characteristics of the signal that each stage receives and generates.

3 Describe the form in which a compact disc stores music (details of recording not required) and how it is played back.

4 Discuss the factors concerned in the transfer of a voltage signal from a radio tuner to a preamplifier.

5 What is meant by a 'crossover network'?

6 Explain why each channel of a stereo system usually has at least two speakers. Describe a typical crossover network.

7 What is a mixer? Describe a simple mixer circuit and explain how it works.

8 Outline the ways in which digital audio signals are stored in a computer and how they may be played out.

Other questions

9 Describe the structure of a hard disk drive, and how it works.

10 Describe the structure of a colour LCD suitable for use in a mobile phone. (Research on the Web required to answer this.)

11 You are asked to plan (a) a public address system at a railway station, (b) a home theatre, (c) a system to provide background music in a supermarket, (d) an audio/video system for your study/leisure room. Select one of the above and name the electronic units you would include in the system. If possible state manufacturers' names, model numbers, and costs. Describe how to install the system (positioning and mounting the units, cables required, power supply, and so on).

12 Digital technology has made it much easier for people to copy music, images and text. This means that composers, performers and authors may receive little or no payment for the work they have done? What are your views on this? What can be done to make sure that creative people are fairly rewarded?

On the Web

Collect information on stereo and other audio systems and on video systems by visiting manufacturers' sites such as:

www.sony.com,
www.creative.com,
www.lg.com,
www.nikon.com,
www.apple.com , (for iPod)
www.roland.co.jp,
www.yamaha.co.jp,
www. canon. com.

There are also sites that review audio equipment: www.audioreview.com.

Use a browser to find out how these systems work.

Multiple choice questions

1   The filter circuits of a stereo system are located at the:

    A power amplifier stage.
    B equaliser stage.
    C preamplifier stage.
    D input stage.

2   A domain on a magnetic disk is:

    A a small magnet.
    B a region of small magnets, all orientated in the same direction.
    C a region in which all the magnetic molecules lie in the same direction.
    D a region in which some of the molecules lie in the same direction.

3   The main advantage of an electret microphone over a crystal microphone is that:

    A its response is linear.
    B it is cheaper.
    C it does not require a power supply.
    D it produces a larger signal.

4   To obtain maximum voltage transfer from a CD player to a preamplifier:

    A the input impedance of the pre-amplifier must be low.
    B the output of the CD player must be high.
    C both the output impedance of the CD player and the input impedance of the preamplifier must be low.
    D the input impedance of the pre-amplifier must be high.

5   The type of amplifier preferred for the output stage of a stereo system is:

    A Class A.
    B Class B.
    C Class AB.
    D common-emitter.

6   The platters of a hard disk drive may be made from:

    A semiconductor.
    B glass.
    C cobalt alloy.
    D steel.

7   The sensor on the reading head of a disk drive is usually:

    A an electromagnetic coil.
    B a piezoelectric device.
    C a magneto-resistive device.
    D an inductive device.

8   A crossover network

    A has a push-pull output.
    B is one that has crossover distortion.
    C is a set of filters.
    D gives a flat-topped frequency response.

On the Web

You have been asked to write a 1500-word article on the latest developments in audio-video equipment for a special-interest magazine. The magazine caters for one of these groups: sportsmen and women, music lovers, female teenagers, male teenagers, computer hobbyists, games players, retired persons.

Search the web for devices likely to be of interest to members of one of these groups. Write an article describing items of interest to them, with technical details if appropriate.

27 Noise

Sources of noise

Noise is any unwanted signal and every kind of circuit is subject to it. When a radio receiver is turned up to full volume, but is not tuned to a station, there is a steady hiss from the speaker. This noise is **noise** in the electronic sense. It may have originated in the circuit of the radio set, in the surrounding district, or in space. But noise does not have to be audible noise. A TV set that is not tuned to a station shows 'snow' on its screen. This too is noise. In digital data transmission, noise is the erratic reception of 1's for 0's or 0's for 1's. Obviously, noise is something that must be reduced or preferably eliminated.

EMI

There are many sources of **electromagnetic interference**:

- **Electrical machines:** These include everything from a domestic refrigerator to heavy industrial machinery. Whenever switch contacts open there is a momentary arcing. If the load is inductive, as it often is (for example, switching an electric motor) the e.m.f. generated between the contacts is high, perhaps hundreds of volts. The arc is a high current in which electrons are being rapidly accelerated. This generates an electromagnetic field, a radio wave which spreads for an appreciable distance from the source. It produces a series of pulses in any conductor that it meets. Motor vehicles come into this category for their ignition systems generate their sparks by inductive circuits, producing electromagnetic radiation at the same time. Electromagnetic noise can be picked up by a circuit, and possibly be amplified by the circuit. Or it may be picked up on mains wiring and enter equipment through the power supply circuit.

- **Mains hum:** In mains-powered equipment the 50 Hz (or 60 Hz in USA) mains frequency may pass through the power supply circuits and appear as **ripple** on the output. In may also enter the circuit by induction in an earth loop (below). Another source of noise arising in the power supply is high-frequency **switching noise**, which is generated in switch-mode power supply units.

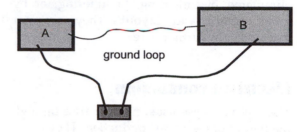

ground loop

Two pieces of equipment are plugged into mains sockets; both are connected to the mains earth line. A shielded cable joins them, and the shield is earthed at both ends. This completes a loop. Any electromagnetic fields passing through this loop generates a current in it, causing interference in the equipment. To prevent interference, break the loop by earthing the connecting cable at only one end.

- **Thunderstorms:** Flashes of lightning are similar to the arcing in electrical machines, though on a larger scale. They produce intense electromagnetic fields that can be heard as noise on a radio receiver many kilometres away.
- **Cosmic electromagnetic action:** Cosmic radiation arriving from Outer Space produces electromagnetic waves when it interacts with the ionosphere, and these are picked up by sensitive electrical equipment.

Mechanical effects

In an amplifier based on thermionic valves, any vibration of the equipment may lead to vibrations of the electrodes within the valves. This makes the characteristics of the valve fluctuate, causing a noise signal to be generated. The effect is known as **microphony**. It is not a common problem nowadays, when valves are seldom used, but a similar effect is found in cables. If the cables are vibrated, the capacitance between the conductors varies, producing a noise signal.

The noise sources described above are more often classified as **interference,** rather than as noise. They can often be reduced or even eliminated by shielding, by filtering, or by good design and layout. They are not discussed any further here.

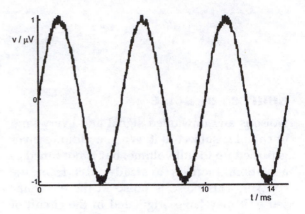

A 200 Hz sinusoid, amplitude 1 µV, has picked up a 60 nV noise signal, mainly due to thermal noise. The sinusoid now has a randomly spiky form and, if this noisy signal is amplified and sent to a speaker, we hear a hissing sound as well as the 200 Hz tone.

Electrical conduction

The types of noise listed next all arise through the nature of electrical conduction. They occur in all circuits and are difficult to reduce:

- **Thermal noise:** This is also known as **Johnson noise.** Electrical conductors all contain charge carriers, most often electrons. Although there is a *mass motion* of carriers when a current is flowing, there is always a *random motion* superimposed on this. Their motion is made random by forces between the carriers and the molecules in the lattice as the carriers pass by. Random motion produces random changes in *voltage*, called thermal noise (right). This is what causes the hissing in the output of an audio circuit. It is similar to the noise made by other random processes such as steam escaping from a steam valve, or water running through shingle on the seashore. The amount of thermal noise depends on the amount of random motion of the charge carriers and this depends on temperature.

- At higher temperatures, the carriers gain extra energy from vibrations of the crystal lattice of the conductor, and so random motion becomes more important.

The noise level depends on the absolute temperature, so a conductor needs to be cooled to absolute zero to eliminate this type of noise. Thermal noise also depends on resistance and on bandwidth. Being completely random, thermal noise has no particular frequency. Or we can take the other viewpoint and say that it contains *all* frequencies at equal amplitudes. Because it contains all frequencies, it is often called **white noise.** If a circuit has a wide bandwidth, a wider range of noise frequencies can exist in it. The total amount of noise in the system is greater. Because thermal noise depends on resistance, it is produced by all components in a circuit, but particularly with those that have high resistance.

- **Shot noise:** With a steady current the *average* number of charge carriers passing a given point in a circuit is constant. But the number passing *at any instant* varies randomly. The random variation of *current* produces another form of white noise called shot noise. Shot noise is a variation of current but the current passes through resistances in the circuit and so becomes converted to variations in voltage. So the effect of shot noise is added to thermal noise.

Like thermal noise, shot noise includes all frequencies, so the amount of it in a circuit increases with bandwidth. If currents are large (A or mA) the random variations in current are relatively small and shot noise is low. But for currents in the picoamp range, shot noise is *relatively* greater, and may have an amplitude that is 10% or more of the signal amplitude. Shot noise occurs most in semiconductors.

- **Flicker noise:** This is also known as **1/f noise** because its amplitude varies inversely with frequency (below). It is only important when signal frequency is less than about 100 Hz, and may be ignored at frequencies greater than 1 kHz. The cause of it is unknown. It occurs in semiconductors, possibly because of imperfections in the material. It also occurs in valves, where it is possibly an effect of irregularities in the surface of the cathode.

Signal processing

Processing a signal does not necessarily produce the exact result intended and this can give rise to noise. An example is **quantisation noise**. When an analogue signal is being converted to digital form, a smoothly varying analogue voltage is converted to a stepped digital voltage. The steps introduce an irregularity into the signal which has the same effect as noise. If the conversion has a resolution of 12 bits or more, the stepping usually can be ignored. One bit in 2^{12}, is 1 in 4096, which is an error of only 0.024%

Signal-to-noise ratio

There are several ways of expressing the amount of noise present in a signal. It is not usually the absolute power of the noise that is important, but rather the power of the noise relative to the power of the signal itself. For this reason, one of the most often used ways of expressing the amount of noise is the **signal–to–noise power ratio**. A ratio of two powers is most conveniently stated on the decibel scale. Because power is proportional to voltage squared, the signal-to-noise ratio is defined as:

$$SNR = 10 \log_{10}(v_s^2/v_n^2) \text{ dB}$$

In this equation, v_s is the rms signal voltage and v_n is the rms noise voltage. The equation has the standard form used for expressing a ratio of powers in decibels.

Example

A signal has an rms value of v_s = 2.4 V. The rms noise level is v_n = 7 nV. Calculate the signal-to-noise ratio.

$$SNR = 10 \log_{10}(2.4^2/(7 \times 10^{-9})^2)$$
$$= 171 \text{ dB}.$$

The SNR is used to calculate the degree to which a circuit, such as an amplifier, introduces additional noise into an already noisy signal. If SNR_{IN} and SNR_{OUT} are the SNRs of the signal at the **input** and output, the **noise figure** of the amplifier is given by:

$$NF = SNR_{IN}/SNR_{OUT}$$

> **Self Test**
>
> If the rms noise level is 15 nV and the rms signal level is 350 mV, how much is the SNR?

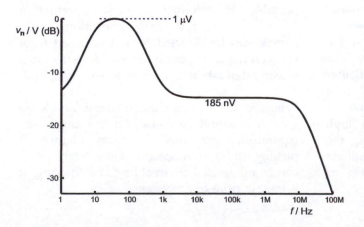

> This graph is an analysis of the common-emitter amplifier on p. 72. It plots the amplitude of the noise signal at the output of the amplifier for frequencies in the range 1 Hz to 100 MHz. Over most of the audio range the noise is 185 nV, due to thermal noise and shot noise. The noise level drops at both very low and very high frequencies because these frequencies are outside the bandwidth of the amplifier, so they are not amplified.

Noise reduction

Noise can not be eliminated from a circuit because noise is an inevitable effect of current flowing through components. But it is possible to reduce noise by using the techniques listed below:

- **Avoid high resistances:** Thermal noise is generated when currents pass through resistance and is proportional to the resistance. For low noise, design the circuit so that resistors are as small as possible.

- **Use low-noise components:** Some components are specially manufactured to reduce their noise levels. The BC109 transistor is the low-noise equivalent of the BC107 and BC108 because it has low flicker noise. FETs have no gate current, so no thermal noise is produced. JFETs have less shot noise and flicker noise than BJTs because there is only a small leakage current through their gates. MOSFETs have no shot noise. Many of them have low resistance, so thermal noise is low too. Using FETs in a circuit before the signal is amplified avoids introducing noise that will be amplified by later stages. Often a two-stage amplifier has a MOSFET for the first stage, followed by a BJT in the second stage. Low-noise versions of op amps are another way of avoiding noise.

- **Use low currents:** These result in low thermal noise so it is best to keep current low in the early stages of an amplifier or similar circuit. A collector current of 1 mA is suitably low for a BJT amplifier stage.

- **Restrict bandwidth:** The graph on p. 241 shows that the noise levels in the amplifier fall off at low and high frequencies because they are outside the bandwidth of the amplifier. Restricting the bandwidth can further reduce noise. In this particular amplifier the upper −3dB point is at 11 MHz. For many applications there is no need to pass signals of such high frequency. In these cases, reducing the high-frequency response will eliminate much of the noise, without affecting its performance in other ways.

The bandwidth of radio frequency amplifiers needs to cover only the carrier wave frequency and the sidebands which occur about 15 kHz to either side of this. Restricting the bandwidth to such a narrow range means that noise is very much reduced. In general, noise may be reduced by restricting the bandwidth of a circuit to just that range of frequencies it needs to do its job.

Noise in telecommunications

In all three modes of telecommunications the signal gets progressively weaker as the distance from the transmitter decreases. We say that it is **attenuated**. Attenuation is usually expressed in decibels. The amount of attenuation depends on the transmission medium.

In transmission lines, signals of higher frequency are attenuated more than those of low frequency. For example, in RG58 coaxial cable attenuation is 0.33 dB per 10 m of cable at 1 MHz, rising to 7.65 dB at 1 GHz. This limits the rate at which data can be transmitted.

Optical fibre has much smaller attenuation, typically 0.15 dB per *kilo*metre.

Radio transmission in free space obeys the inverse square law, so that doubling the distance reduces the signal to quarter strength (−2.5 dB). So much depends on other factors such as absorption in the atmosphere and refraction in the ionosphere that it is not possible to produce reliable attenuation figures. Under favourable conditions radio signals can be directed to and received from spacecraft and can circle the world, so the exact rate of attenuation is not easy to specify.

Whatever the medium, the further it travels the weaker a signal becomes. Yet the longer the communications link the more chance of picking up interference and noise. Interference can be minimised by shielding and filtering. It is least in fibre-optic systems.

True noise is introduced at the transmitter. In many systems the transducer is resistive and generates noise on its output signal. Further noise is introduced by the amplifying and transmitting circuits. The amplifiers at the receiver also introduce noise. Noise may be kept to a minimum by using some of the general electronic techniques listed above. There are also some measures specially applicable to telecommunications systems:

- **Companding:** This term is a combination of the words *compressing* and *expanding*. The principle is to amplify the signal using a special amplifier that amplifies the lower-power signals more than the higher-power signals. In terms of audio, a lower-power signal is the sound of a subdued instrument such as a triangle, and the sound of all instruments in their quietest passages. Also the higher harmonics of all instruments are of low power. In a compander, all of these are boosted relative to other more powerful signals. The loudest instruments remain the loudest and the quietest remain the quietest but the range of power from quietest to loudest is compressed. This takes the lowest powers of the wanted signal up above the general noise level. Such selective amplification produces a distorted signal overall, but the balance is restored at the receiving end by a circuit which has the reverse action. There, the lower-power signals are amplified less than the high-power signals. The range of signal powers is expanded and the original balance of the signal is restored.

- **Frequency modulation:** Noise and interference are voltage signals. If they are superimposed on an amplitude-modulated signal, they affect the amplitude and are still there after the signal has been demodulated. This does not happen with frequency modulation or with phase modulation. Once the carrier has been frequency modulated, random increases or decreases of amplitude have no effect on the *frequency* so any noise acquired after modulation is lost on demodulation.

- **Repeaters and regenerators:** The signal-to-noise ratio decreases with increased transmission distance so it is essential to restore the signal at intervals along the transmission path. For analogue signals there are **repeater** stations at regular intervals along the path. These receive the signal, filter it to remove noise and interference as much as possible, amplify it, and finally re-transmit it. For digital signals there are **regenerator stations**. These receive the signal and have the relatively easy task of distinguishing between the 0's and 1's. Even if the signal is noisy and the pulses are badly distorted, the original sequence of binary digits can usually be recovered. The repeater station then transmits a new stream of perfectly formed pulses exactly representing the original signal. This is one reason why digital data is preferred for long-distance telecommunications.

Questions on noise

1 Name six sources of interference in electronic circuits and state how they may be avoided or reduced.

2 Name three sources of noise in electronic circuits and state their cause.

3 What is quantisation noise?

4 Define the signal-to-noise power ratio. If the rms signal voltage is 120 mA and the rms noise voltage is 2.7 µV, what is the SNR?

5 Describe how to avoid noise in a BJT amplifier.

6 Describe how to avoid noise in a radio-frequency amplifier.

7 Explain the difference between a repeater and a re-generator on a long transmission line.

8 What is companding and how does it help reduce noise?

9 What is the relationship between the bandwidth of a system and the noise generated in it?

Multiple choice questions

1   The noise produced by the random motion of charge carriers is:
     A thermal noise
     B shot noise
     C flicker noise
     D $1/f$ noise.

2   The noise produced by the varying numbers of charge carriers in a very small current is:
     A thermal noise
     B shot noise
     C flicker noise
     D $1/f$ noise.

3   The main sources of shot noise in a circuit are:
     A semiconductors
     B resistors
     C any form of resistance
     D capacitors.

28 Telecommunications

The job of a communications system is to convey information from one place to another place that is some distance away. In this book we are concerned with communications systems that employ electronic devices. These are known as **telecommunications**, and include communication by electrical cable (telegraph and telephone systems), optical fibre and radio. There are many other non-electronic systems of communication such as smoke signals, semaphore and homing pigeons.

This chapter deals with the principles of telecommunications, including the forms in which information is transmitted. Following chapters explain the techniques used for transmitting data by cable (Chapter 29), by optical fibre (Chapter 30) and by radio (Chapter 31).

Telecommunications systems have many features in common, as illustrated in the diagram below.

This is a big simplification because there are hundreds of different communications systems in use. Some do not make use of all the stages shown in the figure. In others the functions of different stages may be combined into one stage. Specialist systems may have additional stages not shown here.

The table overleaf describes the main features of telecommunications systems in more detail. There are some terms in this table that you may not understand. Use the Index to to find where in the book these terms are explained. The term 'transducer' is used loosely in the table. Not all of the devices mentioned are true transducers.

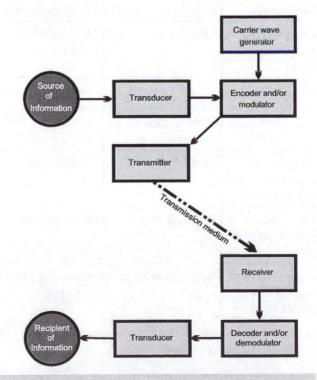

Information passes through a telecommunications system from the source to the recipient, taking various forms along its journey.

Stages in telecommunications systems

Stage	Function	Cable	Optical fibre	Radio
Source of information	Stores or originates information.	Stored: computer memory, magnetic disk, compact disc, human brain, video recording, book. Original: temperature, colour, pH, weight, sound, a new idea in a human mind.		
Transducer	Converts information to an electrical signal.	Microphone, keyboard, photocell, thermistor, electromagnetic head of a disk drive. This stage may include an amplifier.		
Encoder	Makes a signal suitable for transmission.	Companding analogue signals, converting analogue to digital, compressing digital, parity bits, formatting to various protocols (ASCII, e-mail, teletext), conversion to pulse codes. Assembling data into packets. No encoding in telephone connection to local exchange.		
Modulator	Modulates a carrier wave with the signal.	Analogue signals by amplitude, frequency, phase or pulse modulation. Digital signals by frequency shift keying, pulse code modulation. In local telephone circuits: not used for speech, but modems used for fax and Internet.		
Carrier wave generator	Produces sinusoid of fixed frequency and amplitude.	Radio-frequency oscillator.	Laser or light-emitting diode, usually infrared.	Radio-frequency oscillator.
Transmitter	Transfers signal to transmission medium.	Line driver.	Modulated laser or LED.	Radio-frequency amplifier and antenna.
Transmission medium	Carries signal to receiver.	Twisted pair of wires or coaxial cable carries current.	Narrow glass or plastic fibre carries light.	Free space allows propagation of electro-magnetic waves.
Receiver	Takes in signal and converts it to electrical signal.	Line receiver	Photodiode.	Antenna and radio-frequency amplifier.
Demodulator	Extracts signal from modulated carrier.	Reverses the modulation process.		
Decoder	Converts signal to its original electrical form.	Reverses most of the encoding process.		
Transducer	Converts signal to its original physical form.	Speaker, electron gun in TV tube, electric motor in remote-controlled robot.		
Recipient of information	Receives and possibly responds to information.	Data recorder, TV set, computer memory, human mind, printer, fax machine, pager, computer monitor.		

Carriers and modulation

Except for sending audio signals as a varying voltage on a local telephone line, analogue information is more often sent by superimposing it on a signal that has a considerably higher frequency. This process is called **modulation** and the signal of higher frequency is known as the **carrier**. The amplitude, frequency, or phase of the carrier is modified in proportion to the instantaneous value of the analogue signal.

Amplitude modulation (AM) is illustrated below. The signal is a sinusoid (a). There is a short period at the beginning (on the left in the diagram) when there is no signal.

The carrier is also a sinusoid but with a higher frequency, and shows its unmodulated amplitude on the left of curve (b). As soon as the signal begins, the amplitude of the carrier varies according to the signal superimposed on it. This is the **modulated carrier**.

In this example, the signal causes the amplitude of the carrier to vary by as much as ±50%. The **modulation depth** (or **modulation factor**) is 50%. In the figure, the carrier frequency is 12.5 times that of the signal. It is limited to that amount because it would be less easy to draw a diagran with the frequency any higher. In practice, the ratio of frequencies is much greater.

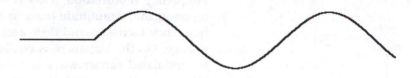

(a) Original signal

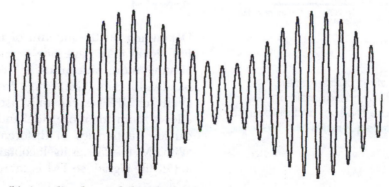

(b) Amplitude modulated carrier

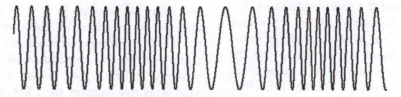

(c) Frequency modulated carrier

Signal (a) is used to modulate the amplitude of a carrier, producing the modulated waveform (b), which has constant frequency. In (c) the signal modulates the frequency and amplitude remains constant.

Example

An AM radio transmitter on the Medium waveband has a frequency of 800 kHz. It is modulated by an audio signal of 1 kHz. With these typical values, the carrier frequency is 800 times the signal frequency.

An unmodulated carrier is of a single frequency, f_c. When it is modulated, the composite signal (carrier plus modulation) consists of many frequencies. These comprise f_C itself and, for each modulating frequency f_m, the sum and difference frequencies f_{c+m} and f_{c-m}. As there are usually many different modulating frequencies present, their sum and difference frequencies occupy two bands above and below the carrier frequency.

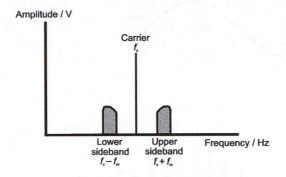

Amplitude modulation of the carrier generates two sidebands, containing frequencies that are the sum and difference of the modulating frequencies and the carrier.

These are the **sidebands**, referred to as the **upper sideband** and the **lower sideband**. The amplitude of the signals in the sidebands is usually less than that of the carrier, as in the figure. Therefore most of the power of the signal is in the carrier and less in the sidebands. But the carrier itself provides no information, so AM is wasteful of power.

If the highest modulating frequency is f_{max}, the modulated signal occupies a band of frequencies extending from $f_c - f_{max}$ to $f_c + f_{max}$. The bandwidth of the signal is $2f_{max}$.

Both sidebands contain identical information so there is no point in transmitting them both. In **single sideband** (SSB) transmission, the carrier and one of the sidebands is suppressed before transmission. The power of the transmitter is devoted to transmitting the single remaining sideband. This is much more efficient. Also in SSB, the bandwidth of the transmission is reduced to f_{max}. This means that transmitters may be spaced more closely in the radio spectrum.

In a telephone system transmitting speech, the highest frequency present is 3.4 kHz. The bandwidth of the modulated signal is 6.8 kHz, but is reduced to 3.4 kHz by SSB transmission.

Frequency modulation (FM) is illustrated at (c) on p. 247. Amplitude remains constant but frequency increases and decreases with signal voltage. On the left, there is no signal and the unmodulated carrier wave is seen. Frequency increases as the signal rises above 0 V and decreases as it falls below 0 V. As in the diagram for AM, the amount of modulation has been exaggerated to make the diagram clearer.

Depending on the amount of modulation, an FM signal may have one or more sidebands. Having more sidebands, it requires greater bandwidth than AM. A high-frequency carrier (VHF or UHF) is generally used because this allows room for more sidebands between the carrier frequencies of different transmitters. With FM, the carrier itself contains information about the signal, so FM is more efficient than AM.

A further advantage of FM is that it is not affected by changes in the amplitude of the signal. If reception conditions are poor, an AM signal varies in amplitude, being received strongly at times, but fading at other times. This does not happen with FM, since changes in amplitude have no effect on frequency.

It is explained on p. 243 that FM is not affected by noise or interference. Both of these can badly affect AM transmissions.

Pulse modulation

Analogue information may also be transmitted by various systems of pulse modulation.

Pulse amplitude modulation (PAM): The analogue signal is sampled at regular intervals and a pulse is transmitted for each sample. The pulse amplitude is related to the sampled value of the signal.

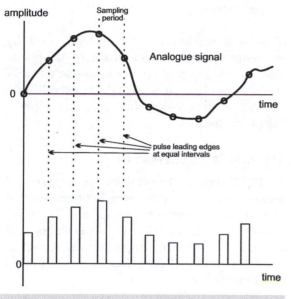

PAM generates a series of pulses, varying in amplitude.

As is usual in sampling of analogue signals, samples must be converted at double the highest frequency of interest. For example, if the highest frequency of interest is 8 kHz, the signal must be sampled at least 16 thousand times per second.

PAM is liable to noise and interference in the same way as AM transmission. It has uses in time multiplexed transmissions (p. 252) for the pulses of other transmissions can be sent in the gaps between pulses. On the whole, PAM is not commonly used.

Pulse width modulation (PWM) or **pulse duration modulation** is the system most commonly used. The analogue signal is sampled at regular intervals as for PAM.

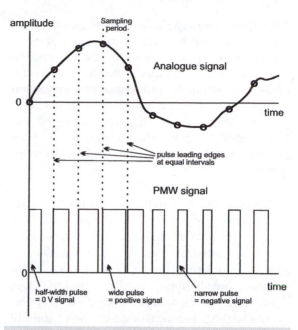

In pulse width modulation, the widths of the pulses are proportional to the instantaneous value of the analogue signal at each sampling.

The PWM signal consists of a series of pulses with their leading edges occurring at the regular sampling times. The width of each pulse is proportional to the amplitude of the analogue signal at sampling. The scaling is such that an instantaneous amplitude of 0 V results in a pulse of half width. The pulse is wider when the signal voltage is positive. The pulse is narrower when the signal voltage is negative.

At the receiving end, the original signal is recovered by sending the pulsed signal through a lowpass filter. A rough description of how this works is to take two extreme conditions. A succession of pulses wider than average results in an increase of voltage from the filter, so reconstructing the original positive-going input voltage. Conversely, a succession of narrow pulses, corresponding to a negative voltage, results in falling output from the filter.

An advantage of PWM is that the circuits for producing PWM pulses and for recovering the original signal are relatively simple.

A further advantage is that spikes and noise on the system affect pulse amplitude but have little effect on pulse length. A disadvantage of PWM is that the transmitter has to be powerful enough to generate full-length pulses yet, on average, it is producing only half-length pulses. This makes inefficient use of the transmitter. In effect, the transmitter operates at only half the power it could otherwise generate.

Pulse position modulation (PPM) uses pulses of constant amplitude and length, but their timing varies. With no signal, the pulses occur at regular intervals, one for each sample. If a sample is positive, the pulse is delayed by an amount proportional to the signal voltage. Conversely, if the sample is negative, the pulse occurs earlier. The pulses make use of the full power of the transmitter, and the transmissions are relatively immune from noise. The draw-backs of this system are that circuits are relatively complicated and that the transmitter and receiver must be synchronised, which is not required for PWM.

Pulse code modulation (PCM) begins like the other systems with the analogue signal being sampled using an analogue-to-digital converter. Each sample is converted to an 8-bit value, which can take any value between 0 and 255. The sampled values are then formatted automatically into larger groups for transmission. Extra bits are added to aid error-checking on reception and to synchronise the receiver with the transmitter.

Data from several sources may be automatically combined, transmitted simultaneously, and separated at the receiving end. Very high transmission rates may be used, the fastest being over 500 megabits per second with the capability of transmitting 7680 channels simultaneously.

PCM is also used for signals that are digital in origin, such as signals sent from one computer or other digital system to another such system. Whether the signal originated in analogue or digital form, the values transmitted are either 0s or 1s.

Frequency shift keying

This is one of several related systems for transmitting digital information. FSK uses two frequencies, one to represent '0' and the other to represent '1'. In one version of FSK, a '0' is represented by a burst of 4 cycles at 1200 Hz, and a '1' is represented by a burst of 8 cycles at 2400 Hz. The two frequencies are easily distinguishable at the receiving end and the original sequence of 0s and 1s is reconstructed.

This is a relatively slow system, used by modems operating over ordinary telephone lines. Many types of fax machine also use FSK with audio frequencies when establishing contact with another fax machine over a telephone line.

Digital transmission

In PCM data is transmitted as a series of pulses. A data stream consists of pulses that are coded to represent the 0s and 1s of the original data. If the original data is analogue it is first converted to digital form, using an analogue-to-digital converter, before coding it for transmission.

There are several ways of coding data for transmission, one of which is illustrated in the diagram below.

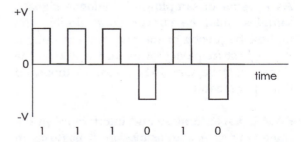

The signal level is positive for '1' and negative for '0', or could be the other way round. The essential feature is that the signal returns to 0 V inbetween each pulse. This is **return-to-zero (RZ)** coding.

Each pulse is separate from its neighbours so it is not necessary to send a clock signal for synchonising transmitter and receiver.

Not requiring a clock signal is an advantage of RZ coding. Its big disadvantage is that returning to zero between pulses means that fewer pulses (and therefore less data) can be transmtted in a given time. In other words, the bandwidth is considerably reduced.

Data transmission rate is higher with non-return-to-zero (NRZ) coding, as illustrated below.

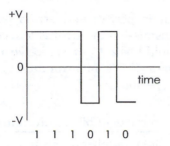

The same data is tranmitted in about half the time. But the receiver will need to receive a synchronising signal so that it will know when each pulse begins and ends.

In the coding above, a bit is represented by a *pulse*. Other methods of NRZ coding may represent a bit by a *change* in level. In the system illustrated below (NRZ Mark), a change in level represents a 1. A 0 is represented by absence of change at the appropriate time.

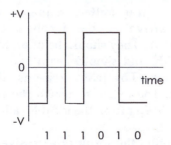

This sequence begins with the signal level low from previously transmitted data. Had it been high, the signal would have been the inverse of this. The actual signal level is not important— it is the *change* in level that conveys the information.

This system of coding is used in recording on magnetic disks, where the change (or no-change) is a change in direction of the polarity of the magnetic domains.

Frequency division multiplexing

Pulses can be transmitted directly as pulses, but more often they are used to modulate a high-frequency carrier signal, using PCM or other coding system.

Modulation of any kind makes it possible to send a large number of different signals simultaneously on *one* channel of transmission. The aim is to take advantage of the full bandwidth of the channel. This is the range of frequencies that the channel can transmit. Usually, this range is much wider than the bandwidth of an individual signal. So a number of different messages may be modulated on to a number of different frequencies, sent down the same channel at the same time, and separated at the receiving end. This is **frequency division multiplexing**.

Example

A voice signal requires a bandwidth of 3.4 kHz to reproduce understandable speech. If such a signal is SSB modulated on to a 64 kHz carrier, the lower sideband extends down to 64 – 3.4 = 60.6 kHz. The frequency space between 60.6 kHz and 64 kHz must be reserved for this transmission.

There is plenty of space above 64 kHz, which can be used for transmitting other signals. The next higher carrier could be at 68 kHz, which, when modulated, occupies from 64.6 kHz to 68 kHz. This does not overlap the band occupied by the signals on 64 kHz. In this way we can stack a group of voice signals on carriers spaced 4 kHz apart. A **group** of 12 speech signals modulated on to carriers 4 kHz apart occupies a 48 kHz bandwidth.

Modulation can be taken a step further. Groups of signals can themselves be modulated on to carriers of even higher frequency. If a number of these carriers are spaced 48 kHz apart, they produce a **super-group**.

This process of stacking signals by modulating them on to groups and super-groups (and even higher groupings) of frequencies makes it possible to transmit hundreds of signals simultaneously along a single channel such as a coaxial cable.

At the receiving end, tuned circuits and demodulators are used to separate out the super-groups, then the groups, then the original modulated signals, and finally the individual analogue signals. These are then routed to appropriate recipients.

Frequency division multiplexing is used also with signals of other kinds. The bandwidth needed depends on the nature of the original signal. Music signals, for instance, require a greater bandwidth than speech. Typically, a musical signal needs 15 kHz bandwidth for a single transmission, so the carriers must be spaced more widely apart. For the highest quality, the bandwidth is increased to 20 kHz. Bandwidths for video transmissions are rated in tens of megahertz.

Time division multiplexing

This is another approach to conveying many signals on a single channel. Signals from several sources are allocated to short time slots (3.9 μs for 1 byte). A total of 30 time slots, plus two for control/synchronising, takes 125 μs, which gives 8000 30-byte samples per second.

At the receiving end the signals are sorted out and routed to their proper recipients The diagram opposite shows how a TDM system could be applied to PCM transmission of several channels carrying analogue signals.

TDM is often used in telephone exchanges and in radio satellite communications.

Transmission rates

There are two different ways of expressing the rate of transmission:

Bit rate: the number of binary bits (0s or 1s) transmitted in 1 second.

- **Baud rate:** the number of 'signal events' transmitted per second. An example of a signal event is a single pulse of given length or amplitude. Another example is a burst of one of the FSK frequencies.

If one signal event represents one binary digit (as with FSK), the bit rate and the Baud rate are equal. In many systems a signal event represents two or more bits. In such cases the baud rate is half of the bit rate or less.

No system is perfect and errors ooccur. In digital transmission the error is to receive a '1' when it should be a '0', or to receive a '0' when it should be a '1'. The **bit error ratio** of a system for a given period is:

$$BER = \frac{\text{number of bits received in error}}{\text{total number of bits transmitted}}$$

This is often expressed as error bits per 10 000.

The BER of a line is tested by using a pattern generator to trasmit a sequence of pulses representing 0s and 1s. The pattern may possibly be all 1s or all 0s to test the system's timing reliability, or it may be a regular mixture of both. At the receiving end is a digital device set to receive the same pattern and to report errors.

Jitter

Ideally, a transmitted sequence of pulses should arrive at the receiver exactly as transmitted. They should be of uniform height and width, and should begin and end at the right times. For many reasons, this rarely happens. The extent to which the pulses of a sequence differ from the ideal is called **jitter**.

Jitter is often the result of excessive electronic noise (internal or external) in the system. This may cause variations in the clock rate. The clock produces pulses of unequal length or pulses that are slightly delayed or advanced. The deviations in timing may be random, or may show bias in one direction. In either case the operation of the system is impaired.

Jitter is avoided or reduced by specially designed ani-jitter circuits. Essentially these accept the signal and process it before sending it on with improved timing. In a device called a dejitteriser the signal stores the signal as it arrives, then passes it out at a regular rate equal to the average rate of the arriving signal.

Pulse shape

In theory, a pulse waveform rises instantly to its full amplitude, stays there for a while, and then falls instantly to zero. This never happens in practice because a voltage can increase or decrease only when a current flows, and the current takes *time* to carry charge from one place to another.

As the graphs on the right show, a pulse rises and decays (falls) in a finite time. By definition, the **rise time** is the time taken for the voltage to rise from 10% to 90% of its final value.

Pulses (a) in theory and (b) in practice. Pulses have a finite rise time t_r and decay time t_d. Decay time is also known as fall time.

In a telecommunications system the shape of the pulse may be modified by the filtering effect of any circuit or cable through which it passes. As explained on pp. 262-263, a transmission line comprises resistive, capacitive, and inductive elements.

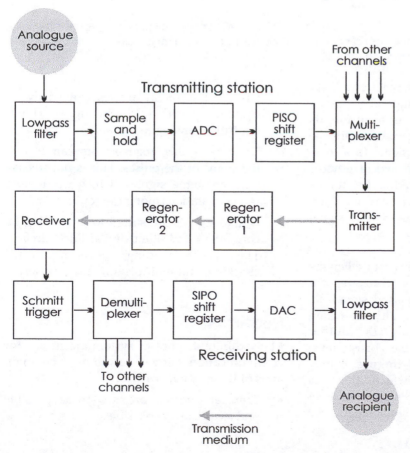

Transmitting analogue data over a long-distance PCM system, using time division multiplexing. The transmission medium could be cable, optical fibre or radio. A system for transmitting digital data would be similar, but would replace the ADC and DAC by a coder and a decoder at each end of the system (for 2-way communication). The coder and decoder may be combined into one device, known as a codec. Instead of a hardware codec, the same functions may be taken over by a codec program running on a computer.

The resistance is that of the conductor. There is capacitance between the conductor and nearby conductors, including those connected to ground. There is the inductance of the conductor, even though it is not coiled. The resistance, capacitance, and inductance act as a filter circuit, usually a lowpass filter.

The graph below shows what happens to a 10 kHz pulse after it has been through a lowpass filter with the –3 dB point at 33 kHz. The pulse has lost its squareness. Both the rise time and the decay time of the signal have been extended.

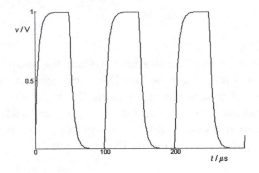

When a pulsed signal passes through a circuit, its shape is distorted by stray capacitance.

The signal that emerges at the other end is far from square in shape. It must be passed through a Schmitt trigger circuit to restore its squareness. If this is not done there could be problems when feeding this signal to a logic circuit. This is because some logical devices require that the rise time should be less than a specified minimum. The device may not operate reliably when presented with a signal like that shown in the drawing.

Another form of distortion may occur when the input to a circuit changes abruptly, for example, when receiving a pulsed signal. Its output may swing too far. This is called **overshoot**. After overshooting, the output oscillates on either side of the correct level before settling. This is called **ringing**. It may occur in any circuit in which there is capacitance and inductance. Small amounts of overshoot and ringing are eliminated by the Schmitt trigger circuit.

Comparing analogue and digital transmission

Digital transmission is the basis of today's 'information explosion' and the main coding system used is PCM. These reasons for its success are:

- Binary coding uses 0s and 1s which are easily distinguishable even in a poorly received signal.

- Digital signals are highly immune from noise. The signal below is noisy but there is no problem in recognising it as a pulsed signal.

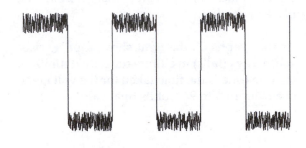

A high level of noise does not prevent the pulses from being readable.

- Distorted or noisy pulses can easily be squared up with a Schmitt trigger. This is not true of analogue signals.

- On long routes, regenerators can restore the shape of the pulses. This is not so easy with analogue signals, which need more repeater stations along the route.

- Digital signals are suited to processing by computers and other digital logic circuits. This makes coding, decoding, error checking, and multiplexing much easier.

Networks

Many different types of system are in use for communication between stations. The main aims of these systems are to:

- Transfer information as soon as possible and as quickly as possible.

- Transfer information without error or loss.

- Make maximum use of the transmission medium, both in terms of *time* (minimising idling time), and *bandwidth* (transmitting many channels simultaneously).

The final point is important because submarine cables, communications satellites and other hardware (see next three chapters) are expensive to build, maintain, and operate.

We will look at ways of connecting information sources with recipients, starting with the simplest systems and going on to world-wide networks. It is essential that communicating stations must 'speak the same language', in the sense of using compatible coding and decoding procedures, compatible multiplexing and demultiplexing procedures and, in general make certain that messages transmitted from one station can be received and interpreted by the recipient. To this end, a wide range of **standards** and **protocols** have been devised by international bodies. We will look at a few typical and widely used examples of these.

RS-232

This is a standard for cable communication between two computers. The term 'computer' in this account includes not only personal computers but any computer-like equipment such as data-recorders and modems.

Data can be output from a computer either in **parallel** form or in **serial** form. A parallel connection has a line for each bit in the data words. For example, in an 8-bit system there are 8 parallel data lines, plus ground (0 V) lines and a few other lines for handshaking between the stations. This kind of link is used for sending signals from a PC to an attached printer. The Centronics interface is an example. The advantage of parallel connection is that a whole byte of information can be transmitted in one operation. The disadvantage is that it is limited to cables a few metres long, Usually, transmitter and receiver are in the same room. The RS232 ('RS' stands for 'Recommended Standard) connection is an example of a serial connection.

The complete RS232 connection uses a 25-way connector, and most of the lines are in use. However, this provides features such as a secondary channel and timing signals that are not generally used. Timing is necessary for **synchronous** data transmission, in which the transmitter and receiver must be kept in step. Most users of RS232 employ **asynchronous** data transmission, as described below.

The simpler RS232 connection, which is the one most commonly used, has a 9-pin connector. The directions in which signals are transmitted are illustrated in the diagram. This shows a typical case, in which a computer is communicating with a modem.

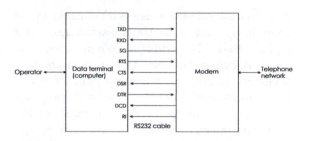

A computer can be connected through an RS232 cable to a modem, and then to the telephone network.

Three of the nine lines are concerned with carrying the data:

TXD (or TD): transmits serial data from computer to modem.

RXD (or RD): receives serial data from modem to computer.

SG: signal ground, the common return path for TXD and RXD signals.

Four of the lines are concerned with handshaking. These are the lines by which the computer and modem coordinate their actions:

RTS: request to send. The level on this line goes high when the computer is ready to transmit data to the modem along the TXD line. This is the first stage of handshaking.

CTS: clear to send. On receiving the RTS signal, and if it is ready to receive data, the modem makes the level on this line high. The computer then starts sending the data.

DTR: data terminal ready. This signals that the terminal is operating properly and that the modem may be connected to the telephone network.

DSR: data set ready. This signals that the modem is correctly connected to the telephone network.

DCD: data carrier detect. This indicates that incoming data is detected on the telephone line.

RI: ring indicator. Goes high when the modem is called from the telephone network.

RS232 is intended for use with a cable up to 15 m long, and with transmission rates not higher than 20 kbits/s. When used for communication between two computers it is necessary to cross over certain pairs of wires. For example the TXD terminal of one computer is to be conected to the RXD terminal of the other computer, and the other way about. The same applies to the RTS/CTS and DSR/DTR pairs.

The standard also comprises electrical specifications. These include logic voltage levels. A 'high' level (logic 1) is between +3 V and +15 V. A 'low' level (logic 0) is between –3 V and –15 V. The hardware to produce these levels are described on pp. 264–265.

ASCII code

This is the **American Standard Code for Information Interchange**. It is a way of coding alphabetic, numeric and punctuation characters into groups of seven bits.

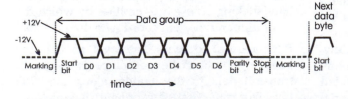

A code group being transmitted on an RS232 cable.

For convenience, the codes are quoted as numbers from 0 to 127. In this code the character 'a' is '65'. In digital form, this is:

$$100\,0001$$

Although RS232 does not specify the coding technique to be used, one of the more widely used coding techniques is illustrated below. The levels in the signals are read from left to right in the diagram. The sequence for transmitting a letter 'a', for example, is:

Marking: waiting for transmission to begin. This is an asynchronous transmission, so the receiver must be ready for it at any time. During marking, the TXD line is at logic 0.

Start bit: A single bit (logic high) indicates the start of a transmission.

Data bits D0 to D6: In this case, these are the seven binary digits coding for 'a', as listed above.

Parity bit: Logic 0 or 1, as explained on p. 178. If this is incorrect, the receiver may automatically signal the transmitter to repeat the group.

Stop bit: Logic low, indicating the end of the transmission.

Marking: Waiting for transmission to resume and the next group to arrive.

Circuit switching

In the case of an RS232 or similar short-distance link, the connection is usually more-or-less permanently plugged together. Switching the connection on or off is not considered. However, when there are a number of stations in a neighbourhood, maybe several thousands, and they may each need to be connected to one of the others only occasionally, some form of switching is required.

A common example is a **local telephone network**. The subscribers are each connected to the **local telephone exchange** by an individual local line. The connection is generally a pair of copper wires, fed into larger cables on their way to the exchange.

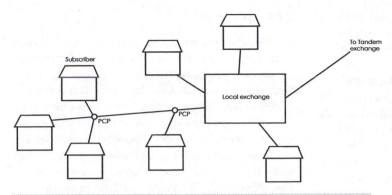

Telephone subscribers cluster around their local telephone exchanges. Each has an individual 2-wire connection to the exchange.

The local exchange provides for communication within in an urban district, or county, or area of similar population size. The cost of local calls is low; they are free in some countries. For subscribers outside the local area the exchange is connected to a Local Tandem exchange. Connection may be be cable or by optical fibre.

A Local Tandem exchange serves a group of adjacent urban areas.

Usually there are a number of main cables running cross-country to the exchange. These carry numerous pairs of conductors. At convenient intervals, the cables pass through **primary connection points**. These are usually roadside cabinets. At these points the paired-wire cables from nearby subscribers are connected to pairs of wires in the main cable and thence to the exchange.

The connections between subscriber and exchanges are unswitched. The subscriber picks up the telephone and is immediately in contact with equipment in the local exchange, usually hearing the 'dialling tone'. All switching is done in the exchange. When the subscriber dials the number of the called subscriber, the switching equipment connects the calling subscriber's line to the called subscriber's line. The ringer sounds and the call is made when the called subscriber picks up the telephone.

It is a feature of conventional circuit switched networks that the call is blocked if the called subscriber is already talking to some other subscriber. There is nothing to be done but to hang up and call again later. However, now that the old-fashioned mechanical switches at the exchange have been replaced by digital switching equipment, some improvements have been developed. These include 'Call Waiting' and 'Message Bank' facilities which are available for extra rental charges.

Local Tandem exchanges are connected to Regional Tandem Exchanges and these in turn are connected to National Tandem Exchanges. A call to a distant part of a country may be routed first to the caller's local exchange, then to the Local Tandem Exchange, followed by the Regional Tandem Exchange and the National Tandem Exchange. From there it is routed down through the hierarcy of exchanges until it reaches the local exchange of the called subscriber. All of this switching is done automatically.

At the top of the system are the International Exchanges, which communicate directly with the National Tandem Exchanges and with each other in different parts of the world.

Message switching

In the circuit switching system, described above, the caller and the called station need to be active at the same time. Whether two people are conversing, or one fax machine is sending a message to another, signals are passing in a continuous stream from one end of the line to the other. Origination and reception are more or less simultaneous. The system operates in *real time*.

The alternative system of communication is **message switching**. In this, the message is stored in memory and then relayed to the recipient at some future time.

Electronics — Circuits and Systems

Message switching requires a digital system, though some messages may be analogue signals such as music and speech that have been converted to digital form. The aim of message switching is to maximise the use of the system, exploit its high speed, increase its efficiency and minimise operating costs. A stored message is transmitted as soon as there is an opportunity to deliver it. If a station or route is busy, any messages that arrive can be queued. They are not blocked (or even lost!) as in a circuit switching system.

One of the more important techniques for message switching is **packet switching**. In this, digital data is made up into packets of fixed length. This is done by a digital device known as a **packer assembler/disassembler** or **PAD**. The packet begins with a block of data or **header** which contains such information as:

Synchronising bits.
Address of origin.
Destination address.
Sequence number.
Error checking code.

The sequence number is used when the message extends over two or more packages. It might happen that the different sections of a message are sent along different routes, so they arrive at their destination out of order. The PAD at the destination uses the sequence numbers to combine the sections of the message into a single message, with all its parts in the right order.

Following the header comes the main message and then the **footer** that concludes the transmission. Header and footer are stripped off at the receiving end and only the message reaches the recipient.

Networks that employ packet switching may have several different configurations. A **bus network** has the terminals spaced out along its length, each communicating with the terminals on either side of it. There are also ring networks and star networks. The transfer of data in the network is performed according to agreed rules, known as a **protocol**.

TCP/IP

A widely used protocol, used on the Internet and many others, is known as **Transport Control Protocol/Internet Protocol**. The Internet is a world-wide network. It has grown enormously since it was first developed in 1969 for military purposes. It has incorporated several other networks and certain networks, such as Compuserve, operate in conjunction with it. The vast number of users attracted by the World Wide Web, which runs on the Internet, has contributed to its expansion.

At present, the Internet uses a protocol known as **IPv4**, which is Version 4 of the Internet Protocol. Its packet-switching format is described in the previous section. One limitation is that the address field is only 32-bits long. This gives 2^{32} possible addresses, which is over 4 thousand million, but this is insufficient to meet demand. In writing the address of a computer on the Internet, the 32 bits are divided into 4 bytes of 8 bits each. These are then written out as 3-digit decimal numbers, each in the range 000 to 255, separated by full-stops, generally known as **dots**. For example, an address might look like this hypothetical example:

123.111.222.001

In the future, if IP Version 6 comes into operation, addresses will have 128 bits.

Transport Control Protocol is concerned with ensuring that a packet arrives at its destination free from error. If a package is lost or arrives in corrupted form, TCP returns a message to the sender asking for a repeat transmission. If a message in broken into several packets, which arrive out of order, TCP places them in the correct order.

TCP also controls the rate of data transmission. If too many packets are being lost, perhaps because of bad reception or overloading, the rate of transmission on the network is reduced to allow the system to cope better.

An address that is a string of 12 digits is not easy to remember. Keying in such an address is prone to error.

258

To overcome this problem and to help make the system more user-friendly, the Internet has introduced a system of **domain** addresses. Each computer on the Internet has, as well as its numerical address, an address in the following format:

newnespress.com

The lower level domain name is 'newnespress', which is owned by the publishers of this book. The upper level domain name is 'com'. This is three-lettered and indicates the nature of Newnes Press, which is a commercial body. Other upper level domain names are 'org' for a non-commercial organisation, 'ac' or 'edu' for an academic organisation such as a university, 'gov' for a governmental organisation, and 'net' for a network. Commercial sites located in certain countries (such as the United Kingdom) use the name 'co' instead of 'com'.

The domain name may also have a two-lettered third part, which indicates the country in which the computer is sited. For computers in the USA, this is generally omitted, but examples for other countries are 'uk' for United Kingdom, 'au' for Australia and 'jp' for Japan.

Sites that are connected on the World Wide Web have their domain names prefaced by 'www'.

As might be expected, there are organisations in each country supervising the issuing of domain names. A table of domain names is updated continually and held on the routing computer (or **routers**) in the network. Given any domain name, the router can use its **look-up table** to find the corresponding numerical address and thus begin the process of forwarding the packet to its destination.

At a router, packets are continually being received and checked using complex error-checking procedures. The packet is then sent on to the next router on its way to its destination. If the shortest route is heavily loaded, the router is able to select an alternative route for the packet

Questions on telecommunications

1 Match the stages of the diagram on p. 245 against the stages of transmission of information by: a smoke signal, semaphore, a video surveillance system in a supermarket, and any other named system of your choice.

2 Name some sources of information that make up the first stage of a telecommunications system. In what form(s) does the information exist?

3 Explain, giving examples, the functions of transducers in information systems.

4 What is modulation? Explain the difference between amplitude modulation and frequency modulation.

5 Compare the advantages and disadvantages of amplitude modulation and frequency modulation.

6 What is a sideband? In what way does the production of sidebands affect the bandwidth of a modulated carrier?

7 Describe how an analogue signal is transmitted and received using pulse width modulation.

8 Compare pulse width modulation with pulse position modulation.

9 Describe how a binary message is sent using frequency shift keying.

10 Explain what is meant by the amplitude, rise time and decay time of a signal pulse.

11 Describe the transfer of information over an RS232 cable with 9 wires.

12 The Association for Science Education gives its Internet address as:

www.ase.org.uk

Explain the meaning of the four parts of this address.

13 What are the meaning of these terms: protocol, router, packet switching.

14 Describe the structure and operation of a local telephone network.

15 What is frequency division multiplexing? What are its advantages?

16 A SSB voice signal requires a bandwidth of 3.4 kHz? If it is modulated on to an 80 kHz carrier, what frequencies does it contain?

17 What is the bit error ratio of a transmission? How is it measured?

18 What is jitter? Name one of its causes. How can it be reduced?

19 Draw diagrams to show the waveforms of these pulse signals: (a) return to zero, (b) non-return to zero, (c) non-return to zero mark.

20 What is the advantage of a RZ system over NRZ systems? What is its disadvantage?

21 Draw a diagram of the wave forn of this signal, encoded as NRZ mark: 001110110.

Multiple choice questions

1 An AM signal has a modulation depth of 20%. This means that:

 A the signal frequency may vary by as much as 20%.
 B the modulated carrier amplitude varies by as much as 20%.
 C the signal frequency is 20% of the carrier frequency.
 D the amplitude of the signal is 20% of the amplitude of the carrier.

2 A modulation system in which there are pulses of constant amplitude and length but with varying timing is known as:

 A PPM.
 B PWM.
 C PCM.
 D PDM.

3 One of the systems used for transmitting binary data is:

 A PCM.
 B PWM.
 C PDM.
 D PAM.

4 Of the systems listed below, the one most subject to interference from spikes and noise is:

 A FSK.
 B AM.
 C FM
 D PPM.

5 When a pulse signal oscillates before settling to its new level, the effect is called:

 A overshoot.
 B the time constant.
 C ringing.
 D oscillations.

6 A carrier with frequency 250 kHz is amplitude modulated with a signal of which the maximum frequency is 8 kHz. The bandwidth of the signal is:

 A 258 kHz.
 B 16 kHz.
 C 8 kHz.
 D 242 kHz.

7 A system operates with even parity. The seven bits of a byte are 0110101. When the parity bit is included the byte is:

 A 10110101.
 B 01101010.
 C 00110101.
 D 01101011.

8 What type of signal is shown by the drawing below?

 A NRZ.
 B RZ.
 C PWM.
 D FSK.

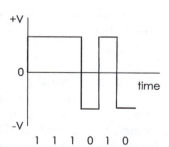

9 If a system has jitter:

 A the timing of the pulses is irregular.
 B the rise time of the pulses is too slow.
 C it has a high bit error ratio.
 D there is a fault in its codec.

29 Cable transmission

Telecommunication by cable is an extension of the passing of signals directly from the circuit to external conductors. In a wired intercom, for example, the sound signal is carried along a pair of wires from the master station to the remote station. When the master station is trasnmitting, the remote station is in essence just a speaker at the end of a long lead. The signal is unmodulated. The distance between transmitter and receiver is limited to a few hundred metres. Communication between a telephone in the home and the local exchange falls into the same category.

Most cable communications differ from the above in one or both of the following ways:

- The distance between transmitter and receiver is greater.

- The signal is modulated on to frequencies much higher than audio frequency.

These factors have important effects on transmission. To understand these we first look at the types of cable used and also examine some of the properties of transmission lines.

Types of cable

A wide range of cables is available, but the two types most commonly used are the twisted pair and coaxial cable.

The **twisted pair** consists of two insulated wires twisted around each other. Because the wires are unshielded (that is, they have no metallic sheath around them), there is capacitance between the two wires and between each of the wires and ground or any other nearby conductors. Because they are twisted around each other, the capacitances of each wire to ground or to external conductors is balanced. Any capacitive effects are common mode and have little effect on the signal. The cable is said to be **balanced**.

> **Common mode**
> Affecting both wires equally.

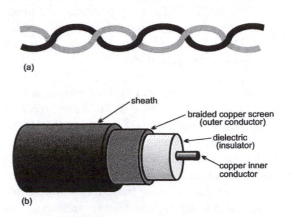

(a)

sheath

braided copper screen (outer conductor)

dielectric (insulator)

copper inner conductor

(b)

The two types of cable most often used are (a) the twisted pair, and (b) coaxial cable.

The induction of signals produced by external magnetic fields is minimised by twisting. In any circuit which contains one or more loops, an external magnetic field passing through each loop induces an e.m.f. in that loop. But in a twisted pair, the direction of induced e.m.f.s in the wires is reversed in adjacent loops (see below) so they cancel out.

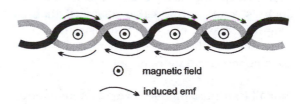

⊙　magnetic field

induced emf

The loops of the twisted pair are threaded by a magnetic field increasing in a direction upward through the paper of the page. By Fleming's Left-hand Rule, this induces an e.m.f. in a clockwise direction in each loop. Because of the twisting, these e.m.f.s cancel out in adjacent segments of the wire. Changes in direction and intensity of the field affect all loops equally.

Coaxial cable is widely used for carrying radio-frequency signals. The signal is carried in the inner conductor, which may be a single copper wire or may consist of a number of narrower copper strands. Stranded wire is used when flexibility is important. The inner conductor is insulated from the outer conductor by the dielectric, a layer of polythene or polyethylene.

The outer conductor is made of copper and is usually braided to give it flexibility. This is connected to ground so it serves to shield the inner conductor from external magnetic fields. It also prevents magnetic fields produced in the inner conductor fron spreading to other nearby circuits. Because there is no capacitance between the inner conductor and external conductors and, because the inner conductor is shielded from external magnetic fields, a coaxial cable is described as **unbalanced**.

Transmission lines

When a twisted pair or coaxial cable is being used for transmitting signals over a large distance, or the signal has high frequency, or both circumstances apply, we have to consider the properties of the medium in more detail. We regard this not as a mere length of conductor but as a **transmission line**. There is no sharp distinction between a simple lead, such as a speaker lead, and a transmission line, such as a submarine cable. It is just that, as length increases, or as transmission frequency increases, the distinctive properties of the line become more and more important.

Not all transmission lines extend for kilometres under the oceans. In microwave transmitters and receivers the frequency is so high (up to tens of gigahertz) that the wavelength of the signal in the conductor is only a few tens of millimetres. Even conductors joining one part of the circuit to another part need to be treated as transmission lines. At the very highest frequencies the mechanism of propagation of the signal becomes rather different. We use **waveguides** (box-like metal channelling or metal strips) instead of ordinary cable.

A tranmission line may be considered to be an electronic circuit in its own right. The diagram below represents a balanced transmission line by its equivalent electronic components. The resistors R represent the resistance of the copper conductors. In reality, the resistance is spread along the whole length of the wires, but here we represent it by individual resistors. The resistors are the **lumped equivalents** of the evenly distributed resistance of the wire.

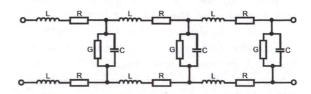

The lumped equivalent of a transmission line. In theory, the line consists of an infinite number of units like these three, spaced equally along the line.

If the cable is a short one carrying audio signals, resistance is the only property of the line that is important. Ohm's Law is all that is needed for analysing the circuit. With short runs of a suitable gauge of copper wire, even resistance may be ignored.

Inductance is another factor that adds to the impedance of the cable. Inductance is usually associated with coils, but even a straight wire has inductance (consider the wire to be part of a one-turn coil of very large radius). In the figure, this too is represented by its lumped equivalent, the two inductors.

Capacitance between the wires is represented by the capacitors C, joining the wires at regular intervals. The size of the capacitance depends on the spacing between the conductors and the material (dielectric) between them. Usually the dielectric is plastic, often PVC in the case of a twisted pair, or polythene in a coaxial cable.

Finally, there is always a small leakage current through the insulation or dielectric, represented by the conductances, G.

Using lumped equivalents reduces the transmission line to a relatively simple 'circuit', which can be analysed by the usual techniques. For instance, we know that the reactance of capacitors becomes very low at high frequencies and it is easy to see from the diagram that capacitance becomes more important than conductance. Conversely, the reactance of inductors becomes very high at high frequencies, so inductance becomes more important than resistance. As a result, conductance and resistance may be ignored at high frequencies.

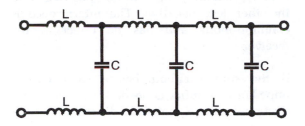

At high frequencies, resistance and conductance may be ignored, leaving only inductance and capacitance.

At high frequencies, the lumped equivalent circuit consists of only capacitance and inductance, as illustrated above. This arrangement of capacitance and inductance has the properties of a lowpass filter. Signals passing along the line lose relatively more of their high-frequency components. This has the effect of making pulses more rounded. They lose their sharp edges. They become more spread out. If this occurs, adjacent pulses may merge, causing errors in the signal at the far end of the line.

The dominance of capacitance and inductance at high frequencies leads to another effect. Inductance and capacitance store and release energy but they do not absorb it (p. 57). *In theory*, there is no loss of energy or power along a transmission line with a high-frequency signal. *In practice*, some of the energy escapes as electromagnetic radiation (in twisted pairs, less in coaxial cable), or as heat generated in the conductors when current passes through them.

Summing up, there is a fall in signal power along the line. This can be ignored in short lines, which we then consider to be **lossless**.

Characteristic impedance

The most important characteristic of a transmission line is its characteristic impedance, Z_0. This is defined as the ratio between the voltage and current (v/i) at the input of a line of infinite length.

If the impedance of the signal source is equal to Z_0, all of the power generated by the source is transferred to the line. Conversely, if there is an impedance Z_0 at the far end of the line, all the power of the signal is transferred to this. It follows that a lossless line transfers all the power from a source to the receiver. But, for this to happen, the source and receiver impedances must be **matched** (equal) to the line impedance.

If the impedance at the far end of the line is not equal to Z_0 (including when it is an open circuit, or when there is a short circuit) part or all of the energy of the signal is reflected back along the line. This causes standing waves to form in the conductors, and less power is transferred to the receiver. For efficient transmission of signals, this is to be avoided by careful matching of impedances.

The characteristic impedance of a cable at high frequencies depends on the inductance and capacitance per unit length of cable. It is independent of frequency, provided that frequency is high. In other words, the cable behaves like a pure resistance, which remains fixed, whatever the frequency.

In practice, cables are made with standard characteristic impedances. Transmission hardware has input or output impedances that match these standards. The most commonly available value of Z_0 for coaxial cable is 50 Ω, another common value being 75 Ω. Twisted pairs have standard impedances of 300 Ω or 600 Ω.

With lower frequencies, the impedance varies with frequency. The range of frequencies that the line can transmit without significant loss of power is called its **bandwidth**. The bandwidth extends from the lowest frequency to the highest frequency at which the output signal is less than 3 dB below the input signal. The graph below shows a simulation of a transmission line consisting of a twisted pair of copper wires with characteristic impedance of 600 Ω and length 200 m.

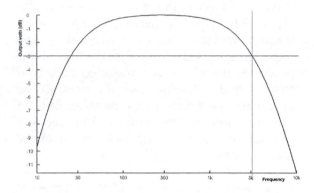

The pass band extends from 25 Hz to 3.0 kHz, The bandwidth is 3000 − 25 = 2975 Hz.

The amplitude of the signal across the load at the receiving end is plotted against frequency. Both are plotted on logarithmic scales, and the amplitude is scaled in decibels. The line at −3dB indicates the half-power level.

Signal strength rises above half power at 25 Hz and falls below it at 3 kHz. The bandwidth of the system is defined as the difference between these two frequencies, which is 2972 Hz.

Line driving

For analogue audio signals, a pair of centre-tapped transformers may couple the transmitter and receiver to the line, as illustrated below. The taps are grounded so that when one line of the twisted pair is positive of ground, the other is negative of ground by exactly the same amount. The line is **balanced**.

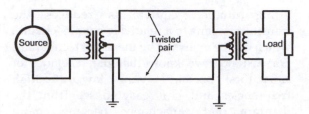

Source and load may be transformer-coupled to a balanced transmission line to minimise EMI.

This minimises electromagnetic interference because such interference will affect both lines equally, making one more positive and the other less negative. The interference is common-mode and has no effect at the receiver.

If the source circuit has a high output impedance, a buffer is needed between it and the transformer to match the source to the line.

Audio signals may be fed directly to the line using a **line driver**. A wide range of these is available as integrated circuits, which usually have several drivers and receivers in the same package. Each driver is a differential amplifier with low output impedance to match the impedance of the line and with high slew rate and high bandwidth. It has two outputs (see below) which are in antiphase for connection to the two lines of a balanced transmission line.

Self Test

What is the meaning of *slew rate*?

The receiver is a differential amplifier and the two wires of the transmission line are connected to its (+) and (−) inputs.

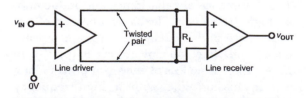

The essentials of using a line driver and receiver to send data along a twisted pair, or any other form of balanced line. The line is terminated in a load resistor equal to the characteristic impedance of the line.

The receiver has high slew rate, high bandwidth and high common-mode rejection. Although most types of driver and receiver are intended for use with balanced lines, they are also able to operate on unbalanced lines, in which one path is grounded and the signal passes along the other.

For digital and pulsed signals, the line driver is usually a logic gate, often NOT or NAND. TTL buffer gates can be used, as in the diagram below, though special line driver gates are preferred, because these have higher slew rates and higher bandwidth. CMOS drivers are used in low-power circuits.

The receiver is also a logic gate, sometimes a NOT gate. It may have Schmitt trigger inputs to 'square up' the distorted incoming pulse and to reduce the effects of noise on the line.

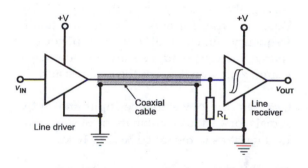

A basic circuit for using a line driver and receiver to send data along an unbalanced line. The line is terminated in a load resistor equal to the characteristic impedance of the line, usually 50 Ω or 75 Ω in the case of a coaxial cable.

A wide range of ICs is available for interfacing logic circuits to RS232 cables. The MC1488P, for example, contains four RS232 line drivers. Three of these function as NAND gates and the fourth as a NOT gate. The IC requires a power supply of up to +15 V and –15 V. It converts input from a TTL circuit into the equivalent RS232 levels. As these are inverting gates, the conversion includes inversion. Thus, an input of 0 V (TTL logic low) produces an output of +15 V, while an input of 5 V (TTL logic high) produces an output of –15 V.

The MC1489P has four NOT gates for use as line receivers. They have Schmitt trigger inputs to square up the incoming pulses. The IC operates on 5 V to 10 V and can accept inputs of up to ±30 V. Its outputs are the corresponding standard TTL levels of 0 V and 5 V.

Because most equipment connected by RS232 operates on TTL voltage levels, the more recent line driving and receiving ICs operate on +5 V and +3 V (many recent computers operate on this lower voltage). There is no need for the ±12 V (or higher) supply that is usually associated with RS232 systems. Instead, the IC contains a circuit unit that generates the higher voltages from the low supply voltage. This greatly simplifies construction of RS232 interfaces.

Crosstalk

When several lines are close together, the current passing along one line may cause current to flow in one or more adjacent lines. This may be the result of capacitance, between the two lines or the effect of inductance. The effect is known as **crosstalk**, and often occurs in multicored cables and in the ribbon cables used for connecting computers to peripherals. It may be eliminated in multicored cables by having each of the cores enclosed in a separate earthed shield, though this makes the cable considerably more expensive.

Peripheral
Equipment such as a monitor, keyboard, printer, modem or CD drive, which is connected to a computer.

Connecting alternate lines to ground helps to reduce crosstalk in ribbon cables. Crosstalk also occurs between adjacent tracks on circuit boards, particularly if the tracks run side-by-side for appreciable distances. It is avoided by running grounded tracks between the active ones.The size of the current induced by a magnetic field is proportional to the *rate of change* of the magnetic flux. This in turn is proportional to the rate of change of the current generating the field. Therefore, induced crosstalk is more serious at higher frequencies.

The effect is the same for capacitative crosstalk because capacitance decreases with frequency and high-frequency signals pass more readily across the capacitative barrier.

In data signalling at high rates, the pulses have very rapid rise-times, and these result in the most serious crosstalk problems. Spikes several volts high may be generated on an adjacent line at the beginning and end of each pulse. The effect can completely disrupt the signal on the adjacent line. With many types of line driver the rise time can be adjusted so that it is no more rapid than necessary, so minimising crosstalk.

Activity — Transmission lines

Set up a circuit in which a sinusoidal voltage source feeds a signal through a source resistor into a transmission line, and there is a load resistor at the far end.

The transmission line could be a 100 m length of twisted pair or coaxial cable. The resistor values are equal to the characteristic impedance of the line, which is usually given in the retailer's catalogue. This circuit may be breadboarded, or set up with simulation software.

Use an oscilloscope or the corresponding software feature to examine voltage signals at the input and output of the line. Measure the amplitudes of the signals and calculate their maximum power, using $P = V_{rms}^2 / R$.

Run the test with various load resistors in the range 10 Ω to 1 kΩ to find which resistor obtains the greatest amount of power from the line. Try this for sinusoidal signals for a range of frequencies from 1 MHz up to 20 MHz.

Here are some hints for setting up the tests. At high frequency, $Z_0 = \sqrt{(LC)}$. In this equation, L and C are the inductance and capacitance per metre. Typical values for a twisted pair are $L = 600$ nH and $C = 50$ pF. Applying the equation, these values make $Z_0 = 109.5$ Ω, independent of the length of the line.

If the simulator requires a value for R, a typical value is 0.17 Ω per metre. Obtain values for other types of line from data sheets.

To calculate the delay time taken for a signal to pass along the line, begin with the velocity of the signal in free space, which is 300×10^6 m/s. The velocity is less than this in a conductor, usually between 0.6 and 0.85 of the free space velocity. A typical delay for a 100 m line is 475 ns.

Replace the sinewave generator with a square-wave generator and investigate the effect of the line on the shape of the pulses at various frequencies, with matched and unmatched loads.

Activity — Transformer coupling

Set up the circuit on p. 260, using 1:1 line-isolating transformers. Apply an audio-frequency sinusoid at one end (through a source resistor) and use an oscilloscope to examine the signal across the load resistor.

Calculate the power of the input and output signals for various combinations of source and load resistors in the range 50 Ω to 10 kΩ.

This circuit can also be run as a simulation.

Activity — Line drivers and receivers

Use data and circuits from the manufacturers' data sheets to set up a 100 m balanced transmission line. For audio signals (sinusoids in the audio frequency range) use drivers such as the SSM2142 and receivers such as the SSM2143. Use an oscilloscope to investigate the transmisson of signals along the line.

Set up a 10 m unbalanced coaxial transmission line and use a pair of RS232 driver and receiver ICs to transmit digital data along the line at up to 20 kHz. The MC1488P and MC 1489P are suitable types. Examine pulse shape at the source and load.

Other pairs of drivers and receivers that can be used include the 75172 line driver and 75173 receiver, and the low-power CMOS pair, the 14C88 line driver and 14C89A receiver. Refer to the data sheets for circuit diagrams.

Questions on cable transmission

1 Compare the structure and properties of a twisted pair with coaxial cable.

2 Explain what is meant by the lumped equivalent of a line carrying a high-frequency signal.

3 What is meant by the characteristic impedance of a transmission line? Why is it important for the source and load to be matched to the line?

4 Explain why, in theory, there is no loss of signal along a transmission line.

5 Describe how a signal source and a load may be transformer-coupled to a balanced transmission line.

6 What is meant by crosstalk and how can it be reduced or avoided?

7 Explain how line driver and receiver ICs are used in the transmission of digital data.

8 Outline the main features and describe the action of an RS232 transmission line.

Multiple choice questions

1 At high frequencies the most important properties of a transmission line are:

    A resistance and capacitance.
    B capacitance and inductance.
    C resistance and conductance.
    D inductance and conductance.

2 When we say that the load on a transmission line is matched we mean that:

    A it has the same capacitance as the line.
    B the end of the line is short-circuited.
    C it has the same impedance as the line.
    D standing waves are formed along the line.

3 A balanced transmission line is one in which:

    A one conductor is grounded.
    B both conductors carry the same signal.
    C the line is shielded against EMI.
    D the signals on the conductors are the inverse of each other.

Calculating Z_0

The Companion Site has a Calculator for working out the characteristic impedance.

On the Web

Refer to manufacturers' or retailers' data sheets to find out more about the latest types of line drivers and receivers.

30 Optical transmission

Optical fibre

When a ray of light meets the boundary between a medium (such as air) and another transparent medium (such as glass), which is optically denser, it may be partly reflected and partly refracted. Ignoring the part that is reflected, the diagram below shows what happens when the ray passes from air to glass. The ray is bent or refracted *toward* the normal.

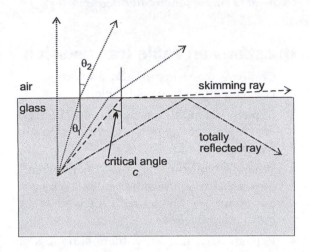

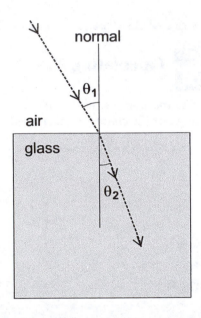

When a ray of light passes from one transparent medium to another, it is refracted. Note that in this and later diagrams the grey tone represents the fact that the glass has a higher refractive index than the air, NOT that the glass is tinted. In fact, the glass used in optical fibres is exceptionally clear.

The drawing on the right shows what happens when the ray passes from the medium with higher n (glass) to the medium with lower n (air). All depends on the angle (θ_1) at which the ray strikes the boundary.

> **n**
> The refractive index of a medium such as air and glass.

If a ray of light is travelling in glass and meets a glass-air boundary at an angle equal to or greater than the critical angle, the ray is totally reflected back into the glass.

If it strikes perpendicularly, it passes straight out of the glass into the air. If it strikes at a small angle to the normal, part of it passes out and is refracted away from the normal. As the angle θ_1 is increased, a point is reached at which the refracted ray *just* skims along the surface ($\theta_2 = 90°$). If θ_1 is increased a little more, the ray can not escape from the glass. The whole of it is reflected. This is called **total reflection**. The value of θ_1 at which this happens is the **critical angle**, c. For a typical glass-air surface, c is about 42°. Total reflection occurs whenever θ_1 is equal to or greater than c.

> **Normal**
> A line perpendicular to the boundary at the point where the ray strikes it.

The angles at which these effects occur depend on the wavelength (colour) of the light. This causes a beam of ordinary white light to spread into its component colours. The diagrams assume that the light is monochromatic so the ray remains narrow.

> **Monochromatic**
> Light of a single wavelength.

In the diagram below, light is passing along inside a narrow cylinder of glass. When it strikes the glass-air surface it is not able to emerge into the air. This is because θ_1 is greater than c. The ray is totally reflected. It passes along inside the cylinder and may be totally reflected many more times before it emerges from the other end. This is the basis of the **optical fibre**. Even if the fibre is curved, as in the drawing, the light travels from one end to the other without emerging.

A ray of light travelling in a glass fibre is totally reflected when it strikes the surface of the glass. If the fibre has slight bends in it, the ray is conducted along the fibre without loss.

It may sometimes happen that a fibre is too sharply curved. The θ_1 is less than c and the light emerges. Light also emerges if there are scratches in the surface of the glass, producing local regions where θ_1 is less than c.

A practical fibre optic cable consists of a narrow fibre of glass, the **core**, surrounded by a layer of glass of lower optical density, the **cladding**.

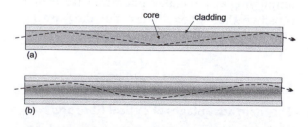

Optical fibre usually has (a) a core of high refractive index surrounded by a cladding of low refractive index, or (b) the refractive index decreases gradually from the centre out. In (a) the ray is totally reflected every time it strikes the boundary. In (b) the ray is gradually redirected back to the centre of the core whenever it deviates into the outer region.

Note that the reflecting surface is the inner surface of the cladding, so it is protected from becoming scratched.

In **graded index** fibre, the refractive index decreases gradually from the centre of the core outward. The ray is bent gradually as it enters the outer regions and is gradually returned to the central region. Its path along the fibre is smoothly curved instead of being a sharp zigzag. This type of cable produces less distortion of light pulses and is better for long distance communication.

The glass used for fibre optic cable is of very high purity and clarity, prepared in a clean environment. The loss of signal due to absorption in the glass is about 0.15 dB (3.5%) per kilometre. Cheaper cables made from polymer are used for short-distance runs (up to 100 m). They have a higher absorption rate. Fibres of either type are enclosed in a tough plastic sheath for protection.

Light sources

The light signal is provided by a light-emitting diode or a solid-state laser, both of which produce monochromatic light. The source is enclosed in a metal housing which clamps it firmly against the highly polished end of the fibre. LEDs are inexpensive and reliable but their light output is low. The light output is proportional to the current. The intensity (amplitude) of the light signal is easily modulated and this is the most commonly used modulation system for LEDs.

Solid-state lasers are more expensive than LEDs and require more complicated circuits to drive them, but their power is greater. Their output may be modulated in intensity, frequency or phase, and also by the angle of polarisation. Like LEDs, they are mounted in a metal housing.

At present most of the LEDs and lasers operate in the infrared region but, with further research, operation is being extended into the visible spectrum.

Because of the very high frequency of light waves, it is possible to modulate the light signal at very high rates, measured in terahertz (1 THz = 10^{12} Hz). Consequently, the bandwidth of optical channels is around 100 times wider than that of radio channels. Such a wide frequency range provides enormous bandwidth for stacking simultaneous transmissions on a single channel. There is room to carry 100 000 voice channels over a single optical fibre channel. This is one reason why optical fibre is becoming more and more widely used for telecommunications.

Light receivers

The devices used for detecting the signal and converting it to an electrical signal need to have very high sensitivity and the ability to operate at high frequencies. The most commonly used receivers are avalanche photodiodes, PIN photodiodes and phototransistors.

Avalanche photodiodes (APDs) are operated with reverse bias. When stimulated by light, a few electron-hole pairs are generated. These move in the electric field, striking other atoms in the lattice and generating more electron-hole pairs. The effect is similar to that in an avalanche Zener diode. The result is a rapid increase in the number of electron-hole pairs and a rapid increase in the current passing through the diode. The response time is about 200 ps, so the device is suited to high-frequency operation.

PIN photodiodes are so-called because they have a layer of intrinsic (i-type) material between the n-type and p-type material. This widens the gap between the n-type and p-type so that the capacitance between them is reduced. With reduced capacitance, the PIN diodes are more suited to operate at high frequencies. The wide gap also increases the absorption of light at the junction, so increasing sensitivity.

Intrinsic material

Pure, undoped semiconductor, with low conductivity.

Self Test

What charge carriers are present in the depletion region of a reverse-biased diode?

PIN diodes are reverse-biased in operation, and light causes the generation of electron-hole pairs. In a circuit such as that below, the increased current passes through the resistor and produces an increased voltage across it. This is amplified to produce the electrical signal.

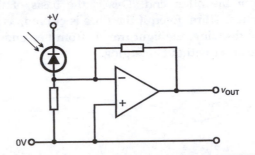

A basic circuit for converting an AM optical signal into an AM electrical signal.

A **phototransistor** is enclosed in a transparent package or in a metal can with a transparent lens-like cap. It is connected into a circuit in the same way as an ordinary npn transistor, except that it is not necessary to make any connection to the base. When light falls on the transistor it generates electron-hole pairs in the base layer. A current then flows from base to emitter, which is equivalent to the base current of an ordinary BJT.

Because of the transistor action, a much larger current flows from collector to emitter. Their amplifying action makes phototransistors more sensitive than photodiodes, but they are less able to operate at high frequencies.

Advantages of optical fibre

The main advantages of optical fibre are:

* **Bandwidth**: optical systems have a higher bandwidth than any other system of telecommunication. Many more signals can be frequency multiplexed or time multiplexed on to a single channel. This allows the development and implementation of highly complex stacking systems.

- **Cable cost**: the cost of optical fibre is greater than that of copper cable, but it is falling. Offset this against the fact that, on long distances, repeater stations can be up to 160 km apart. Often they are much closer than this because the power for the repeater unit is supplied along a pair of copper wires running in the cable alongside the fibres. This limits the distance between repeaters to 10 km. Local power supplies are used if the distance between repeaters is greater than this. Even so, optical fibres give longer runs compared with copper cables, which need repeaters every few kilometres.

- **Immunity from EMI**: unlike copper cable, optical fibre is immune from the effects of external electromagnetic fields. It is particularly suited to industrial applications where electrical machinery and heavy switching generate strong interference. There is no crosstalk between fibres bundled in the same cable.

- **Safety**: faults in copper cabling may generate excess heat which may lead to fire. This is much less likely to occur with optical fibre.

- **Security**: surveillance equipment can be used to read messages being transmitted on copper cables. This is not possible with optical fibre.

- **Corrosion**: unlike copper cable, optical fibre does not corrode.

Questions on optical transmission

1 Describe what happens when a ray of light travelling in air meets an air-to-glass surface.

2 Describe what is meant by critical angle.

3 What is graduated index optical fibre and how does it work?

4 Describe how an LED is used as the light source for optical fibre.

5 Describe the action of an avalanche photo-diode as an optical fibre receiver.

6 What are the advantages of optical fibre as a medium of telecommunication?

Multiple choice questions

1 Total reflection occurs when a ray of light:

    A passes from air to glass
    B is refracted away from the normal
    C strikes the surface at an angle less than the critical angle
    D strikes the surface at an angle greater than the critical angle.

2 It is important not to bend an optical fibre too sharply because:

    A it may break
    B it alters the refractive index
    C light may escape at the bend
    D the sheathing plastic may split.

3 The type of receiver most suited for high-frequency reception from an optical fibre is:

    A a PIN photodiode
    B an avalanche photodiode
    C a laser diode
    D a phototransistor.

On the Web

Find out more about optical fibre, light sources and receivers. Details to look for include transmission frequencies, and bandwidths. Try keywords and phrases such as optical fibre, infrared, and infrared laser.

31 Radio transmission

The electromagnetic spectrum

When an electron is accelerated, it radiates energy in the form of **electromagnetic waves**. We can think of the waves spreading through space like ripples on a pond, but in three dimensions. The wave fronts are spherical and continuously expanding.

Electromagnetic waves include (in order of increasing wavelength): gamma rays, X-rays, ultraviolet, visible light, infrared, microwaves, and radio waves. All are the same type of radiation and have similar properties. For example, they all travel through free space with the same velocity, 300×10^6 m/s. Their wavelength (λ) is related to frequency (f) by the equation:

$$\lambda = \frac{3 \times 10^8}{f}$$

They can be reflected, refracted, and diffracted, for example, and they can be absorbed by the medium through which they are travelling.

There are differences. For example, gamma rays and X-rays can penetrate dense matter such as metals, but radiation of longer wavelength can not. These radiations and also ultraviolet are ionising radiations. They cause electrons to be knocked free of the atoms of materials as the radiation passes through. They can also be very damaging, possibly lethal, to living tissues. They produce the ionised layers of the upper atmosphere that are so important in radio transmission.

Radio waves have the longest wavelengths of all types of electromagnetic radiation. The shortest wavelengths (100 mm to 1 mm) are found in the Extremely High Frequency band (30 GHz to 300 GHz). The longest wavelengths (1000 km to 10 000 km) are in the Extremely Low frequency band (30 Hz to 300 Hz). All are used for telecommunications of various kinds.

The two other wavebands used in telecommunications are visible light and infrared. They are used in optical fibre systems. It might be mentioned that, in the days before electronics, visible light was used for communicating by visual signals. Techniques included smoke signals, beacons, semaphore, heliograph and signal flags. All techniques were line of sight transmissions.

Among the many systems using electromagnetic radiation, we list a few examples:

TV remote controllers: The source is an infrared LED radiating pulse coded signals.

Car engine immobilisers: Emit radio signal.

Remote control for toy vehicles and robots uses pulsed infrared or radio. The 315 MHz and 433 MHz frequencies are available for use by hobbyists to use, unlicensed.

Optical mice (for computer control) use LEDs to transmit 1500 images per second of the surface below the mouse. Digital signal processors in the computer interpret the change in the images as the mouse is moved.

RFID radio-frequency tags are passive devices that can be attached to almost anything, and even implanted in the human body. When the tag detects the magnetic field generated by the receiver, it emits an identifying radio signal that the receiver can interpret. In a security system it identifies a wafer card, allowing authorised holders to pass through. In a stock room it identfies how many of each kind of stocks are present and sends this information to a computer for making an inventory, At the entrance to a freeway it can identify each car as it passes by and automatically charge the freeway fee to the owner's account.

The satellites of the global positioning system use the 3990-1550 MHz UHF band. This band is also used for communicating with other satellites.

Description	Abbreviation	Frequency range	Wavelength range	Propagation method	Applications
Gamma rays	γ-ray	$> 1 \times 10^{20}$ Hz	$< 1 \times 10^{-12}$ m	Penetrative Ionising	Nuclear physics
X-rays	HX and SX	$> 3 \times 10^{16}$ Hz	< 10 nm	Penetrative Ionising	X-ray astronomy
Ultraviolet	UV	750 THz - 3×10^{16} Hz	400 - 10 nm	Ionising	
Visible light		400 - 750 THz	750 - 400 nm	Line of sight Optical fibre	Telecomms
Infrared	IR	300 GHz - 400 THz	1 mm - 750 nm	Line of sight Optical fibre	Telecomms
Extremely high frequency	EHF	30 - 300 GHz	10 - 1 mm	Line of sight	Microwave links, experimental, amateurs
Super high frequency	SHF	3 - 30 GHz	100 - 10 mm	Line of sight	Microwave links, radar, amateurs
Ultra high frequency	UHF	300 MHz - 3GHz	10 - 1 mm	Line of sight Skywaves	TV, telecomms, radar, mobile phones, citizen band
Very high frequency	VHF	30 - 300 MHz	1 m - 10 mm	Line of sight Skywaves	Telecomms, TV, FM radio, amateurs, control
High frequency	HF	3 - 30 MHz	100 - 10 mm	Line of sight Sky waves	Telecomms
Medium frequency	MF	300 kHz - 3 MHz	1 km - 100 m	Ground waves	Telemetry. control, telecomms
Low frequency	LF	30 - 300 kHz	10 - 1 km	Ground waves	Telecomms
Very low frequency	VLF	3 - 30 kHz	100 - 10 km	Ground waves	Experimental, telecomms
Voice frequency	VF	300 Hz - 3 kHz	1000 - 100 km	Ground waves	AM radio, marine, aircraft
Extremely low frequency	ELF	30 - 300 Hz	10000 - 1000 km	Ground waves	

The electromagnetic spectrum. Radio wavebands on light grey background. Frequencies are allocated by national and international organisations. Some of the applications are allocated only a small range of frequencies within the bands. For instance, the Global Positioning System uses only two frequencies within the L-Band (390-1550 MHz) which is aasub-band of the UHF band.

Broadcast radio

Many kinds of radio transmitter have an antenna that radiates the waves in all directions with more-or-less equal strength. It is **broadcast** to everyone within range. This is a feature of radio communication that other forms of telecommunication do not normally have. A radio signal spreads through free space. We can communicate with many people at once, as when politicians or disc jockeys speaks to their audiences by radio.

Another benefit of radio is that we can broadcast messages to particular individuals without having to know exactly where they are. Ship-to-shore radio, mobile telephones and paging systems rely on radio to convey messages to receivers that are not in a known location.

Ground waves

A signal that is broadcast in all directions can travel from the transmitter to the receiver in two ways. If the frequency is less than 500 kHz, it spreads close to the surface of the Earth as a **ground wave** (see overleaf). It is not able to pass out of the atmosphere because of the ionised layers referred to earlier. These have a lower refractive index than the atmosphere, so the radio waves of lower frequency are reflected back toward the Earth's surface.

Ground waves of lower frequency are more affected by the nature of the surface, particularly when it consists of moist soil or water. The wave fronts nearest the surface are slightly held back so that, instead of being vertical, they are tilted forward. The result of this is that the waves follow round the curvature of the Earth. VLF transmissions can travel for thousands of kilometres in this way and can reach almost any part of the world.

Unfortunately, the surface of the Earth is not completely smooth. Ranges of hills can block the path of ground waves so that the signal does not reach the area just beyond the hills. This limits the range of transmission of ground waves in many terrains. However, if the range has a sharp edge, the waves may be diffracted and then stations immediately behind the hills are able to receive the signals.

Sky waves

As frequency increases above 500 kHz the radio waves are able to penetrate the lower ionised layers. As they pass up through the layers the refractive index is decreasing. The effect is similar to that of light passing along a graded index optical fibre. The waves are refracted along a curved path and eventually directed back to Earth again.

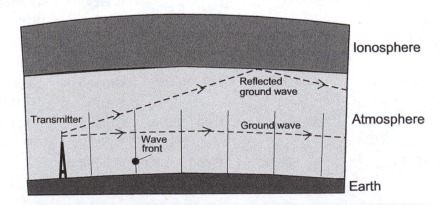

Ground waves travel in a direction more or less parallel to the earth. Although low frequency ground waves can encircle the Earth, those of higher frequency have relatively short range and are often blocked by high ground. This drawing is not to scale.

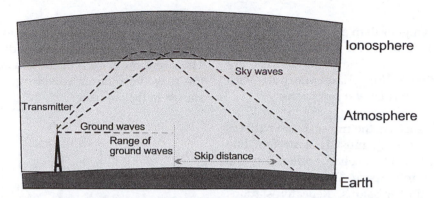

Radio waves with frequencies between about 500 kHz and 25 MHz are refracted back to Earth in the ionosphere, because of the lower refractive index of the ionised layers. They then become sky waves, which are important for long-distance communication. This drawing is not to scale.

These are known as **sky waves**. They reach the Earth at a considerable distance from the transmitter and are important for long-distance communication. The main drawback is that reception of sky waves varies in strength. This is because it is the ultraviolet radiation from the Sun that produces the ionised layers. The layers are not permanent, varying with the time of day, the season, and the number of sunspots.

At a certain distance from the transmitter there may be a point that is too far away for the ground waves to reach, but is too near for the sky waves to reach. The signal is not received at this distance, which is known as the **skip distance.**

Because of the curvature of the Earth and the ionised layers, some of the sky waves skim past the Earth's surface and once more head up toward the ionised layers. Again, they are refracted back toward the surface. In this way, the signal from a sufficiently powerful transmitter may hop around the world.

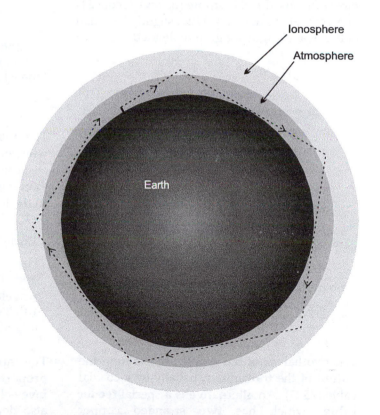

In good conditions, sky waves may be repeatedly refracted in the ionosphere and circulate round the Earth several times. This produces annoying echoes, repeated at intervals of about 1/7 second. This drawing is not to scale.

At frequencies higher than about 25 MHz, radio waves are able to penetrate the ionosphere and pass out into Space. These frequencies are therefore suitable for communication with spacecraft and interplanetary missions.

On the earth's surface, the transmission path for high-frequency signals must be a straight line from transmitter to receiver, otherwise the signals are lost into Space. These are known as **line-of-sight** transmissions. Repeater stations are located on high ground to relay the transmission to areas out of sight of the main transmitter.

Antennas

There are many types of antenna, but the one most often used is the **dipole** (below). It consists of a vertical metal rod or wire, slightly less than half a wavelength long, and divided into two sections.

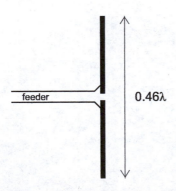

The motion of the electons in the standing waves of a half-wave dipole antenna generates radio waves.

The two halves of the dipole are fed from the output of the transmitter, usually by a coaxial cable (75 Ω). An alternative is a special **feeder cable**, which has two stranded copper conductors running parallel with each other and spaced about 9 mm apart. The halves of the dipole are insulated from each other, so it acts as an open circuit at the end of a transmission line. Standing waves appear in the dipole and in the connecting cable.

An antinode of the voltage waves is located at the ends of the dipole and a node at its centre. The situation in each of the rods is like that in the terminal quarter wavelength of the cable shown in (b) opposite.

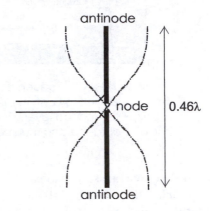

A standing wave approximately λ/4 long is developed in each half of a dipole, so that the whole dipole is approximately λ/2 long.

The total length of the dipole would be half a wavelength. However, the velocity of electromagnetic waves in the conductor is less than it would be in free space, so the total length is 0.46λ, not 0.5λ.

Example

For a UHF transmitter operating at 554 MHz, the wavelength is λ = 3 × $10^8/554\times 10^6$ = 0.54 m. The dipole length is 0.54 × 0.46 = 25 cm.

Self Test

What is the length of a dipole suitable for use with a VHF transmitter operating at 205 MHz.

The input impedance of a dipole antenna is proportional to the length of the rod and inversely proportional to the wavelength. It is also dependent on other factors. This means that it is possible to design an antenna so that its input impedance matches the characteristic impedance of the feeder. A typical dipole antenna has an input impedance of 75 Ω and needs a feeder cable of the same impedance for maximum power transfer.

Standing waves

With high-frequency transmissions, the cable is several wavelengths long. A succession of waves enters the cable from the source and is carried to the far end. What happens there depends on what is connected there between the two conductors:

Matched load circuit (having an impedance equal to Z_0): All of the power of the signal is transferred to the load.

Short circuit: No power is dissipated in this so the power is returned to the cable in the form of waves travelling back toward the source. Where the two conductors are short-circuited, they must be at equal voltage. The amplitude of the wave is zero at this point. Interference between the forward and returning waves sets up a pattern of standing waves as shown at (a). There are alternating nodes (zero amplitude) and antinodes (maximum amplitude). The nodes are spaced $\lambda/2$ apart, with an antinode at the source end (representing the alternating voltage signal supplied by the source) and a node at the far end (because it is short-circuited).

Open circuit: As with a short circuit, no power is dissipated at the far end, so the waves are reflected back along the cable (b). The far end is a point of maximum voltage change, so an antinode occurs at this point. As with a short circuit, there is an antinode at the source end.

What happens when a signal is sent along a transmission line depends on the loading at the far end. If this is (a) a short circuit or (b) an open circuit, a pattern of standing waves is set up along the line.

Unmatched receiving circuit (having an impedance that is not equal to Z_0): Some of the power is transferred to the load, but some is reflected back along the cable. This produces standing waves of smaller amplitude.

The dipole is an example of a **standing wave antenna**, in contrast to some other types, which are **travelling wave antennas**. The formation of standing waves means that the dipole is resonating at the frequency of the carrier wave, and radiating part of the energy fed to it as electromagnetic radiation.

At a receiver, a dipole antenna intercepts arriving electromagnetic radiation, which induces currents in it.

If the frequency is correct, the antenna resonates and standing waves are formed in it.

The antenna receives signals at its resonant frequency much more strongly than those of other frequencies. This helps to make the receiver more **selective**. Although a dipole transmits and receives best at its resonant frequency, it is also able to operate reasonably well at frequencies that are within ±10% of this.

Directional transmissions

The vertical dipole radiates equally strongly in all horizontal directions, so is suitable for broadcasting. But it is wasteful of power to broadcast a signal in all directions when it is intended for only one particular receiver.

The dipole may be made directional by adding further elements to it. The drawing opposite shows a dipole with two **parasitic elements**, known as the **director** and the **reflector**. These have about the same lengths as the dipole elements. In an antenna being used on a transmitter, the parasitic elements pick up the signal from the dipole, resonate at the signal frequency, and re-radiate the signal.

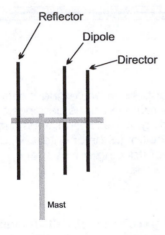

Reflector
Dipole
Director
Mast

Addition of a reflector and director makes the dipole antenna directional and increases its gain.

They act as two signal sources, and are spaced certain distances apart so that they reinforce the signal being radiated from the dipole. The signal is stronger than that radiated from the simple dipole, and is concentrated in one direction.

The array may have further directors added to it to increase its gain and make it even more directional. Multiple directional arrays, which may have up to 18 elements are known as **Yagi arrays**.

As a receiving antenna, a dipole with parasitic elements is used for picking up signals from weak or distant stations. A common example is the antenna used in fringe TV reception areas. The electromagnetic field causes electrons in the elements to oscillate to and fro. If the frequency of the incoming transmission is correct, the elements all resonate to the incoming signal and re-radiate it. At the dipole, their radiations are added to the incoming signal, so increasing the strength of the signal passing to the receiving circuit. Standing waves are set up in the dipole and the parasitic elements and some of their energy passes to the receiving circuits.

The use of directional dipoles makes it possible to set up microwave links with repeater stations at intervals. Theoretically, the stations at either end of a link should be able 'see' each other, for this is line-of-sight transmission. The range depends mainly on terrain. Repeater stations are generally located on hill-tops or on mountain-tops when available.

Other local conditions may extend the range slightly above line-of-sight range. In practice, the maximum range is about 50 km, though it may be as far as 100 km in some localities. The number of repeater stations required on a long route is far fewer than for cable or optical fibre links.

The use of a resonant standing wave antenna is important in many types of radio communication. It promotes reception of weak signals from distant transmitters or, conversely, allows us to use transmitters of lower power.

In some applications, antennas half a wavelength long are too long for convenience. For example, for radio pagers in the VHF band the wavelength is at least 1 m, so a half-wave dipole would have to be at least 50 cm long. This is not convenient for a device that has to be carried in a jacket pocket. A much shorter antenna is used is such equipment, but tuning components are added to the receiver circuit to produce the equivalent of a resonating antenna.

Radio waves are reflected by a metallic surface, but the surface does not have to be a continuous sheet. Perforated metal or a wire mesh may be used for cheapness and lightness. The only requirement for reflection to occur is that the dimensions of the reflector or dish must be several times greater than the wavelength.

Transmissions are most directional when the antenna is mounted in front of a reflector which is parabolic in section. If the antenna is placed at the focus of the reflector, all radiation incident on the reflector is reflected as a parallel-sided beam.

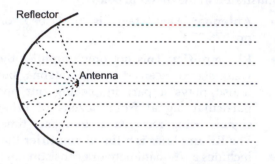

For frequencies in the megahertz bands and above, a parabolic dish concentrates the energy from the antenna into a narrow beam. Conversely, a dish may concentrate a weak signal on to the antenna to enable a distant transmitter to be heard.

Conversely, all rays arriving from a distant transmitter, being parallel with each other, are concentrated on an antenna placed at the focus. Radio-telescopes have very large dishes and similar dishes are found at the ground stations of satellite communications systems. Smaller reflectors are used for line-of-sight communications links and for receiving TV programmes from satellite transmitters at home.

Transmitting circuits

The main stages of a typical radio transmitter are illustrated in the drawings opposite. It normally comprises the following stages:

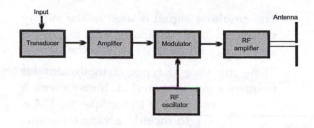

A block diagram of a rado transmitter. The modulator mixes the signal with the oscillator output to produce a modulated carrier wave.

- **Input**: This is the source of information that is to be transmitted. Examples are listed under 'Sources of information' on p. 246.

- **Transducer**: Examples are listed under this heading on p. 246.

- **Amplifier:** The signal from the transducer may have insufficient amplitude, so it needs amplification at this stage. Usually this stage is an audio frequency amplifier, but an amplifier with wider bandwidth is needed for video transmissions and for high speed digital data links. The diagram does not show an encoder, but there might be one at this stage.

- **Modulator:** This takes the amplified signal and modulates it on to the output of the radio frequency oscillator. In an amplitude modulated (AM) transmitter, the modulator often consists of a special multiplying IC. This has two inputs. One input (the carrier) is multiplied by a varying amount determined by the instantaneous voltage at the other input (the signal). The output of the IC is the amplitude-modulated carrier. In a simpler system of amplitude modulation there is no modulator as such, as shown below.

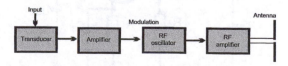

In simpler designs of transmitter the signal from the amplifier acts directly on the oscillator circuit to modulate its output.

The amplifier signal is used as the supply voltage of the RF oscillator. This makes the amplitude of the carrier vary as the signal.

There are several types of modulator for frequency modulation (FM) transmitters. A simple technnique for FM is to include a varactor in the tuned network of the RF oscillator. The signal is applied to the varactor so as to vary the capacitance of the network. This makes the resonant frequency of the network vary according to the signal. The output from the network is the frequency-modulated carrier.

> **Varactor**
>
> A diode-like device the capacitance of which depends on the voltage across it. Also known as a **varicap diode**.

- **Oscillator**: This is based on an amplifier but includes a resonant network (p. 366) or a crystal to set the frequency. Feedback from the amplifier provides energy to keep the network or crystal oscillating.

- **RF amplifier**: An RF amplifier is used to increase the output power of the transmitter. The drawing below shows a common-emitter BJT amplifier (p. 72) modified as an RF amplifier. The transistor specified has a good response at high frequencies. The input and output capacitors are only 1 nF so that they can readily pass high frequencies. There is an inductor instead of the usual collector resistor.

The inductor has high impedance at high frequencies so a large voltage develops across it, boosting gain. Therefore, the RF amplifier has only a narrow bandwidth, centered on the carrier frequency and extending on one or both sides to include the sidebands.

- **Antenna**: This stage is described on pp. 276-279.

Receivers

In many ways a receiver is like a transmitter in reverse, but there are differences. The stages of a **tuned radio frequency receiver** (TRF) are illustrated in the diagram below:

- **Antenna**: This stage is described on pp. 276-279.

- **RF amplifier**: This not only amplifies the weak signal received by the antenna, but it also plays a part in picking out one particular signal from dozens of other signals being received at the same time. The RF amplifier is a **tuned amplifier** that includes a resonant network consisting of a capacitor and inductor. It amplifies the frequency to which it is tuned and the sidebands of the transmission, but not lower or higher frequencies. Usually the signal passes through 2 or 3 such tuned amplifiers as this gives a better separation between the wanted and unwanted signals.

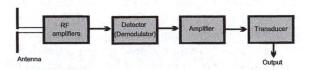

A block diagram of a TRF radio receiver. The RF amplifier is tunable to pick up transmissions of a particular carrier frequency.

- **Detector**: This stage demodulates the signal but, in radio receivers is usually given the name 'detector'.

Compare this RF amplifier with the common-emitter amplifier on p. 72.

V_{CC} 15 V

R_1 18k Ω

L_1 5m

C_2 1nF

C_1 1nF

Q1 2N2222A

V_{OUT}

V_{IN}

R_2 12kΩ

R_3 1kΩ

C_3

0V

In an AM system the detector is essentially a rectifier diode, followed by a lowpass filter. The diagram below illustrates a simple detector circuit.

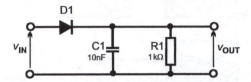

A simple diode detector circuit demodulates AM transmissions. Because of its low forward voltage drop, a germanium diode is generally used in this circuit.

Below is a demonstration of its action. The modulated carrier (grey) is rectified by the diode. Only the positive-going half of the signal can pass through it. C1 and R1 then filter this signal. The output of the circuit (black) shows that the low-frequency component (the modulating signal) passes through the filter, but the high-frequency component (the carrier) is almost eliminated.

- **Amplifier**: In the block diagram on p. 229, the stages we have described so far are all represented by the 'radio tuner' box.

  The demodulated signal may need amplification. An amplifier is used that has a suitable bandwidth. For example, an audio frequency amplifying system is used for speech and music.

- **Transducer**: In an audio system, this is a speaker, but other transducers may be used as listed in the table on p. 246.

- **Output**: This is the recipient of the information that has been received. Examples are listed under 'Recipients of information' in the table on p. 246.

The block diagrams of the transmitter and receiver do not include encoders or decoders (codecs), which form part of certain digital radio communications systems.

Superhet receivers

The **superheterodyne** (or '**superhet**') receiver has a performance that is superior to that of the TRF receiver. Its main feature is that it is much more selective than the TRF receiver.

Selectivity

The ability to pick out a transmission on one particular frequency from a number of other transmissions on slightly different frequencies.

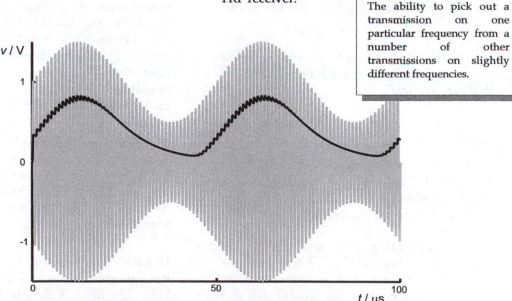

In this simulation of the action of the diode detector the signal from the RF amplifier (grey) is a 1 MHz carrier, amplitude modulated at 20 kHz. The demodulated signal is in black.

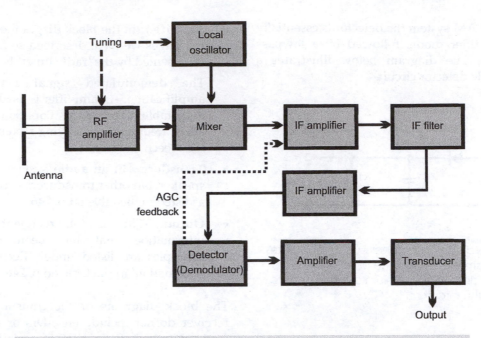

The superhet receiver converts the carrier to an intermediate frequency so that it can be processed by stages designed to operate best at that one frequency.

One of the problems with the TRF receiver is that 2 or 3 tuned RF amplifiers are needed to provide good selectivity. It is difficult to design and control a series of RF amplifiers that will all tune to precisely the same frequency over the whole range of frequencies that the set can receive.

This problem does not arise with the superhet because there is only one RF tuned amplifier stage (see overleaf).

The operation of the superhet receiver is as follows:

- **Antenna**: This stage is described on pp. 276-279.
- **RF amplifier**: A single tuner RF amplifier. Its output is the modulated carrier wave signal.
- **Local oscillator**: This is a tunable RF oscillator. The RF amplifier and the oscillator are tuned at the same time, so that the frequency of the tunable oscillator is always a *fixed* amount above that of the amplifier. In many superhets the difference is 455 kHz.

- **Mixer**: This combines the signals from the amplifier and oscillator. They produce **beats** (see box, opposite) at their frequency difference. This is known as the **intermediate frequency** (IF) and is often 455 kHz. This becomes the new carrier frequency, and has the signal amplitude modulated on to it, just as it was on the signal from the RF amplifier.

> **Heterodyne**
> Producing a lower frequency from the combination of two almost equal higher frequencies.

- **IF amplifier**: This amplifies the signal. Because the intermediate frequency is 455 kHz, whatever the frequency of the original transmission, this amplifier is designed to give its best performance (maximum gain, minimum distortion) at 455 kHz.

- **IF filter**: This is a bandpass filter with a very narrow pass-band centred on 455 kHz. The filter is designed very exactly with sharp cut-off on either side of 455 kHz, so as to cut out signals on nearby frequencies. The narrowness of the

- **IF amplifier**: Further amplification with an amplifier designed to work best at 455 kHz.

- **Detector, amplifier, transducer and output**: As for the TRF receiver, but there is negative feedback from the detector to the first IF amplifier. If the detected signal is weak, the feedback increases the gain of the IF amplifier. This **automatic gain control** compensates for variations in signal strength, either when receiving from a weak station or when the signal from a strong station fades temporarily.

Self test

If a superhet is receiving a station at 750 kHz, and the IF is 455 kHz, what is the frequency of (a) the local oscillator, and (b) the signal passing through the IF filter?

Beats — Extension Box 22

When two signals of almost equal frequency and amplitude are mixed together, the resulting signal changes regularly in amplitude. We say that it **beats**. The frequency of the beats equals the difference in frequencies of the two signals.

The upper drawing below shows two signals with equal amplitudes and with frequencies of 200 Hz and 220 Hz. When these are mixed we obtain the signal in the lower drawing, which beats at 20 Hz.

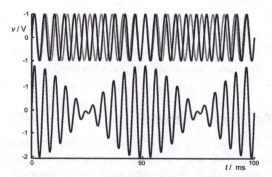

Beats occur with sound waves too. For instance, if two motors are running *almost but not exactly* at the same speed, their combined 'hum' pulsates regularly and slowly.

Mobile telephones

One of the most obvious of the changes that electronics has made to our lifestyle is the invention of the mobile telephone. It has made life easier in so many ways. If you want to phone a plumber and he is out on a job, there is no problem in talking to him instantly. Prior to the development of mobile phones you would probably have had to wait until he came home in the evening. A big advantage is the ability to contact a person without having to know where they are.

Offset against these instances and many benefits like them are the nuisances caused by people loudly using their phone in restaurants and, worse still, thoughtlessly leaving them switched on during concerts and theatrical performances. There have also been cases when criminals have made use of mobile phones to aid them in committing crimes.

Earlier mobile telephone systems were based on analogue technology but in Europe these have been replaced by digital instruments operating under the **Global System Mobile**.

The essential feature of a mobile system is that, wherever a subscriber may be, there must be easy radio connection to the public telephone network. A mobile phone is a small low-power device and can have only a small antenna. Therefore it can operate only at a relatively small distance from the nearest base station. Base stations must be close together so that the station and the phone are always in range.

The systems works on the principle of dividing the country into **cells**, which gives us the alternative term **cellular phone**. Cells are grouped into clusters (usually of 4, 7 or 13 cells) each cluster being based on a **mobile switching centre**. Cells are from 3 km to 35 km apart, depending on local conditions. In GSM900, the system is full duplex. Base stations transmit in the 935-960 MHz band. Mobiles transmit in the 890-915 MHz band. GSM1800, GSM1900 and GSM2100 are similar systems operating in the 1800, 1900, and 2100 MHz bands.

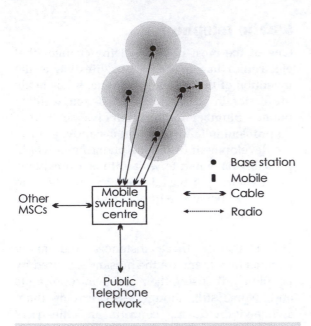

●	Base station
▮	Mobile
←→	Cable
←·····→	Radio

A mobile switching centre connects mobile telephones to the public telephone network.

Within the 25 MHz transmission bands there are 125 carrier frequencies, spaced 200 kHz apart. These 125 channels are further subdivided by **time division multiplexing**, into 8 time slots of 540 μs. This gives a total of 1000 channels per cell.

However, a bandwidth of 200 kHz is insufficient for the high-frequency pulse-coded signals so tranmissions in adjacent cells must be at least 400 kHz apart. The allocation of frequencies within a cluster is carefully planned so that there is no interference with transmissions in adjacent clusters. This is an example of **frequency division multiplexing**.

More distant cells can be allowed to operate on the same frequencies. This is known as **frequency re-use**. Frequency re-use is one technique for accommodating large numbers of subscribers on the system.

The frequency on which a given mobile transmits and receives is controlled by signals from the computer at the mobile switching centre. The MSC may often change frequencies as long as the mobile is switched on, and without the user being aware of the fact.

The strength of the signal from the mobile is constantly being monitored by the nearest base station. As the mobile moves out of a cell, its signal strength decreases. This is detected by the base station and frequencies are changed to put the mobile in contact with the base station of the new cell into which it has moved. This is done under computer control, unknown to the user, and is known as **seamless hand-over**.

Like all digital transmissions, the signals include error-checking codes and are corrected where necessary.

Bandwidth is saved by the system of coding used. Speech signals are digitised and then compressed. On reception, they are decompressed, using a system that produces speech of a reasonable quality.

The computers in the system need to know the location of all the mobiles so that incoming calls may be correctly routed to the nearest base station. They keep a register of all the mobiles and where they are. This includes a register for mobiles entering the system from other countries. All this plus the routines needed for timing and charging calls, and for producing the eventual telephone bill, are the task of one of the more complex electronic systems yet devised.

Activity — Radio receivers

Use an oscilloscope to trace the signal through a commercially-built battery-powered AM or FM radio receiver. Identify the demodulation stage and examine its input and output.

Activity — Beats

Feed the output from two audio frequency generators into a mixer circuit. If you do not have a mixer, build one from an op amp as on p. 101. Set the generators to produce sinusoidal signals of equal amplitude but slightly different frequencies. Use an oscilloscope to observe the output of the mixer. Measure the beat frequency.

Questions on radio transmission

1 Describe and compare the propagation of (a) ground waves and (b) sky waves.

2 Give an example of (a) reflection, (b) refraction, and (c) diffraction of radio waves.

3 Describe the structure and action of a half-wave dipole antenna.

4 Give examples of the use of radio for (a) broadcasting and (b) person-to-person communication.

5 Describe how a parabolic reflector is used for directional radio transmission.

6 List and describe the action of the main stages in a radio transmitter. Illustrate your answer with a block diagram.

7 Describe how a diode detector works.

8 List and describe the main stages of a tuned radio frequency receiver. Illustrate your answer with a block diagram.

9 Explain the principle of the superhet receiver and show why it gives superior performance.

10 What is meant by the *selectivity* of a radio receiver? Explain how the selectivity of a radio receiver may be increased.

11 Outline the structure of a cellular telephone system, explaining how a subscriber may remain in contact with the system while travelling within a country.

12 What techniques are used to maximise the number of subscribers who may use the GSM simultaneously.

Multiple choice questions

1 The electromagnetic waves of longest wavelength are:

    A infrared
    B radio
    C gamma rays
    D ultraviolet.

2 Sky waves are refracted in:

    A the ionosphere
    B the atmosphere
    C the stratosphere
    D free space.

3 The frequency of ground waves is:

    A up to 500 kHz
    B over 500 kHz
    C above 25 MHz
    D below 30 kHz.

4 The demodulation stage of a radio receiver is often called:

    A the IF amplifier
    B a lowpass filter
    C the detector
    D a rectifer.

5 A varactor is a:

    A variable-capacitance diode
    B variable capacitor
    C variable-frequency tuned amplifier
    D a type of triac.

6 The frequency of the local oscillator of a superhet is:

    A different from that of the carrier
    B the intermediate frequency
    C the same as the carrier
    D twice that of the carrier.

7 Beats are produced by mixing:

    A two signals of equal amplitude
    B two signals of almost equal amplitude and frequency
    C the carrier frequency with the intermediate frequency
    D signals of equal frequency.

8 A mobile telephone communicates directly with:

    A the base station
    B other mobile telephones
    C the public telephone network
    D is nearest mobile switching centre.

9 To make maximum use of the allocated bandwidth, the mobile telephone system makes use of:

    A TDM
    B cellular broadcasting
    C TDM and FDM
    D multiplexing.

On the Web
There is a lot more to find about radio.

32 Instrumentation systems

This chapter refers back to several previous topics, which readers should revise in detail.

The purpose of an instrumentation system is to measure a physical quantity, such as mass, temperature, force, or length. In this book we are concerned only with electronic instrumentation systems. Like many other systems, a typical instrumentation system comprises three stages, input, processing and output. The input stage is usually a sensor and the output stage is usually a display.

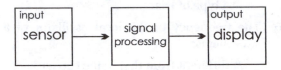

The sensor may be a thermistor, a strain gauge, a light-dependent resistor, or any other device that can be used to convert a physical quantity (temperature, light intensity, force) into an electrical quantity (particularly potential difference).

The display is often the dial of a meter but could be a computer display. The signal processing stage depends on the nature of the sensor and of the display. It may include amplification where the signal from the sensor is weak. It may include filtering to remove unwanted signals such as noise. It may be wholly analogue, or it may be partly digital.

Telemetry

This is measurement at a distance, where the sensor must necessarily be a distance away from the display. For example, in a nuclear reactor there are radiation and temperature sensors to monitor conditions inside the reactor. It is not safe for humans to enter these areas to take readings directly, so the displays are outside the reactor in a shielded control room. As another example, take a balloon sent up by meteorologists to measure atmospheric temperature at high altitudes.

Telemetry is used where it is dangerous, difficult, or simply inconvenient for a person to make the measurement directly. The system diagram of a telemetric instrumentation system therefore has some extra stages.

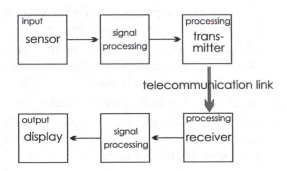

The telecommunication link may be by cable, optical fibre or radio, as explained in Chapters 28-31. It may be analogue or digital. If it is digital, it will include stages for analogue-to-digital conversion, error checking and correction, encoding and decoding. However, these extra stages make no difference to the essential function of converting a physical quantity to a displayed electrical quantity.

Note that telemetry is not to be confused with **remote sensing**. An example of remote sensing is the use of a satellite to investigate soil properties over a wide area of land. The sensor is directed at a remote site. There is a considerable distance between the site and the sensor. An example on a smaller scale is when the temperature of a furnace is measured by directing a radiation pyrometer at the open doorway of the furnace to measure the thermal radiation being emitted. Again, the sensor is at a distance from the site of measurement, the reason being that it would be damaged by bringing it closer.

> **Radiation pyrometer**
>
> An instrument based on a thermistor, thermopile, or other thermally sensitive device, designed to be used at a distance from the site where temperature is being measured.

Measuring voltage

The simplest instrumentation systems are those that measure voltage differences. The **measurand** is an electrical quantity, so there is no need for conversion. The medical instrumentaton systems that produce EECs and ECGs (p. 103) are examples. The activity of the brain or heart produces voltage differences within the brain or body. We measure these voltage differences between a pair of electrodes placed on the skin.

> **Measurand**
> Quantity to be measured.

The voltage differences produced by the brain or heart are small. One way to amplify and measure a small voltage difference, is to use a differential amplifier. This can be based on a pair of BJTs or on an op amp. The op amp provides the more sensitive circuit but, like the BJT differential amplifier, it has the disadvantage that its input resistance is relatively low. For example, the input resistance of the circuit on p. 102 is 10 kΩ at both inputs. Measurements are imprecise if the voltage source has high impedance.

An **instrumentation amplifier (or in amp)** is used where greater sensitivity and precision are required. As shown below, this consists of three op amps. It can be built from separate op amps but it is more convenient if the op amps and resistors (except for R_G) are fabricated ready-connected on a single chip. All the on-chip resistors are exactly equal, which provides an accurately balanced circuit.

Both inputs of the instrumentation amplifier go only to the op amp inputs and not to any grounded resistors (compare with p. 102). As a result, the input impedance of the in amp is very high (in the order of 10^{12} Ω), resulting in precise measurements of voltage differences.

The resistor R_G is external to the IC, and may be connected to two of its pins. If the resistor is omitted and the pins are left unconnected, the amplifier has unity gain. Different gains may be obtained by connecting a resistor of suitable value at this point.

For example, in the NA118P instrumentation amplifier IC, the gain is given by:
$$\text{gain} = 1 + 50\,000/R_G$$
By using a suitable resistor, the gain can be set anywhere between 1 and 10 000. This amplifier operates on ±1.35 V to ±15 V, and has a gain-bandwidth product of 500 MHz.

Some other types of instrumentation amplifier have a set of resistors provided on the chip. By connecting appropriate pairs of pins, the user can select a gain of 1, 10, 100, or 1000.

For accurate measurements, free from the effects of EMI, it is important for a differential amplifier to have a high common mode rejection ratio. The NA118P has a CMRR of 110 dB. In other words, the differential mode gain is over 300 000 times greater than the common mode gain. This gives it high immunity to electromagnetic interference.

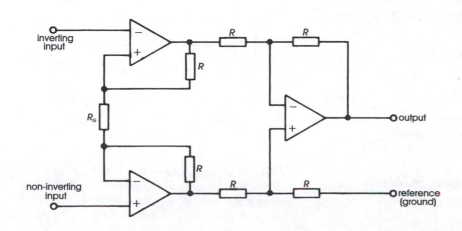

An instrumentation amplifier consists of three op amps; all resistors except R_G are equal.

Sensors

To measure quantities other than voltage it is necessary to employ a sensor. By suitably interfacing the sensor to the processing stage, we obtain a measurable voltage difference that is related in size to the measurand. Ideally, a sensor has these characteristics:

- **Selectivity**: It is affected by the measurand but unaffected by other quantities. For example, a light sensor should be unaffected by changes of temperature.
- **Sensitivity**: A small change in the measurand produces a relatively large change in the response of the sensor.
- **Operating range**: It operates over a wide range of values of the measurand.
- **Linearity**: Equal changes in the measurand result in equal changes in the response of the sensor.
- **Response time**: In some applications a very short response time is essential.
- **Stability**: Its response does not change with age, mistreatment, or other effects of use.
- **Output**: This is easy to measure electronically.

Emf–generating sensors

These generate an emf proportional to the measurand. Two common examples are:

Photovoltaic cell: This consists of a block of n-type silicon, with a thin layer of p-type material diffused into one surface. This layer is thin enough to be transparent to light. As in a diode, the energy of light falling on the depletion region causes electron-hole pairs to be produced. The result is a virtual cell, with the p-type layer 0.7 V positive of the n-type.

If the p-type and n-type layers are connected to an external circuit, the emf between the layers produces a current through the circuit. Electrons flow through the circuit from the n-type layer and, on returning to the p-type layer, combine with the holes. For as long as light falls in the cell, the supply of electrons and holes is renewed and current flows. The size of the emf depends on the amount of light.

A photovoltaic cell converts one form of energy (light) into another (emf) so it is a **transducer**. It has the advantage that its response is linear. The cell can be made with a large area to give it high sensitivity. Its response time is in the order of 50 ns. This is much shorter than that of most other light sensors such as LDRs (350 ms), most photodiodes (up to 250 ns) and phototransistors (up to 15 μs).

Photovoltaic cells are used as sensors in optical instruments, and as receivers in fibre-optical equipment. Their rapid response time suits them for these applications. Larger photovoltaic cells are used as solar cells for electricity generation.

Thermocouple: A thermcouple consists of two junctions between dissimilar metals or alloys (below). The junctions are usually made by twisting together and soldering wires of the metals or alloys. In the example illustrated, the two alloys are nickel-chromium and nickel-aluminium. This pair of alloys is often used and is known as a Type K thermocouple.

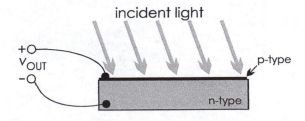

A photovoltaic cell is essentially a pn diode.

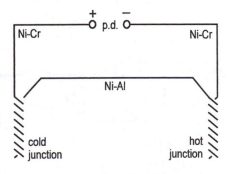

A thermocouple generates a potential difference when there is a temperature difference between its junctions.

One of the junctions, the **cold junction,** is kept at a steady reference temperature. It might simply be located inside the case of the measuring instrument. The other junction, the **hot junction,** is placed where the temperature is to be measured. Owing to the Seebeck effect, a potential difference is generated across the pair of junctions, linearly proportional to the difference of their temperatures.

The voltage output of a thermocouple is small (about 40 µV/°C), so amplifiers are required. For greatest precision, a bridge circuit is used. It usually includes a dummy lead in one arm to compensate for the Seebeck effect where the thermocouple is joined to the different metal (copper) of the leads. Thermocouples can be made only a millimetre or so in diameter so their response time is short. Types made from metal foil are even faster (10 µs). Thermocouples are robust and are frequently used in industrial applications for temperatures between –200°C to 2500°C.

A **thermopile,** consists of several thermocouples in series, with their hot junctions grouped together, and their cold junctions separately grouped together. Since they are in series, they produce a larger output voltage than a thermocouple.

Resistive sensors

There are few sensors that produce an emf directly, but there are many that show a change of another electrical property, resistance, in response to a change of measurand.

The sensors that we have studied in earlier chapters belong to this category. These are the photodiode, the light-dependent resistor, the thermistor and the strain gauge. In Chapter 2, the first three sensors listed above are being used to trigger a transistor switch. The circuit is triggered when the resistance reaches a given level. In a measurement system, the varying resistance produced by a varying measurand is to be converted into a varying voltage. The circuit at top right shows a way of doing this.

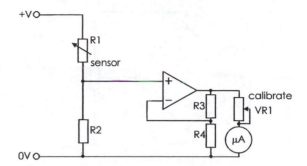

A basic measurement system for resistive sensors.

The varying resistance may be measured by connecting the sensor in series with a resistor as part of a potential divider (Chapter 2). We then measure the voltage across it, or across the resistor. A change of measurand is thus converted into change in voltage, which is displayed by using a meter.

In the case of the photodiode, the circuit is as above, with the photodiode reverse biased. In fact the reverse voltage drop across the diode is not quite the same thing as a voltage drop across a resistor but the practical effect is the same.

For greater precision, a resistive sensor is connected as one of more of the arms of a bridge.

Gas sensor: A gas sensor is used for detecting and measuring small quantities of gases in the air. Depending on the type, it detects carbon monoxide, nitrogen oxide, fuel gases, or the vapours of certain organic solvents. They can also detect insecticides and cigarette smoke. Gas sensors are useful for monitoring air pollution and for warning of the danger of explosions.

The sensing element is a coil of fine platinum wire coated with special mixtures of metal oxides and catalysts. There is also a dummy element identical to the sensor, except that the coating mixture does not contain the catalysts.

Both elements are enclosed in a dome of stainless steel mesh, about 1 cm in diameter.

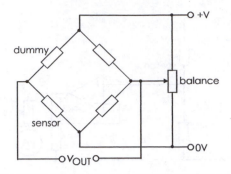

A gas sensor uses a bridge to compare the resistances of the sensor coil with that of a dummy coil.

In the detector for inflammable fuel gases, a current is passed through the coils to heat them to about 350°C. The bridge is balanced so that v_{OUT} is 0 V when no gases are present. When gases are present, they become adsorbed on to the coating of the coils. The catalysts on the sensor coil cause the gases to be oxidised, generating extra heat. The sensor coil becomes hotter than the dummy coil, which has no catalysts. The steel mesh prevents the combustion of inflammable gases from spreading and causing an explosion. The increased temperature of the sensor coil increases the resistivity of the platinum, and the bridge goes out of balance. v_{OUT} becomes positive, and this is detected by an amplifying circuit, which triggers an alarm.

This sensor is an example of indirect measurement. The measurand generates heat, the heat increases the resistivity of the platinum wire, the increased resistivity throws the bridge out of balance, which causes an increased output voltage.

Capacitative sensors

In these, changes in the measurand causes a change of capacitance. An example is an electret microphone. As the diaphragm vibrates, the distance between the plates of the capacitor varies and this varies the capacitance. Another way of varying capacitance is to vary the nature of the dielectric. This is the principle on which some humidity sensors work.

The dielectric is ofen a thin layer of aluminium oxide, which is a hygroscopic substance. One side of the layer is coated with a thin gold film which is conductive but will allow water vapour to pass through. There are two thicker metal films on the other side of the dielectric layer.

> **Hygoscopic**
>
> Absorbs water vapour from damp air: releases it to dry air.

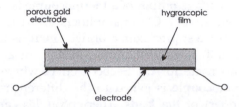

This humidity sensor is the equivalent of two capacitors in series.

Changes in humidity bring about changes in capacitance. This can be measured electronically. One way is to make the capacitance part of an oscillator circuit, then measure the frequency of the oscillations.

Capacitance can also be measured by bridge techniques. If the bridge is powered by an alternating current, the reactances of the sensor and the balancing capacitor act is a similar way to the resistances of a conventional resistance bridge.

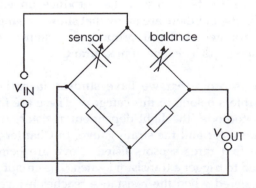

The simple capacitance bridge produces an alternating output with amplitude varying according to capacitance.

The capacitance is measured by balancing the bridge, the knob of the balancing variable capacitor being calibrated to give the readout. Alternatively, the bridge is balanced to give zero output when the measurand is at some suitable base level. Any change in the measurand causes an AC output of measurable amplitude. This is measured by rectifying it and measuring the amplitude of the resulting DC signal with a voltmeter.

Inductive sensors

A **linear inductive position sensor** (LIPS) is used to measure position over a range from 1 mm to 1.5 m. It depends on variation in the self inductance of coils as a rod of magnetic material slides in or out of them. The rod, or **target** (see below) is linked to a moving object by a plastic or other non-inductive link.

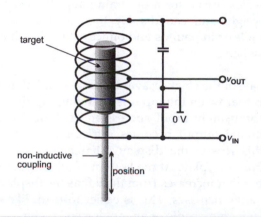

The amplitude of the output of the LIPS depends on the position of the target.

The circuit is a bridge circuit, with the two sections of the coil in two of the arms and two capacitors in the other two arms. The bridge is fed with a 1 MHz signal that keeps it resonating. The amplitude of the alternating signal at v_{OUT} depends on the relative self inductance of the two sections of the coil. The target is conductive (usually aluminium) but not magnetic, so the alternating field of the coil induces eddy currents in it. These affect the inductance of the coils, by varying amounts, depending on what fraction of the target is within each of the coils.

The output is sampled at the same stage in each cycle to give a DC signal. The DC voltage is related to the position of the target and hence to the position of the object to which the target is coupled.

The interface of this sensor is an integrated circuit running on a 5 V supply. The output ranges from 0 V to 5 V as the target is moved over its full range. As in previous examples, variations in the measurand have been converted to variations in a voltage which can be measured, displayed, or recorded in a variety of ways.

Signal processing and output

Some of the ways in which signals from the sensor are processed have been described in the previous examples. In most of the cases, processing involves a potential divider or (for greater precision) a bridge circuit. This may be followed by amplification.

At this stage the signal may be converted into digital form for transmission, conversion, and eventual storage. A dual slope converter gives the greatest precision.

In industry, hand-held data loggers are used by staff to keep a check on stock. The storekeeper is able to move around the warehouse, keying in details of stock held or stock being taken into or removed from store. Information about the current condition of the stock may also be keyed in. For example, information on the present water content of stores of timber.

The data recorded in the data logger is automatically transmitted by radio to a central computer located in the offices of the company. Conversely, the storekeeper can key in the batch number and retrieve from the central computer all stored data on the history of any item of stock.

Inventories are prepared and new stock is ordered automatically by the central computer. This is but one example of the extensive processing of data in industrial plant.

Measurement displays

Two kinds of measurement display are illustrated in the photograph below. On the left is a **moving coil multimeter**, sometimes known as a **analogue multimeter** because its display is a needle moving over a scale. It can settle at any position on the scale. This type of meter can be read with a reasonable degree of precision, provided that the user avoids parallax when taking the reading. This means looking straight at the scale, not from the left or right.

Moving coil meters give a quick visual indication of the value of the measurand and it is easy to follow rapid changes in its value. The main disadvantage is that the meter coil takes an appreciable current, so it draws current from the circuit under test. This may result in an erroneously low voltage reading. This disadvantage can be eliminated by feeding the signal to a voltage follower before sending it to the meter.

Low-cost multimeters cover most of the measurement requirements of the electronic engineer.

Simple moving coil voltmeters or ammeters are often used in displays. Instead of being calibrated in volts or ohms, their scales are marked to indicate the value of the measurand. For example, moving coil meters are used on automobile dashboards to display measurands such as engine temperature and fuel level.

A **digital multimeter** (right, in the photograph) converts the analogue input to digital form before displaying it. The data is processed digitally and displayed on a 7-segment or similar LCD. The value can be read more precisely (assuming that the measuring circuits of the meter have precision appropriate to the number of digits). The inputs of the measuring circuits have high impedance so they do not drain appreciable current from the circuit under test. This gives improved accuracy compared with moving coil meters.

Since the data is processed by digital circuits it is easy to arrange for additional processing, even in an inexpensive meter. The meter in the photo can display not only the current reading but the maximum, the minimum and the average reading since a measurement session was begun. Similarly an electronic kitchen scales uses a form of strain gauge to measure weight and can display the result in metric units or in pounds and ounces at the touch of a switch.

One of the chief disavantages of a digital meter is that when the input is changing rapidly, the confusingly rapid succession of digits on the digital output becomes difficult to read. For this reason the display often includes a bar graph display. At the bottom of the display is a bar that increases from the left as the displayed value increases. This is easier to read, like the swinging needle of an moving coil meter.

A bargraph display can be built using a row of LEDs. There are special ICs that, given an input voltage, light the appropriate LEDs. The display can be programmed to display either a thermometer-like bar, or a moving spot of light. Bargraphs are ofen used to indicate the output power of audio amplifiers.

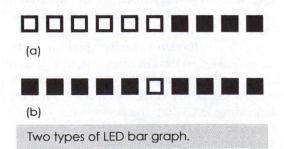

Two types of LED bar graph.

If the data is processed by a computer or micro-controller, there is almost no limit to the range of displays that become feasible. One of the more elaborate is illustrated in the photograph. It shows the monitors in the control room of the Ironbridge Power Station, Shropshire. The data from sensors located at dozens of points in the plant is processed and displayed on these screens.

Monitors in the control room of an electricity generating station display the outputs from numerous measurement systems.

The engineers can read temperatures, steam pressures, rates of revolution, voltages and other measurands relating to electric power generation. The display givens a complete picture of what is happening in all parts of the power station, providing the engineers with all information they need to control the turbines and generators.

Activity — Instrumentation

Study an instrumentation system that you have seen in operation. Draw a block diagram of the system. Describe the sensor used with as much detail as you can obtain from manuals and data sheets. Outline the electronic circuits used for processing the data. What type of display is used in this system and what are its advantages and disadvantages?

Questions on instrumentation

1 What are the desirable features of a sensor used in an instrumentation system?

2 Name two sensors. Describe how they work and explain how their response is processed in named instrumentation systems.

3 What is a thermocouple? How does it work? Design a bridge circuit to interface a thermocouple to a millivoltmeter.

4 Design a system for measuring humidity based on the varying resistance of a hygroscopic substance.

5 Design a system to measure the rate of rotation of a flywheel.

6 What kind of sensor would you use in a weighbridge to measure the weight of loaded vehicles? How would you process the output from the sensor?

33 Electronic control systems

The simplest kind of electronic control system comprises three stages: input, processing, output. Examples are given on pp. 11 and 12. A general diagram is drawn below. These are called **open-loop** systems.

There are also **closed-loop** systems, in which *part* of the signal from the output stage is fed back to the input stage. For example, in a thermostat (opposite), a small part of the heat from the heater warms the thermistor sensor.

Feedback is an important concept in control systems. With **negative feedback**, the output is made to have a negative effect on the input. If the temperature of a room is too high, the heater is turned off. If it is too low, the heater is turned on. This action makes a system stable — the room temperature is held constant.

Typically, there is a sensor (such as a thermistor) which produces a voltage signal that is proportional to the output. A feedback circuit operates by comparing this voltage with a fixed voltage, the **set point** or **reference signal**. Often a comparator (perhaps an op amp) receives the two voltages and compares them. It produces an output voltage called the **error signal**. The error signal is proportional to the *difference* between the feedback and set point signals. If they are equal, their difference is zero and the error signal is zero. No further action occurs. If they are different, the error signal acts on the input stage of the system until they become equal.

Chapter 34 looks more closely at the theory of control systems, such as are used in industry for process control.

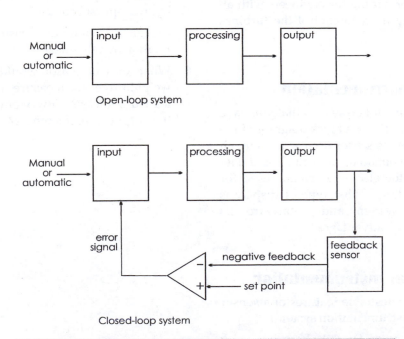

Open-loop system

Closed-loop system

The main components of an open-loop control system and a closed-loop sysyem with negative feedback.

Regulators and servos

Many control systems can be described either as regulator systems or as servo systems.

A **regulator system** holds the value of a given quantity (temperature, speed, position) at a fixed level, the set point. Examples are a thermostat and the system for keeping a CD spinning at constant speed. In some regulator systems (for example, an incubator for growing bacterial cultures) the set point may be fixed. In others it may be altered occasionally (for example, a room heater).

In a **servo system** the set point is often changed and the system is continually adjusting to new values of the set point. Examples are steering systems in vehicles, in large vessels, and in aeroplanes (including flying model planes, p. 299). Other examples are the systems for controlling the position of robot arms.

The systems described in this chapter are analogue systems. In Part 3 of the book we describe several other kinds of control system, which are based on digital electronics. The regulator and servo systems described in this chapter can all be realised as digital circuits, usually with a microcontroller in charge of the processing.

Temperature control

A system which controls an electric heater to hold the temperature of a room at a fixed level is an example of a regulator system. The **set point**, can be adjusted from time to time to suit current conditions.

The circuit at top right is a fully automatic system. It could be used to control the temperature of an oven, a refrigerator, an incubator, a room or a building.

The temperature sensor (R1) is a thermistor. This is part of a potential divider and the feedback signal is the voltages at the junction of R1 and R2. In the diagram, this is labelled t_0. It goes to the inverting input of the operational amplifier IC1, which is connected as a comparator

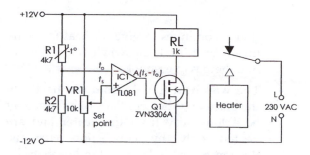

A practical thermostat circuit, the output voltage t_0 from the thermistor/resistor chain depends on the resistance of the thermistor, R1. As temperature increases the resistance of R1 increases and t_0 falls. This is compared with t_s the fixed voltage of the potential divider based on VR1.

Feeding t_0 to the inverting input gives it a negative effect — it is negative feedback.

The set point voltage t_S comes from the wiper terminal of VR1, and goes to the non-inverting input of the op amp. This makes the circuit the equivalent of the comparator shown in the diagram opposite.

The op amp is wired as an open-loop amplifier so has very high gain (A). The output of the op amp is:

$$A(t_S - t_0)$$

This is the **error signal**. Because the open-loop gain of the op amp is so high, the error signal swings fully toward the positive or negative rails, turning the transistor switch fully on or fully off, depending on whether t_0 is less than or greater than t_S.

Feedback in a thermostat

As well as the action of the electronic circuit itself, feedback in a closed-loop system depends on the physical nature of the different parts of the system. Important factors include:

- **The heating element:** a thin wire heats up quickly, so the system responds quickly. We say that it has a rapid **response time**. By contrast a heavy-duty element enclosed in a thick metal sheath heats up slowly.

- It also continues to give off heat for some time *after* it has been switched off. We say that there is **overshoot**.

- **The way heat travels from the heater** to the place to be heated: Parts of a room may be a long way from the heater and are heated only by convection currents. Response time is long. A fan heater has a shorter response time.

- **The response time of the sensor:** a thermistor may be a relatively large disc with long response time, or it may be a minute bead with short response time. The sensor may be in contact with the air or enclosed in a glass bulb.

The effects of these factors produce hysteresis. The heater is switched on at a higher temperature and off at a lower one. It is usually better if the system has hysteresis. It avoids the system being frequently switched on and off when it is operating close to the set point.

In a heating system with hysteresis, the heater goes off when the temperature reaches a given higher level t_U, the **upper threshold**. With the heater off, the temperature falls but the heater does not come on again until the temperature falls below t_L, the **lower threshold**. Then the heater stays on until the temperature reaches the higher threshold. The result is that the temperature is not absolutely constant, but rises and falls over a restricted range.

Rotational speed control

Varying the current through the coils of a DC permanent-magnet electric motor controls its speed. There are several ways of doing this:

- **Manual control:** A variable resistor is wired in series with the coils.

- **Transistor control:** The speed of the motor is controlled by varying the base current to the transistor. Like the manual control technique, this is open-loop control. It can not guarantee that the speed of the motor is exactly related to the setting of the variable resistor or to the size of the base current.

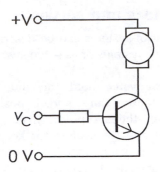

Changing the control voltage v_C varies the speed of the motor. This is an open loop, so speed depends partly on the mechanical load on the motor.

- The reason for this is that the mechanical load placed on it affects motor speed. This may vary for several reasons, for example: the motor is driving a vehicle which starts to go up a slope, the motor is lifting a load which is suddenly decreased, the motor is affected by uneven friction in the mechanism. A slowly rotating motor may stall completely if its load is suddenly increased.

- **Op amp closed loop control:** The circuit opposite has negative feedback. The op amp monitors the voltage across the motor and acts to keep it constant, equal to the control voltage, v_c. This ensures constant current through the coils, keeping it turning at constant speed in spite of variations in the load.

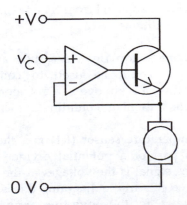

An op amp with negative feedback. Varying v_C sets the speed of the motor.

- **Switched mode control:** Instead of controlling the amount of current supplied to the motor, the current is supplied in pulses of constant amplitude but with varying mark-space ratio.

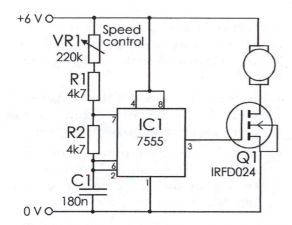

A pulse generator with variable mark-space ratio is used to control the rate of rotation of the motor.

The speed of the motor depends on the *average* current being delivered to it. In the diagram above, pulses are generated by a timer IC. The timer runs at 50 Hz or more, producing a stream of pulses. The inertia of the motor keeps it turning at a steady speed during the periods between pulses.

Adjusting VR1, varies the pulse length and thus the mark-space ratio of the output to Q1. With the values given in the diagram, the mark-space ratio ranges from 2 to 50. This gives a 1 to 25 variation in the amount of current supplied to the motor.

This technique gives smooth control of motor speed over a wide range. There is less tendency to stall at low speeds because the pulses switch the transistor fully on and the motor is driven with full power. It is possible for the pulses to be generated by a microprocessor for fully automatic speed control. The main disadvantage of the method is that it is an open loop. The actual speed attained depends on the load.

- **Stepper motor:** A typical stepper motor has four sets of coils, arranged so that the rotor is turned from one position to the next as the coils are energised in a fixed sequence. This is listed in the table below. The sequence then repeats after this. At every step, two coils are on and two are off.

Step no.	Coil 1	Coil 2	Coil 3	Coil 4
0	On	Off	On	Off
1	Off	On	On	Off
2	Off	On	Off	On
3	On	Off	Off	On

To get from step to step, first coils 1 and 2 change state, then 3 and 4 change state, then 1 and 2, and so on. The result of this is to produce clockwise turning of the rotor, 15° at a time. If the sequence is run in reverse, the rotor turns anticlockwise. Six times through the sequence causes the rotor to turn one complete revolution.

At any step, the rotor can be held in a fixed position by halting the sequence. A pulse generator containing the logic to produce four outputs according to the sequence in the table controls the motor.

The four outputs are connected to four transistors, which switch the current through each coil on or off. The pulses that switch the transistors are provided either by a special IC or by a programmed microcontroller.

The rate of rotation is controlled by the frequency of pulses generated by the microcontroller, one revolution for every 24 pulses. The microcontroller can be programmed to produce the pulses in the reverse sequence, so reversing the direction of rotation. It can also deliver any required number of pulses to control the angle turned.

The advantage of the stepper motor is that its speed is controllable with the same precision as the system clock. It is not affected by the load on the motor, except perhaps an excessive load, which might completely prevent the motor from turning. The stepper motor may be made to turn 7.5° per step by using a slightly different switching pattern. Motors with a 1.8° step angle are also produced.

Position control

There are two types of position that may be controlled:

- **Linear position:** the location of an object (such as a cutting tool) along a straight path, measured from a point on the path.
- **Angular position:** the direction of a lever-like object (such as an arm of a robot), measured as an angle from some reference direction.

If an object, such as a sliding door, or a part of a machine, is to be moved along a straight path, the motive force may be provided by a solenoid, an electric motor, or a hydraulic cylinder and piston. The rotary motion of the motor may be converted into linear motion by a mechanism such as a worm gear or a rack and pinion. A closed-loop control system can supply power to a solenoid, a motor or the valves of a hydraulic system, but it needs a sensor to detect that the moving object has reached the required position. Sensors used for this purpose include limit switches, magnetic proximity sensors, linear inductive proximity sensors, and various resistive and optical sensors.

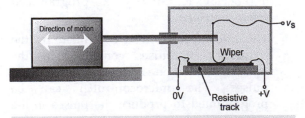

As the object moves, the wiper is moved along the resistive track. The output voltage varies from 0 V up to +V as the object moves from left to right.

A simple resistive linear position sensor is illustrated below. The output of this is proportional to the position of the object. This is compared with v_C, the control voltage, set by the operator or perhaps generated by a computer.

The object is moved along its track by an electric motor and a linear mechanism. The control systems is illustrated below. There are three situations:

- $v_C > v_S$: The output of the comparator is positive. The control circuit supplies current to the motor, so as to increase v_S.

- $v_C < v_S$: The output of the comparator is negative. The control circuit supplies current to the motor in the opposite direction, so as to decrease v_S.

$v_C = v_S$: The output of the comparator is zero. The control circuit supplies no current to the motor. The object stays in the same position.

This is an example of a **servo system**. It is an on-off system, with the extra feature that the motor is not just switched on and off but may be made to turn in either direction. The control circuit is built from op amps and transistors, or it may be a special IC designed for servo driving. If the object is not at its intended position, the motor rotates one way or the other until $v_C = v_S$.

One of the possible snags of such as system is that the system overshoots. Instead of homing steadily on the intended position, the object is moved *past* that position. It is then moved toward the position from the opposite side, and may overshoot again. This may continue indefinitely, causing the object to oscillate about the position instead of settling there. This is called **hunting**. To avoid hunting, most servo systems provide a **dead band** on both sides of the intended position. The motor is switched off as soon as the difference between v_C and v_S becomes less than a given small amount. The object stops a little way short of the intended position but the error is small enough to be ignored.

A **servomotor** is designed to move to a given angular position. The motor has three connections to the control circuit. Two of these are the positive and 0V supply lines. The third connection carries the control signal from the control circuit, which may be a microprocessor.

The rotor of the motor has limited ability to turn. Generally it can turn 60-90° on either side of its central position.

The control signal is a series of pulses transmitted at intervals of about 18 ms, or 50 pulses per second.

The angle of turn is controlled by the pulse length:

- 1 ms: turn as far as possible to the left.
- 1.5 ms: turn to central position.
- 2 ms: turn as far as possible to the right.

Intermediate pulse lengths give intermediate positions.

A small servomotor of the kind used in flying model aircraft and robots. The 'horns' (white levers) are used for connecting the motor to the mechanisms that it drives.

Activities — Control systems

Set up demonstration control such as:

- a thermostat.
- an automatic electric fan for cooling.
- motor speed control.
- lamp brightness control.

- stepper motor driver based on a walking ring counter.
- servomotor driver based on a 555 timer running at 50 Hz, that drives an edge-triggered delay to produces pulses varying in length from 1 ms to 2 ms.

These circuits should be powered from batteries or low-voltage PSUs, not from the mains. Some of these systems could be controlled by a microcontroller.

Questions on control systems

1 Draw a block diagram of a switched-mode closed-loop circuit for controlling the speed of an electric motor. Describe the action of each part of the system.

2 Explain how a thermostat operates.

3 What are the advantages of a closed-loop system when compared with an open-loop system? Illustrate your answer with examples.

4 Give examples to explain what is meant by (a) hysteresis, (b), overshoot, and (c) dead band.

5 What is a stepper motor? How can it be used for controlling (a) rotational speed, and (b) angular position?

6 What are the differences between stepper motors and servomotors? For each type, give an example of control circuit in which you would use it.

7 List and describe in outline the control systems used in three examples of domestic equipment.

 The companion site has several animated diagrams of circuits that are used in control systems.

Set up some of these on a breadboard, or build them into a permanent project.

34 Process control systems

This chapter looks more closely at the structure and behaviour of control systems, particularly those used in industrial processing.

We are concerned only with closed loop systems as open-loop systems are unable to operate as regulators or as servos.

The simplest closed-loop systems are like the thermostat described on p. 295. In this the heater is either on or off. This type of system, often known as an 'on-off' or 'bang-bang' system, is relatively easy to implement. It operates well enough in many applications.

Many industrial applications demand more elaborate control systems. There are three main types:

- Proportional (P) systems.
- Proportional-integral (PI) systems.
- Proportional-integral-derivative (PID) systems.

In this chapter, the systems operate with analogue voltages. However, the same control functions are possible in digital systems. Signals are digital quantities and many of the circuits are replaced by software programs.

Proportional control

The diagram below shows the main functional blocks of a proportional control system and the analogue voltage signals that flow from one block to another.

The system is shown controlling an industrial processing plant. It might, for example be a chemical plant in which several solutions are mixed in a vat to react together. But the same principles apply to other systems such as the control system of an underwater vessel used for repairing an oil well platform.

In the diagram below the function of comparing set point and output is shown divided into two stages:

- A **subtractor** to measure the difference between the actual output and the set point (or desired output), x.
- An **amplifier** to amplify the difference.

The amplified signal drives the actuators, the motors, heaters, stirrers and other units in the processing plant. There may be further amplification in the plant.

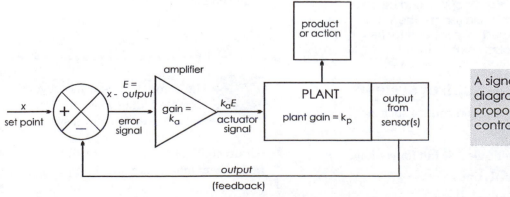

A signal flow diagram of a proportional control system.

In most controllers the plant introduces a time factor, usually a delay. For example, a controller may fill a vat with a solvent. The system operates a valve and the level of solvent in the vat is measured by a float mechanism.

The valve is controlled by the signal y. The amount of solvent in the vat is detected by the float mechanism. The output from this sensor is used as feedback. The feedback is used to calculate the value of y.

Filling the vat to the set point level is a process that takes time — perhaps several minutes or even hours. If the vat is empty, y is large and the valve is opened wide to start filling rapidly. As the vat fills and the level of solvent approaches the set point, y decreases and the valve is gradually closed. It is shut off ($y = 0$) when the vat is full.

The screen shot shows the input and output of a proportional control system, simulated on a computer and plotted against time. The set point voltage is plotted in grey. The black curve shows output voltage. Output is plotted as a voltage that represents the activity of the plant — the level of solvent in the vat or, in other systems, the temperature of a furnace, the position of a ship's rudder, or the angle of elevation of a radio telescope.

In the simulation the set point is suddenly increased, then held constant. The output curves slopes steeply upward at first, then gradually becomes less steep as it approaches the set point.

To begin with, the error E is large, the valve is wide open and the rate of increase of output (increase in the solvent level in the vat) is high. In time the error becomes less and the rate of increase is less. The valve gradually closes, solvent flow diminishes and is eventually cut off. In other words, the rate of increase in output is **proportional** to error.

The advantage of this system is that it saves time by opening the valve wider at the beginning. It may be essential for the operation of the process to deliver as much solvent as possible as quickly as possible.

The next point to notice is that the output approaches the set point, but never reaches it. There is a certain amount of **offset**. The reason for this is explained later.

The screen shot overleaf shows the output curves for four different gain settings. The higher the gain, the smaller the offset. But offset is always there.

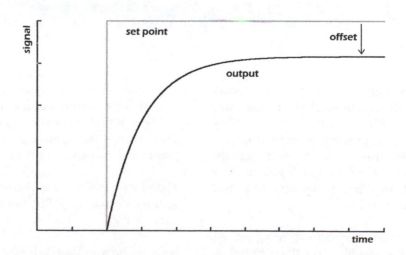

A computer simulation of a proportional control system. A sudden increase in the set point (grey line) results in a rise in output (black).

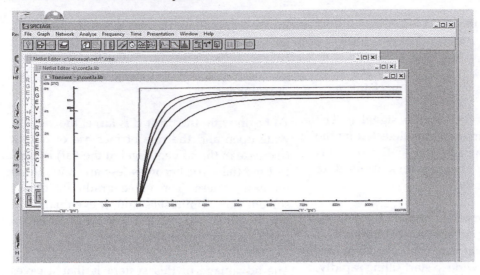

Output is plotted for four different amounts of gain. Increasing gain results in reduction of offset, but does not eliminate it.

This plot shows what happens when gain is further increased. The lowest curve has the same gain as the lowest curve in the screen shot above. The other two curves have higher gains, and the one with highest gain overshoots and oscillates,

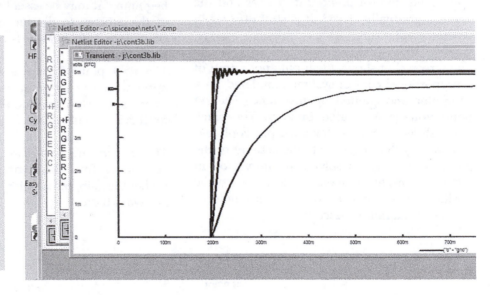

As the screen shot shows, high gain may result in **overshoot** and **oscillation**. This may not be acceptable behaviour for systems involving chemical processing, for example. In mechanical control systems, such as the control of robot arms or the flight surfaces (elevators) of an aircraft, overshooting may damage the mechanism.

When a proportional system is being set up, the gain of the amplifier must be **tuned** to produce acceptable responses. Usually the gain is set at a level that gives the minimum offset with no overshoot or oscillation.

The most important quantity in the control system is the output, but it is interesting to look at the y signal (actuator signal × plant gain). This controls the operating part or parts of the plant, such as valves, motors and heaters.

The plot (opposite) was obtained with the same settings as that on p. 301 but now includes the curve for y.

It shows how y takes its highest value when the set point is increased and the error signal is a maximum. It switches the actuators fully on for maximum effect.

As time passes and the output increases (the vat becomes fuller), E is reduced, resulting in decreasing y. The curves for y and output gradually meet at the same level, a little below the set point. The system is stable.

The cause of the offset error in proportional control systems is due to the feedback of the error signal. The output of the system is $y = k_p k_a E$ (see diagram on p. 300).

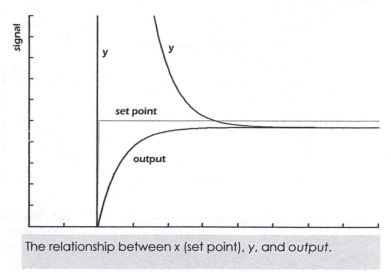

The relationship between x (set point), y, and *output*.

To simplify the equations that follow, put $k_p k_a = K$. K is the **open loop gain** of the system and:

$$y = KE.$$

But E is the difference between x (set point) and y:

$$y = K(x - y)$$

From this we get:

$$y = Kx/(1 + K)$$

As K is increased, the fraction $K/(K+1)$ comes closer and closer to 1, making $y \approx x$, and the offset $(y - x)$ approaches zero.

One way of reducing offset is to add an adjustment signal to the error signal. The size of this signal can be controlled manually. A better technique is used in the proportional-integral system described in the next section.

Proportional-integral control

A proportional-integral (PI) controller uses an integrating block to produce a signal that is the integral over time of the error signal. The output of the integrator is zero to begin with.

For as long as the input to the integrator is positive its output rises at a rate depending on the input.

When E is zero, the input to the integrator is zero and its output remains at the same level. Its output falls if a negative voltage is applied to it.

The two plots overleaf show how the integrator corrects for the output offset.

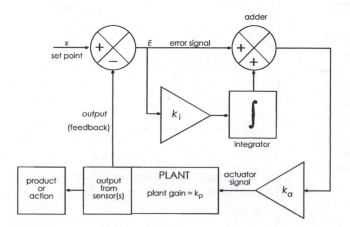

A PI controller (left) adds in the time integral of the error signal, multiplied by the gain, k_i.

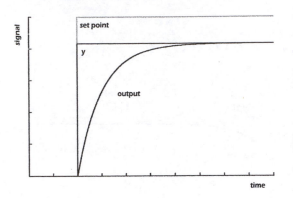

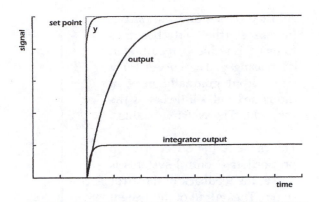

Output of a P-I controller with the integrator switched off, so that it acts like a P controller.

With the integrator switched on, its output rises until the error signal (set point - y) is zero. After that, it remains at a constant level.

On the left above the simulated PI controller is run with its integrator switched off. It then behaves like a P controller. The control voltage y never reaches the set point and therefore the output has a significant offset.

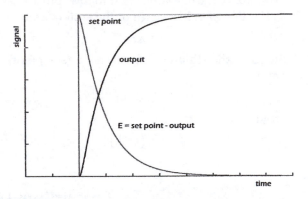

Switching on the integrator adds its output to y. The integrator output rises sharply at first then more slowly. The value of y is boosted by this so that it reaches the set point. At that stage the error E is zero so the integrator output remains fixed. The output rises more slowly because of the load but finally reaches the set point. The vat is full to the required level, the error signal is zero and the valve is closed.

The error signal E is greatest when the vat is empty, so the valve is wide open. As the vat fills, E is reduced, until it becomes zero when the set point is reached.

In a regulator system based on a PI controller, the set point is adjusted to a fixed level and stays there. The output reaches the set point slightly later. In a servo system, in which the set point is continually changing, the output may not be able to keep up with the changes.

The plot on the right shows what may happen. This is the same simulated system as used in the previous plots, but the set point is now programmed as a triangular wave. This could be the system controlling an oscillating motion like that of the heater-dryer fans in a car-wash. The output changes with the same frequency as the set point, but never catches up with it. The next section shows how the response can be speeded up.

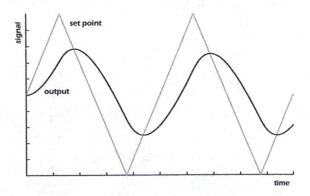

If the set point changes too rapidly, the output does not have enough time to respond.

PID control

PID, or proportional-integral-derivative control introduces a third term into the equation for y. The signal flow diagram below shows that a differentiator block is used to generate a signal proportional to the *rate of change* of the error signal. This **derivative** is added in with the proportional and integral signals and applied to the plant.

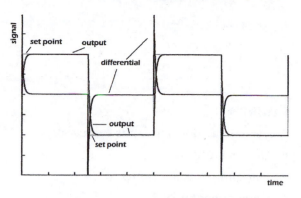

The output of the PID simulation follows the set point almost exactly. The plot also shows the derivative signal produced by the differentiator.

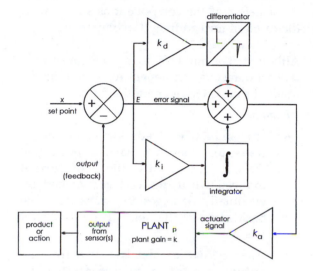

A PID, or 3-term control system is able to follow rapid changes of the set point.

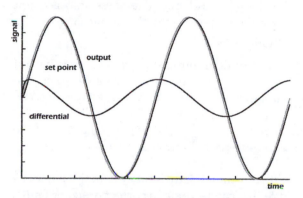

When the set point is a sine curve, the derivative follows a cosine curve because $d(\sin x)/dx = \cos x$.

The effect of adding in the derivative is to increase the rate of change of the output when the set point is changing rapidly. If the set point is steady or changing slowly, the derivative is zero or close to zero, and so has little effect on the output.

A simulation of a square wave set point is illustrated at top right. A square wave changes very rapidly as it swings up or down, but remains constant between swings. The derivative therefore shows sharp upward and downward peaks, but is zero between peaks. The effect of this is to make the output equal to the set point at almost evertystage. Only at each swing does the plot of the output fail to follow the set point, but it catches up again.

Activities — 3-term controllers

Investigate the action of a PID controller, tuning it with its P function only, then switch on the I function, and tune it again. Refer to the user manual for details. Finally, run and tune it with all three functions switched on.

Use a computer simulation of a controller, or build up a controller from function blocks in circuit simulator software. Investigate the effects of setting the values of the gains of the various stages.

35 Systems with faults

Causes

A system may develop faults because of:

- Design faults
- Handling or usage faults
- Component faults

Most systems work properly for most of the time. But even a system that has been properly designed may show faults when it is operated under unusual or extreme conditions. Possibly the designer did not take these into account when the system was in the design stage.

Example

An amplifier normally works correctly, but becomes unstable when the ambient temperature exceeds 35°C. The heat sinks are too small or the ventilation of the enclosure is not adequate.

> **Ambient temperature**
> The temperature of the surroundings.

With a commercially built system there may be little that can be done to correct designer faults, except to complain to the manufacturers. With a system you have built yourself, it is a matter of 'back to the drawing board'.

Handling or usage faults occur when the system is used in a way that was not intended by the designer.

Examples

- A circuit board is connected to a supply voltage of 12 V when it is intended to operate on a maximum of 5 V.
- A circuit is connected to a supply of the reverse polarity.
- This type of fault can also occur accidentally, for example, a spike on the mains supply due to a lightning strike.

Defective components are rarely the *cause* of a fault. Provided that new components have been purchased from a reputable supplier, and are connected into a well-designed system, they should seldom fail.

If components are not new but have been used in other circuits, they may have been damaged previously by overvoltage or excessive current, by incorrect polarity, or they may have been damaged mechanically. Unless precautions are taken, CMOS and similar devices may be damaged by static charges while being handled, prior to being use in a circuit. This is not a defect of the component as such, but is the result of mishandling the component.

Although it is unlikely to be the cause of a fault, a defective component may often be the *result* of a circuit fault of another kind.

Examples

- An inductive load switched by a transistor does not have a protective diode. Consequently, the high current induced when the load is switched off eventually damages the transistor. The basic cause of the defect is faulty circuit design.
- An excessive current is drawn from an output pin of a microcontroller by connecting a low-impedance load to it. The output circuit is damaged and no further output is obtainable from that pin. The basic cause is incorrect usage.

Fault report

The first stage in diagnosing a fault is to state exactly what is wrong with the system. It is not sufficient to report simply that 'It does not work'. The initial report should give answers to the question:

What does it NOT do that it should do?

This can be answered by one or more statements such as:

- The pilot LED does not come on when power is switched on.
- There is no sound from the loudspeaker.

- There is no picture on the screen.
- The heater does not switch off when the set temperature is exceeded.
- Nothing happens when a control button is pressed.

To be able to provide such answers, the engineer must be very clear about what the system should do. Read the User Manual and consult it often.

With some systems it may also be informative to answer this question:

What does it do that it should NOT do?

Typical answers could be:

- There is a smell of scorching.
- The indicator LED flickers irregularly.
- There is a loud buzz from the loudspeaker.
- The servomotor never settles to a fixed position.

Another question that should be asked is:

WHEN does it show the fault?

It may be necessary to run the system for several hours to answer this fully, but typical answers are:

- All the time.
- Not often. It becomes faulty and then recovers.
- Not often, but after the fault develops it stays.
- When it is first switched on, but not later.
- Not when it is first switched on, but starts about 10 minutes later.
- Only when the gain is set high.

Inspection

With the fault report in mind, the next stage is to inspect the circuit, looking for obvious faults. These visible faults, if any, should be carefully listed.

Look for wires that have broken away from the circuit board or the terminals of panel-mounted components, signs of mechanical damage to the circuit board and components, signs or smells of scorching.

At the same time it is important to note what parts of the system can not be inspected and why. In some systems there are complicated components (such as a microcontroller, or a RAM chip) that can not be tested without special equipment. These could be looked at later if the initial inspection fails to deliver any solution to the problem.

Mains supply

Turn off the mains supply and remove the plug from the socket before inspecting or testing mains-powered equipment.

Unless you are fully experienced you should never test or operate mains-powered equipment with the case open. You should never attempt to test the circuit while it is plugged into a mains socket.

Preliminary diagnosis

When the inspection is complete it may already be possible to decide what the cause of the fault may be. The next stage is to carry out some elementary practical tests. It is not possible to provide a set of routines here because so much depends on the type of system and the nature of the fault. Some manufacturers provide checklists of faults and remedies applicable to their equipment. If such help is available, it should be followed.

In the absence of advice from the manufacturer, or even when advice is available, it is essential that the person looking for the fault fully *understands* the system and how it works. A schematic diagram is almost essential, preferably marked with typical voltage levels at key test points.

With circuit diagrams and other data to hand, and with a good understanding of the system, it is usually possible to eliminate certain possibilities and to concentrate on locating particular kinds of defect.

Examples

- A total failure of the equipment to do anything at all suggests that there may be a failure of its power supply. With mains equipment, first check that the mains socket provides mains current when switched on (try it with some other piece of equipment plugged in).

- With a battery-powered circuit, check the battery voltage, remembering that the voltage from a nickel-cadmium rechargeable battery drops sharply as it reaches full discharge. It might have been correct yesterday, but is it still delivering full voltage today?

- From the circuit diagram, pick out points in the circuit that should be at full positive supply voltage, and check these with a meter. Such points include the positive supply pins of all integrated circuits and all other pins (such as unused reset pins) that are wired permanently to the positive supply.

- If the supply voltage is low at any of these points, place an ammeter in series with the power supply and the equipment and measure the current it is taking. If the current is excessive, it suggests that there may be a partial or total short-circuit between the positive supply rail and the ground (0 V) rail. This could be due to a defective component.

- Use a continuity tester to check that all points that should be connected directly to the 0 V rail are in fact connected. This includes 0 V pins of all ICs and all other pins that are wired permanently to 0 V.

If a list of expected voltage levels is not available, try to work out what they should be.

Example

If the base-emitter voltage of a BJT is more than about 0.7 V the transistor should be on.

This means that its collector voltage should be relatively low, perhaps not much greater than its base voltage. Conversely, if the base-emitter voltage is less than 0.7 V, the transistor should be off and its collector voltage may be equal to the positive supply voltage.

Capacitors

In many types of circuit, the capacitors retain their charge for a long time, perhaps for several hours, *after* the power supply has been switched off. They can deliver a very unpleasant and possibly lethal shock.

Treat all capacitors with respect and make sure they are discharged before touching them or their terminal wires. Capacitors of large capacity should always be stored with their leads twisted together.

Common faults

There are some very basic faults that occur more often than faults of other kinds. It is best to look for these first. Several of them are related to the circuit board and construction techniques. Quality control should eliminate most such faults from commercially-built systems, but self-built systems should be checked *before* power is first switched on:

- There are short-circuits caused by threads of solder bridging the gaps between adjacent tracks on the pcb. A hair-thin thread can result in a fault, so it is *essential* to use a magnifier when inspecting a circuit for this kind of fault.

- There are hairline gaps in the tracks. Breaks in tracks may result from flexing due to mechanical stress in mounting the board. These gaps are difficult to see and it may happen that they only open up after the board has been warmed from some time. Faults of this kind may develop after the circuit has been switched on for several minutes.

- Cold-soldered joints are the result of a lead or the pad (or both) being too cold when the solder was applied. It may also be the result of a greasy or corroded surface. The molten solder does not wet both surfaces properly. Often the solder gathers into a ball on the track or lead and the fact that the joint is dry can be seen under a magnifier. Suspected cold-soldered joints should be resoldered.

- Components may be soldered on to the board with the wrong polarity. This applies particularly to diodes, LEDs, electrolytic capacitors and tantalum capacitors. It is also possible (and damaging) to insert an IC the wrong way round in its socket.

- It is possible that a wrong IC has been used. Check all type numbers.

- An IC may be inserted in its socket so that one or more pins is bent under the IC and does not enter the socket. This type of fault is hard to see. The only method is to remove the IC from the socket and inspect its pins.

- Resistors and capacitors may be mechanically damaged by rough handling so that leads lose their electrical connection with the device without actually becoming detached. With ICs, excess pressure on the IC may flex the silicon chip inside so that connections are broken.

- Devices such as diodes and transistors may be totally damaged or their characteristics may be altered by excessive heat. This usually happens when the device is being soldered in, but may also be the result of excess current. The resistance of resistors may also be altered by excessive heat.

Analogue circuits

The items listed above may lead directly to the cause of the fault, which can then be rectified. If this fails, a more detailed inspection is required. In this section we deal with analogue circuits. Suggestions for digital circuits are listed in the next section.

Analogue equipment is checked by further use of a testmeter and with more elaborate equipment such as an oscilloscope. Signal generator and signal tracers are often used with audio equipment. Detailed instructions for using these are found in their Users' Manuals.

The probe of an oscilloscope has high input impedance so the instrument can be used in the same way as a voltmeter for checking voltage levels at key points. In addition, it is possible to detect rapid changes in voltage to discover, for example, than an apparently steady voltage is in fact alternating at a high rate. In other words, the circuit is oscillating. Conversely, we may discover that a circuit which is supposed to be oscillating is static. Clues such as these, combined with the understanding of circuit operation, may point to the cause of the fault. In particular, the switching of transistors may be examined, to see that the base (or gate) voltages go through their intended cycles and that the collector and emitter (or source and drain) voltages respond accordingly. A dual-trace oscilloscope is invaluable for checking transistor operation.

A signal generator may be used for introducing an audio-frequency signal into an audio system. In a properly functioning system, a tone is then heard from the speaker. Beginning at the power output stage, we work backward through the system, applying the signal at each stage and noting whether it appears at the speaker and whether it is noticeably distorted. An oscilloscope can be used to check the waveform, particularly to look for clipping, ringing and other distortions. If we get back to a stage from which no signal reaches the speaker, or distortion is severe, the fault must be located in this stage. Then we examine this stage more closely. Alternatively, we can work in the opposite direction. We supply the input of the system with an audio signal from a radio tuner or disc player. Then we use a signal tracer (a simple audio-frequency amplifier with a speaker) to follow the signal through the system from input to output. Again we note the stage at which the signal is lost and investigate this stage in more detail.

Sometimes it is helpful to use both techniques, working in from both ends of the system to arrive at the site of the trouble.

Digital circuits

The preliminary techniques, such as checking power supplies and circuit boards, are the same for digital circuits as for analogue circuits. Cold soldered joints or cracks in the circuit board tracks can cause a loss of logic signal between the output of one gate and the input of the next.

A common fault is for an input to be unconnected to the power rails or to a logic output. This can happen because of incorrect wiring, a bent IC pin, or a dry solder joint. An unconnected CMOS input will then 'float', staying in one state for some time but occasionally changing polarity. Output from the IC will vary according to whether the disconnected input is temporarily high or low.

With TTL, an unconnected input usually acts as if it is receiving a high logic level. If outputs of a logic IC are not firmly low or high it may be because of lack of connection on the IC power input pins to the power lines. If there is no connection to the 0 V line, output voltages tend to be high or intermediate in level. If there is no connection to the positive line, the IC may still appear to be working because it is able to obtain power through one of its other input terminals which has high logic level applied to it. But output logic levels will not be firmly high or low, and the performance of the IC will be affected. This situation can lead to damage to the IC.

The more detailed checks require the use of special equipment. A **logic probe** is the simplest and most often used of these. It is usually a hand-held device, with a metal probe at one end for touching against terminals or the pins of the logic ics. The probe is usually powered from the positive and 0 V lines of the circuit under test. In this way the probe is working on the same voltage as the test circuit and is able to recognise the logic levels correctly. The simplest logic probes have three LEDs to indicate logic high, logic low and pulsing.

The third function is very important because the level at a given pin may be rapidly alternating between high and low. With an ordinary testmeter the reading obtained is somewhere between the two levels. We can not be certain whether this indicates a steady voltage at some intermediate level (which probably indicates a fault) or whether it is rapidly alternating between high and low logical levels (which probably means that it is functioning correctly).

With a logic probe, the 'high' and 'low' LEDs may glow slightly, but this again could mean either an intermediate level or rapid alternation. The 'pulsing' LED indicates clearly that the voltage is alternating.

Some probes also have a **pulse stretcher** LED. Many circuits rely on high or low triggering pulses that are far too short to produce a visible flash on the 'high' or 'low' LEDs. The pulse stretcher flashes for an appreciable time (say 0.5 s) whenever the probe detects a pulse, even if if lasts only a few milliseconds.

A related device is the **logic pulser**, which produces very short pulses that effectively override the input from other logic gates. They can be used in the similar way to an audio signal generator to follow the passage through a system of a change in logic level.

A **logic analyser** is used for analysing faults in complex logic systems. These can accept input from several (often up to 32) points in the system and display them in various combinations and on a range of time-scales. It can detect and display **glitches** in the system as short as 5 ns duration. The existence of a glitch indicates that there is an unexpected **race** in the system, producing incorrect logic levels.

Glitch
An unintended short pulse resulting from timing errors in logic circuits

Race
When two or more signals are passing through a logical system, one or more of them may be slightly delayed, perhaps because a certain device has a longer delay time than the others. Logic levels do not change at precisely expected times and the result is often a glitch.

There are also input circuits triggered by various input combinations, making it possible to identify particular events in the system and examine circuit behaviour as they occur. A logic analyser requires expertise on the part of the operator, both in using the equipment and in interpreting its results.

Signature analysis is a technique for checking that each part of a logical circuit is acting correctly. When the system has been assembled and found to operate correctly, its inputs are fed with a repeating cycle of input signals. The voltage levels at each key point in the system are monitored and recorded for each stage in the cycle. These are the **signatures** for each point.

If the system develops a fault, its inputs are put through the same cycle and the signatures monitored at each key point. If the inputs of a device have the correct signatures but its outputs show incorrect signatures it indicates that the device is faulty. This technique is one that can be made automatic, so that circuit boards may be tested by computer. Very complicated systems can be put through the testing cycle in a very short time. But the input sequence must be expertly planned, or there may be errors that the test fails to reveal.

Software faults

When a system is controlled by a microprocessor or a microcontroller it may happen that the fault lies not in the hardware but in the software. The tests described above fail to show anything wrong with the power supplies, the circuit board, the wiring or the individual components, but still the system operates wrongly. It may be that there is a **bug** in the program, even one that was there from the beginning. In many programs there are routines that are seldom used and it is relatively easy for a bug to lurk there while the rest of the program functions correctly.

> **Bug**
> A software fault which causes the microprocessor to perform the wrong action.

Eventually the faulty routine is called and the bug has its effect. The same thing can happen as a result of faulty copying of a program or if one of the elements in a RAM is faulty. It is necessary only for a single '0' to change to a '1' or the other way about. Then the machine code is altered, with possibly dramatic effects.

Testing software

Thorough testing of programs at the development stage is essential if all bugs in a program are to be found and eliminated. Although programs can be developed using assembler or a high-level language on a computer, the only way that it can be completely tested is to run it on the system itself.

This is often done by using an **in-circuit emulator**. The emulator is connected into the system by plugging its connector into the socket that the system's microprocessor would normally occupy. The emulator acts in the same way as the microprocessor would act, but it has the advantage that it can easily be reprogrammed if the program does not control the system correctly.

The program can be run and stopped at any stage to allow readings to be taken or to examine the content of memory and of registers in the microprocessor. When the program has been perfected, it is burnt into a PROM for use in the system, and the actual microprocessor chip replaces the emulator.

An emulator is an expensive piece of equipment so its use is limited to industries in which relatively complex logical systems are being produced on the large scale.

Emulators are usually able to simulate a number of different popular microprocessors, which saves having to have a different one for every processor used. But when a new processor is produced, there may not be an emulator to use for testing circuits based on the new processor. This limits the use of in-circuit emulation to the well-tried processors.

Questions on systems with faults

1 Select one of the circuits described below and list all the things that can possibly go wrong with it. For each possible fault, describe the symptoms you would expect the circuit to show, state what tests you would perform to confirm that the fault exists, and briefly state how you would attempt to cure it.

(a) common-emitter amplifier.

(b) common-drain amplifier.

(c) thermostat circuit using a thermistor, a BJT and a relay.

(d) BJT differential amplifier.

(e) tuned amplifier.

(f) non-inverting op amp circuit.

(g) op amp adder.

(h) precision half-wave rectifier.

(i) BJT push-pull power amplifier.

(j) triac lamp dimmer.

(k) half adder.

(l) magnitude comparator.

(m) pulse-generator based on NAND gates.

(n) astable based on 7555 ic.

(o) microcontroller receiving input from a light-dependent resistor, and sending output to a MOSFET, which switches on a lamp in dark conditions.

2 What are the main questions to ask when presented with a faulty circuit or system?

3 List the common faults that should be looked for first when checking a faulty circuit. Explain how to test for them.

4 What simple items of equipment can be used to test an analogue circuit? Explain how to use each item.

5 What simple items of equipment can be used to look for faults in a digital circuit? Explain how to use each item.

Explain briefly (a) the main uses of an oscilloscope, (b) signature analysis, and (c) in-circuit emulation.

36 Input and output

Microelectronic systems are also called **computer systems** or **intelligent systems**. All use a computer-like circuit to perform their processing. None are truly intelligent, though the complexity of the things they do often seems like intelligence.

In Part 3, the emphasis is on **microelectronic control systems**. Analogue control systems are described in Chapters 33 and 34. Here we look at their digital equivalents — systems that can do all that the analogue systems can do and much more besides.

The system diagram of a typical microelectronic system has the standard three stages: input, processing, and output. Some have a few more stages than this — it depends on to what depth we analyse their action.

The processing stage of a microelectronic system consists of a microcontroller or a microprocessor. Both are VLSI integrated circuits. The essential difference

> **VLSI**
> Very large scale integration.

is that a microprocessor has a greater processing capacity and operates in conjunction with other ICs such as memory, a system clock, and input-output ports. The microcontroller has all these and more on the same chip.

Because Part 3 is about the control applications of microelectronics, we are mainly concerned with microcontrollers, which are used in the majority of control systems. From now on, we refer to microcontrollers as 'controllers', because it is a shorter word. 'Controller' also includes systems based on microprocessors, because these can do the same sorts of things.

The main difference as far as input and output are concerned is that external circuits are connected directly to the pins of a microcontroller, but a microprocessor is connected through an **input/output port** IC.

One-bit input

The simplest input device is a switch. This category of input includes push-buttons, limit switches on industrial machinery, keyboard switches, pressure mats in security systems, microswitches like those of a 'mouse', and tilt switches.

An input device usually needs a circuit to link it to the controller. The circuit is known as an **interface**. The basic interface circuits below can be used with all switches. Circuit (a)

> **'Controller'**
> Remember that in Part 3 this term includes microcontrollers and microprocessors.

normally gives a low (logic 0) input but gives a high input (1) when the switch is closed. Circuit (b) works in the opposite sense. By reading the voltage at the pin, the controller can discover which of the two states the switch is in. This is **binary input**.

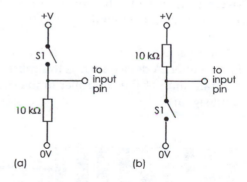

(a)   (b)

Two frequently-used ways of providing 1-bit input to a controller. The value of +V must not exceed the operating voltage of the controller.

A potential divider and transistor switch are used to interface a resistive sensor to a controller. The diagram overleaf shows a circuit for interfacing a thermistor.

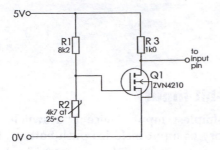

When the temperature is below the set point, the MOSFET is on and the input to the microcontroller is logic 0. The set point is adjusted by varying R1.

This circuit is used to indicate one of two states: the temperature is either *below* the set point (logic 0) or it is *above* it (logic 1). If the output of the circuit is likely to hover at the midway point, the output is fed to a Schmitt trigger circuit before being sent to the controller. A Schmitt trigger is also used to eliminate the effects of noise in the sensor circuit. The transistor switches of Chapter 2 can all be adapted to provide 1-bit input.

Set point

In this case the temperature at which the transistor is to switch on or off (compare p. 269).

A 1-bit input is used to measure linear position in a **bar encoder**. A device such as this relies on the programming of the controller to interpret the input signal.

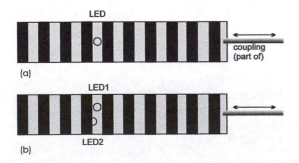

A bar encoder has a plate with alternate transparent and opaque bars. At (a) a single LED allows the distance and speed to be measured. At (b) a pair of LEDs allow direction to be measured as well.

The barred plate moves between an LED (usually infrared) and a photodiode. The photodiode is interfaced to the controller. In (a) the controller alternately receives '0' and '1' as the plate moves. If the controller is programmed to count the 0 to 1 transitions, the position of the plate (and of the machine part to which is is coupled) is known. By timing the transitions the controller can also calculate the speed of motion.

With two LEDs and photodiodes and a 2-bit input, as in diagram (b), the direction of motion can be read as well. The diagram below shows the signals.

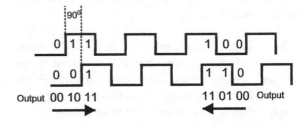

The output of a 2-bit bar encoder consists of two pulsed signals, one shifted 90° with respect to the other.

When the plate moves to the right, input 2 (upper plot) changes state before input 1 (lower plot). The change from 00 to 11 passes through 10. Conversely, as the plate moves to the left, input 1 changes before input 2. The change from 00 to 11 passes through 01. The controller is programmed to read these changes, so detecting the direction of motion.

The same principle is used to read the rotary motion of a shaft. The shaft bears a circular plate marked with radial bars. The controller can detect the angle turned, the rate of rotation and (with 2 diodes) the direction of rotation.

Multi-bit input

An alternative device for position sensing is shown in the diagram opposite. The diagram shows only 4-bit input for simplicity, but input may be 8-bit or more to give greater precision.

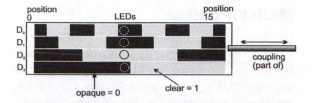

opaque = 0   clear = 1

When the encoder is at position 7, as in the drawing, the code read by the optical system is 0100. If the encoder moves to position 8, only one digit, D3 changes and the code becomes 1100.

The plate is transparent but marked with an opaque pattern. There are four LEDs on one side of the plate and four photodiodes on the other side. The pattern is not based on the binary number system but on a sequence of codes known as the **Gray Code**.

The principle of the Gray code is that only one digit changes at a time. As seen from the diagram, only one digit changes as the encoder moves from position 7 to position 8. If the encloder were marked with the binary equivalents, the change would be from 0111 to 1000 — *all* the digits would change. The difficulty is that, owing to slight errors in the alignment of the LEDs and photodiodes, they would not all change at *exactly* the same instant. Going from 0111 to 1000, they would probably change one at a time. We might get a sequence such as 0111, 0101, 0100, 0000, and 1000. This reads as 7, 5, 4, 0, 8. Using the Gray Code makes this type of error impossible.

The Gray Code is also used for shaft encoding, using a circular plate marked in segments.

If a sensor has analogue output, an ADC is often used to convert this to multi-bit input. Each output of the converter is connected to a single data line. One or two output lines from the controller may be used to instruct the ADC when to begin a conversion and to latch the converted value. There may also be a connection from an 'end of conversion' output of the ADC to tell the controller that conversion is completed and that the converted data is on the lines waiting to be read. This is another example of **handshaking**.

Isolated inputs

In industrial applications it is generally advisable to isolate the controller from external circuits. The voltages used in industrial plant are often higher than those used for powering a controller. Also heavy current switching in industrial circuits produces large voltage spikes that must not be allowed to reach the controller.

The input terminals of programmable logic controllers (PLCs, pp. 324–325) are isolated from the external circuits by means of a circuit like that shown below.

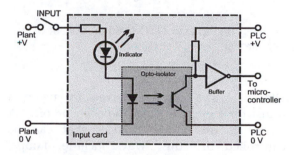

One of the 8 or more channels of a typical DC digital input card of a PLC. The supply from the plant may be at a relatively high voltage, such as 24 V, often unregulated. That used by the controller is often at 3 V or 5 V and is regulated.

The signal from the plant is fed through a dropper resistor and indicator LED to an opto-isolator. When the input is at logic high (say 24 V) the indicator LED is on, and so is the LED in the opto-isolator. The light from this switches on the transistor. The voltage at its collector falls to logic low. This is inverted by the NOT buffer gate, giving a logic high that is sent to the controller.

PLC input cards are also made that operate with an AC input from the plant. An AC voltage is equivalent to logic 1 and zero voltage to logic 0. The AC voltage is passed through an indicator lamp, then rectified by a diode bridge. The resulting DC signal goes to an opto-isolator as in the diagram above.

One-bit output

Low-current devices, such as an LED can be driven directly from an output pin of a controller. A series resistor is needed to limit the current to a safe level.

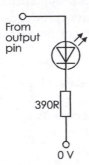

From output pin

390R

0 V

Driving an indicator LED from an output of a controller.

Other devices can be driven by using a transistor switch to turn them on or off. A circuit is on p. 18. Devices that can be controlled in this way include lamps, solenoids, motors, relays, and speakers.

The circuit on p. 297 shows a switched mode motor speed control. In this, the transistor is switched by the output from a 7555 astable. Instead of this, it could be switched by the pulsed output from a controller. The controller is programmed to make one of its output pins go alternately high and low. The program can vary the mark-space ratio and thus control the speed of the motor. The brightness of a lamp can be controlled in a similar way.

PLCs have output cards to interface the controller to the higher voltages and currents met with in the external circuits. The signal from the controller goes to an opto-isolator. The output from the transistor in this then switches on a power transistor connected to the external circuitry.

For controlling heavy loads operating on AC, pulses from a controller are used to trigger a thyristor or a triac. The circuit used is similar to that shown on p. 142

Multi-bit output

Display driving requires several 1-bit outputs, for driving LEDs directly or switching lamps by means of transistors. The LEDs may be the segments of a seven-segment display or the pixels of a dot-matrix display. An example of a program for driving a seven-segment display is given on pp. 338-339. This illustrates the way in which programming replaces hard-wired circuitry or special function ICs for performing a complex activity.

A DAC (p. 222) is used to convert a multibit output to its analogue equivalent. The op amp adder circuit can take a 4-bit output and convert it into a voltage range from 0 V up to the maximum in 15 steps.

Activities

Refer to the programming activities suggested in Chapters 37 and 38.

Questions on input and output

1 Design a 1-bit input circuit to signal to a controller that:

   (a) a pedestrian is waiting to cross the road (Note: the pedestrian is not required to press a button).

   (b) on a bottle-filling machine, a bottle is full of fruit juice.

   (c) the barrier at a car park exit is fully raised to allow the car to leave.

   (d) the number of cars in a car park equals the number of parking spaces.

   (e) the temperature of a greenhouse is almost down to freezing point.

2 What is an opto-isolator? Give an example of the use of an opto-isolator in the input circuits of a control system, with a diagram of the circuit.

3 Draw a block diagram to show how the weight of a packet of tea can be measured with 8-bit accuracy on a packet-filling machine and the weight read by a controller (program details not required).

4 Describe the action of a solenoid. Describe how a solenoid may be driven from a 1-bit controller output.

5 Draw a diagram of an interface for:

(a) controlling a stepper motor,

(b) controlling the speed of a 6 V DC motor.

6 Design a system for controlling a lift that operates between two storeys. Draw a block diagram (Note: programming is not required).

7 List three output devices that give a visual signal and suggest one use for each.

8 Design a controller system to keep a tank topped up with water for a greenhouse irrigation installation. (Note: programming is not required).

9 Explain what is meant by a Gray Code and give the reason for using it.

Multiple choice questions

1 The important thing about the Gray Code is that:

    A all its values are binary.
    B consecutive values differ in one bit.
    C it is completely random.
    D each value contains an even number of '0's.

2 A microcontroller is an example of:

    A VLSI.
    B MSI.
    C IC.
    A EMI.

3 One way to protect a microcontroller from the large voltages and currents often found in industrial plant is to use a:

    A transistor.
    B opto-isolator.
    C protective diode.
    D PLC.

4 In the diagram at the top of the next column, closing the switch makes the input go to:

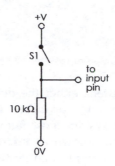

    A logic high.
    B 0 V.
    C an unknown voltage.
    D 5 V.

5 In the circuit below, the control input comes from a microprocessor. The speed of the motor is varied by varying the:

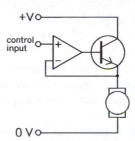

    A operating voltage, +V.
    B frequency of the pulses from the microprocressor.
    C voltage of the input.
    D mark-space ratio of the pulses.

37 Processing

Personal computers, either the desktop or the laptop versions, are probably the kind of microelectronic system most familiar to most people. They are known as **microcomputers.** There are more complicated systems, such as the **workstations** used by geologists, for example, to analyse seismic data when prospecting for oil. These come somewhere between microcomputers and the **minicomputers** used by medium-scale businesses. The most complex systems are the **mainframe** computers used by the largest businesses such as banks and airlines, or by researchers in highly technological applications such as aircraft design. But the distinctions between types of computers based on size are rapidly becoming out of date. The present day personal computer (PC) has far more power than the mainframes of, say, a decade ago. Also a number of smaller computers can be networked to work together with the effective power of a **super-computer**.

At the other extreme to the mainframe is the 'computer on a chip', or **microcontroller**. This is a single IC capable of performing all the functions of a computer but lacking some of the connections to the outside world. It may have no monitor, or only a 16-key keypad instead of a full typewriter-style keyboard. Often it has a fixed program built into it, to perform one particular task. Such microelectronic systems are found in hundreds of different devices, from microwave ovens and smart cards to washing machines and video games machines.

The important thing about all these microelectronic systems is that they all have very similar structure (architecture) and all work in much the same way. Some may lack certain stages and others may have specialised additions, but basically they are all very similar. In the descriptions that follow we refer mainly to microcontrollers, the most widely-used of the microelectronic systems.

From bits to terabytes

Some of the terms mentioned in this section have been defined earlier in the book, but these are the fundamental units of data for microelectronic systems so they are repeated here, from the smallest to the currently largest:

- **Bit:** the unit of information. Takes one of two binary values, 0 or 1. In logical terms, these may mean TRUE and FALSE. In positive logic, 1 is TRUE and 0 is FALSE, and this is the system most commonly used. In electronic terms, 0 means low voltage (usually very close to 0 V) and 1 means high voltage (usually close to the supply voltage, which is +5 V in TTL systems.).

- **Nybble:** The upper or lower four bits of a byte.

- **Byte:** A group of eight bits. Bytes are used in very many microelectronic systems as the basic unit of data.

- **Word:** the unit of data used in a particular system. In an 8-bit system the word is 8 bits long, so a word is a byte. In a 16-bit system, a word is 16 bits long, or two bytes. In a 32-bit system a word is four bytes.

- **Kilobyte (kB):** According to the metric system, this would be 1000 bytes but in digital electronics this term means 1024 bytes, or 2^{10} bytes. This term is often used when describing the data storage capacity of small devices such as the memory of microcontrollers, and the lengths of typical data files.

- **Megabyte (MB):** This is 1024 kilobytes, or 2^{20} bytes, or 1 048 576 bytes. It is *just over* a million bytes. This unit is often used in describing the data storage capacity of devices such as floppy disks (1.44 MB), memory chips, CD-ROMs (650 MB or more), and long data files (especially those containing photographs with over 16 million different colours).

Gigabyte (GB): This is 1024 megabytes or 2^{30} bytes. It is just over a thousand million bytes. The capacity of the hard disk drives of PCs is reckoned in gigabytes. DVDs store 4.7 GB. This is the largest unit in common use at present but a professional photographer, for example, may use several terabytes (= 1000 GB) for storing a collection of high-resolution digital images.

Busses

The block diagram below shows the main features of a microcomputer. It consists of a number of units connected to each other by busses. A bus may be serial or parallel:

- **Serial bus**: a single conductor along which the bits are signalled one at a time by making the line voltage high or low.

- **Parallel bus**: a set of conductors, often 8 or 16 running side by side. At any given moment the buses each carry one bit of a word, so the bus is able to carry a whole byte or double-byte or more of data.

A serial bus is simpler to wire up or to etch on a printed circuit board, but a parallel bus transfers data much more quickly.

Most busses in a microcomputer are parallel busses and are divided into three types:

Data bus: used for carrying data between different parts of the system. The data may be numerical values, coded text or coded instructions.

Address bus: used by the microprocessor for carrying addresses of devices (such as the hard disk drive) or locations in memory. When the microprocessor calls a device into action (to receive data, for example) it places the address of that device on the bus. The addressed device then registers the data or instruction that is on the data bus at that time.

Control bus: used by the microprocessor for controlling the other devices in the system.

Only the microprocessor can place signals on the control and address busses, but the data bus can have data put on to it from several sources. These include the microprocessor, the memories, and the keyboard. In such a system there may be times when several devices may be trying to put data on the bus at the same time. This is referred to as **bus contention** and must not be allowed to happen. To prevent bus contention, each device that may need to put data on the bus has **three-state outputs**.

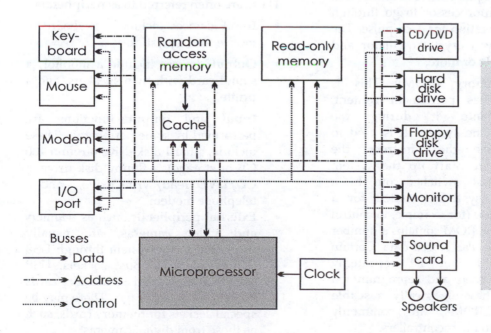

Busses
→ Data
---→ Address
·····→ Control

A microcomputer system consists of a microprocessor, a system clock, memory, and a number of devices to interface the system to the outside world. The parts of the system communicate with each other by placing signals on the busses.

These outputs are under the control of the microprocessor (through the control bus). At any given instant, a three-state output may be in *one* of the following three states:

- Enabled, with low output.
- Enabled, with high output.
- Disabled, with high-impedance output; in effect the output is disconnected from the bus so that it can put no data on it.

The microprocessor allows only *one* device to be connected at any one time. Only that device is able to place data on the bus.

Elements of a microcomputer system

The devices that make up a typical microcomputer system are:

- **Microprocessor:** this is a highly complicated VLSI device and is the heart of the system. Its function is to receive data from certain parts of the system, act upon it according to instructions presented to it in a program, and to output the results. It is described in more detail later.

- **Clock:** This is a square-wave oscillator that causes the microprocessor to go through one cycle of operation for each pulse. The clock is usually a crystal oscillator and operates at 2 GHz or more.

  Read-only memory (ROM): This is memory that has permanent content, programmed into it during the manufacture of the system. It is used to store programs that instruct the microprocessor to start up the system when it is first switched on. In a microcontroller system intended for a particular purpose (for example, to control a dishwasher) the ROM contains a number of programs and also stored data. Certain types of ROM can be repeatedly programmed, erased and reprogrammed electronically. These **electrically eraseable ROMs**, or **EEROMs**, are commonly incorporated into microcontrollers.

- **Random-access memory (RAM):** This is a temporary data store, used for storing the whole or parts of the program that is currently being run, tables of data, text files, and any other information which needs to be registered temporarily. Typical ROM size is 512 MB.

- **Cache memory:** Certain systems, particularly those required to run at high speed, include a small amount of special RAM. This consists of chips that operate at very high speed. It would be too expensive for the whole of RAM to be built from such chips. There may also be a small cache memory built in to the microprocessor.

  Data that is often required is stored in the cache so that it is quickly accessible. Usually the cache is operated so that data on its way to or from RAM is held there for a while. This data is later overwritten. In some systems it is overwritten if it is infrequently used. In other systems, the overwritten data is that which has been there for the longest time. Either way, the data held in the cache is the data most likely to be needed in a hurry.

- **Input/output devices**: To the left and right of the diagram on p. 319 are the units that provide input and/or output to the system. These are often referred to as **peripherals**.

  * **Input** devices include a keyboard, a mouse, and a joystick.

  * **Output** devices include a monitor, a sound card with twin speakers, and a printer.

  * **Input and output** functions are performed by some peripherals. These include, a hard disk drive (around 180 GB of storage), a floppy disk drive, a CD/DVD read/write drive, and a telephone modem.

- **External peripherals** such as scanners and video cameras, are usually connected to the system through USB (Universal Serial Bus) sockets. USB sockets also accept flash memory cards for off-line data storage. There may be special sockets for memory cards, such as those from digital cameras.

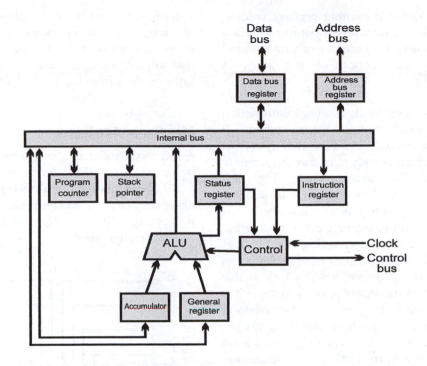

A microprocessor contains the control unit, the ALU and a number of registers (more than shown here) all connected by one or more internal busses, and all on one silicon chip.

Microprocessors

There are many types of microprocessor, each with its own particular features and architecture. The diagram above shows the main parts of an imaginary microprocessor that has most of the features typical of all microprocessors. The heart of the microprocessor is the **control unit**, which finds out what the microprocessor must do next and oversees the doing of it. It goes through a repeated cycle of operations, the timing of which is regulated by the **system clock**. Control lines run from it to all other parts of the microprocessor. It also has connections to certain external parts of the microelectronic system through the control bus.

The microprocessor is also connected to the data and address busses of the system. The **data register** holds outgoing and incoming data and, under the supervision of the control unit, transfers data between the external data bus and the internal bus of the microprocessor.

Similarly, the **address register** transfers addresses from the internal bus to the external address bus. It transfers addresses in only one direction, but the data register has two-way transfer.

The **program counter** (PC) is a register holding an address that indicates where the microprocessor has got to in memory, following its way through the program that is stored there. In short, it is like a bookmark. At each cycle, the microprocessor reads a byte from the address stored in the PC. Then the address is automatically increased by 1, so that it 'points to' the next address to be read. Sometimes the microprocessor needs to jump to a different part of memory and continue from a new address. The control unit puts the new address in the PC, to direct the microprocessor to the different part of the program.

> **Pointer**
>
> A register holding an important address, such as the next address to be read.

The **stack pointer** is another register holding an address. Here the address points to a place in memory where very important data is stored temporarily. The action of the stack is described on p. 344.

The **status register** holds essential information about the operation of the microprocessor. It usually has a capacity of one or two bytes, but the information is mainly stored as the separate bits within the bytes. These bits are individually set to 1 or reset to 0 to indicate certain events. Such bits are usually called **flags**. For example, every time a calculation gives a zero result, the **zero flag** is set to 1. If we want to know if the most recent calculation gave a zero result or not, we can look at this particular bit in the status register to see if it is 1 or 0. Another flag may be set when a calculation gives a positive result. A further example of a flag is the **carry flag**, which is set when there is a carry-out from a calculation, that is, when the result is too big to be stored in the register.

The **instruction register** holds the instructions recently read from memory. These tell the control unit what to do next.

Closely associated with the control unit is the **arithmetic/logic unit** (ALU) which performs all the calculations and logical operations. Most of these operations involve the **accumulator**. A value placed in the accumulator is operated on in one way or another by the ALU, according to the instructions it has read from memory. Then the result is placed on the internal bus, and is often circulated back to the accumulator to replace the value that was there before. At the same time, one or more of the flags in the status register are set or reset, depending on the result of the operation. Some operations involve two values, one stored in the accumulator and one in a general register. The diagram shows only one general register, but most microprocessors have several of these to make operations more flexible. The role of the accumulator and the way in which it is concerned in almost everything the microprocessor does is made clearer by examples given in Chapter 38.

The diagram and the description above apply to most microprocessors and to most microcontrollers too. The main difference is that microcontrollers have memory and other units on the same chip as the CPU.

Microcontrollers

A microcontroller is a single VLSI chip which holds not only the **central processing unit** (CPU, the equivalent of a microprocessor), but also memory, a system clock and other processing units. The diagram below shows a typical arrangement.

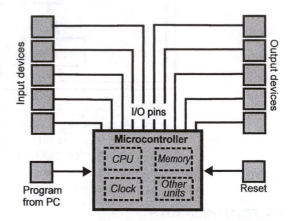

A microcontroller communicates with the rest of a microelectronic system through a number of 1-bit input/output pins.

The input devices in the diagram might be any of the 1-bit inputs described in Chapter 36, such as push-buttons, limit switches, or opto-isolators. Or a group of pins may receive multi-bit input from an input device such as a Gray encoder, or an ADC. However, many microcontroller chips have one or two on-board ADCs.

The output devices might be any of the 1-bit outputs described in Chapter 36. Usually it is possible to drive an indicator LED directly, for a typical output pin can source up to 20 mA, or sink up to 25 mA. Devices such as solenoids and motors are driven by using a transistor switch, either a BJT or an FET, as explained in Chapter 2.

This member of the PIC family of controllers is in a surface-mount package that has 28 pins spaced 1.27 mm apart. Twenty of the pins are for input/output. In addition to the units shown in the diagram opposite, the chip carries two digital timers, an 8-bit real time clock, and a watchdog timer. It runs on a 2.5 V to 6.25 V supply, at up to 20 MHz.

The diagram shows a reset input; a low level on this pin causes the controller to start running its program from the beginning. There is also a temporary connection to a computer, which is used to download the program into the memory of the controller.

The photo (right) shows an Atmel controller being programmed while plugged into a socket on a programming board.

Writing the binary code of the program into the ROM of the controller is under the control of computer software (see the screen at bottom of the page). The person creating the program, types in instructions telling the controller which of its inputs to check on and what to do if the inputs are at logic high or low (depending on the sensors attached to the inputs). The program tells the controller what to do (that is what outputs to activate) when it detects certain combinations of inputs.

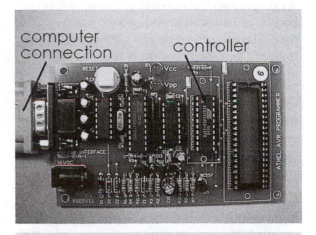

A 20-pin Atmel controller on a programming board. The large socket on the right is for 40-pin members of the Atmel AVR family. There is a connection to the computer's serial port.

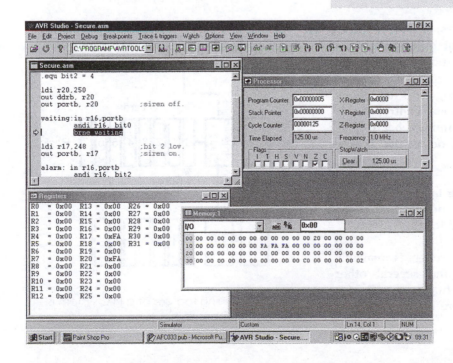

A screen shot showing a program being written to control a security system. The program is being written in assembler (pp. 346-349) in the top left window. Other windows show the state of the registers in the processor, the 32 data registers and the input and output registers. The program is run and tested on the computer, using this software to simulate the processor. When ready, it is downloaded into the controller.

When programming software is run, the computer screen looks something like that at the bottom of the previous page. The operator is typing the program into the window at top left. The operator uses a special 'language' to tell the controller what to do (more about that in Chapter 39). The other windows tell the operator about the state of various parts of the controller. This includes a list of all the input and output pins and the present state of each of these. Using software such as this, the operator creates a program and tests it before actually writing it into the controller. The program can be run at its usual speed or just one step at a time. Stepping through gives the operator time to check that the program is working correctly at each step.The operator can also decide on key stages of the program at which the run should stop. When the program stops at the breakpoint, the operator can study the contents of memory and registers, and the state of the outputs at the end of each stage.

Up to this stage, the controller itself has not been programmed. Everything is being done in the operator's computer. When the operator is satisfied that the program works correctly, it is downloaded into the ROM of the controller. While doing this, the software in the computer converts the instructions keyed in by the operator into a binary code that the processor can 'understand'.

If the program is still not quite right, it is corrected on the computer and the new version downloaded into ROM, to replace the original version. When the program has been thoroughly checked, the controller is transferred from the programming board to its socket on the board of the device that it is to control.

From now on, the controller takes command of the system and is entirely independent of the computer that programmed it.

A very wide range of controllers is available from Atmel, Microchip and several other manufacturers. When selecting a controller for use in a given project there are a number of features to be considered:

- **Size of program memory**: 1 kB or 2 kB is enough for controlling the simpler systems.

- **Size of RAM**: 64 bytes to 128 bytes is typical.

- **Number of data registers**: these are used for storing and manipulating data.

- **Clock rate**: most controllers are easily fast enough for most applications.

- **Supply voltage**: 2 V to 15 V; most controllers can be battery-powered.

- **Number of input/output pins**: from 6 to 20 or more.

- **Instruction set**: number of instructions. A few instructions (as few as 33 in some cases) make it easier to learn to program the controller.

- **Special on-chip units** (if needed): may include analogue comparator (1 or more), ADCs (1 or more), DACs, op amps, voltage reference, and display driver.

Programmable logic controllers

PLC systems are built up from a number of ready-made units that can be mounted on a rack and connected together in various ways. These allow great flexibility in building a system for a particular application.

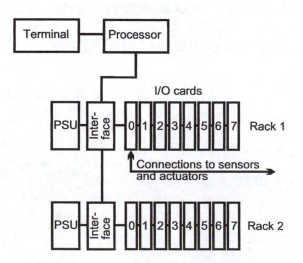

The distinguishing feature of a PLC system is the large number of inputs and outputs.

The door of this PLC cabinet is open to reveal three PLCs on the right, with numerous connections to the sensors and actuators of the colour-printing press that they control. Relays on the left are used for switching.

A **programmable logic controller** is a simple computer-like device especially designed for industrial control systems. PLCs are often used for other purposes too, such as controlling the automatic microwave systems that open and close shop doors. A typical PLC is based on a specialised microprocessor. There is a terminal which usually has a few keys for operating the PLC and an LED or LCD display to show the current status of the system. The terminal also contains the program and data memory. The control sequences and logic are set up on a PC or a special programming unit which is not part of the PLC. When the program has been written it is downloaded into the memory of the PLC, which then runs independently.

The most noticeable feature of a PLC is the large number of inputs and outputs. Each input and output 'card' communicates with the outside world through a number of ports, usually 8 or 16 ports for each card. There may be several racks of cards, because controlling a large industrial unit such as a production line or a printing press involves numerous sensory inputs and a corresponding number of outputs. However, for elementary tasks there are smaller PLCs that have only about 6 inputs and outputs, each complete with interface circuits and housed as a single unit.

Addressing

It is important for the processor to be able to communicate with any other part of the system. This is done by giving every part of the system an **address.** The address is in binary, a set of low and high voltage levels on the lines of the address bus, but we nearly always write it in hex to make it easier to read and understand. The address bus has a line for every binary digit of the address. In a microcontroller system the address bus is internal, that is, it is on the controller chip.

> **Hex**
> The hexadecimal numbering system.

Example
A system has 11 address lines. The lowest possible address is 000 0000 0000 in binary. All lines of the address bus carry a low voltage. The highest possible address is 111 1111 1111 in binary. All lines carry a high voltage. This is the equivalent of addresses from 0 to 2047 in decimal or from 0 to 7FFF in hex. It is a 2-kilobyte system.

> **Note**
> If you find these numbers difficult to understand, turn to the topic on 'Number systems' on p. 330.

Sixteen address lines is a common number for microprocessor systems, but some have 20 address lines (1 MB of addresses) or more. Microcontroller systems need fewer lines, and the address bus is internal, on the chip.

The way that the memory space of a system is allocated to different functions is set out as a **memory map**. A typical memory map of a 64 kB computer system is shown overleaf. Different parts of the system are allocated their own addresses within this memory space. At the lowest addresses in memory there is a ROM chip, containing instructions to the microprocessor on what to do when the system is being started up. The microprocessor has 0000 in its program register when it is first powered up so it always begins at address 0000 and goes there first to find out what to do. There it is given instructions to jump to some other address, perhaps in another ROM, where it finds instructions for starting up the system.

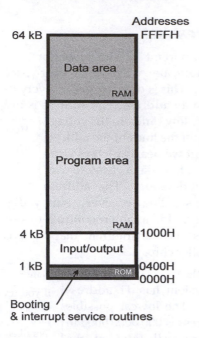

The amount of space in a memory map depends on the number of address lines that the system has. The computer system shown here has 16 address lines. The memory space is allocated to RAM, ROM and all the input and output devices of the system. Not all of the memory space is necessarily taken up; there are ranges of spare addresses where future additions to the system may be located. The numbers on the right are starting addresses of blocks of memory, the H indicating that they are written in hex.

The microprocessor may also go to this low memory when it has received an interrupt (see later), and is then made to jump to one of several interrupt service routines.

The next area of memory may be used for devices such as sound cards and other peripherals. Often these occupy a small range of addresses and the card has logical circuits to decode addresses within its range and to take action accordingly. In general terms, this is referred to as the input/output area of memory space. The devices with addresses within this range are not memory, but they are **memory-mapped**. That is, they are accessed by the microprocessor by using addresses, just as if they were actual memory.

RAM for holding programs and data takes up the main part of the address space. Some addresses in RAM are reserved for temporary storage of data. One example is the stack, which is discussed later.

The memory map of a microcontroller has fewer addresses but is similar in principle. There is a program area of a few kilobytes, but this is ROM, not RAM. In the controllers used in program development this ROM is electrically eraseable, as already explained. However, once a design has been perfected and is to be mass-produced, a version of the controller with a **one-time programmable ROM** is used. The program is burned permanently into this and can not be altered.

The controller also has a small area of RAM for the data area and for the input/output ports.

The memory map of the processor unit of a PLC is organised in a similar way, with EEROM for the program area, and RAM for a large input area, a large output area and a data storage area.

Processing

Having looked at the main devices forming a microelectronic system, we will now see how they operate. This description is based on the operation of a typical controller, such as the PIC or Atmel ICs.

- A controller can do nothing without instructions. The instructions are stored in the program memory as a sequence of operations codes or **op codes**. The central processing unit (**CPU**) of the controller reads and obeys this sequence of instructions one at a time. The stages of the operation are:

- the CPU takes the address stored in the program counter (**PC**),

- it puts that address on the address bus,

- the addressed unit of memory puts its stored data on the data bus,

- the CPU loads the data into its instruction register,
- The CPU does what the instruction says, usually with the help of its arithmetic-logic unit (ALU) and the accumulator.

This process is known as the **fetch-execute cycle**. It is timed by the system clock. The program counter is incremented by 1 every time an instruction has been executed. So the controller goes to the next address in memory to fetch the next instruction. In this way the controller works its way through the program, instruction by instruction, from beginning to end.

To ensure that the controller always starts at the beginning, the PC is reset to 0000H when the power is first switched on or the reset button is pressed. As we shall see, there may be instructions that tell the controller to jump to other parts of the program instead of steadily working its way through it. In such cases the PC is loaded with the address to which the contoller is to jump.

Once an instruction is in the instruction register it may be executed straight away or its execution may require further fetch-execute cycles. The CPU may need additional data to work on. However, most controllers execute the program in single cycles.

As an example of a fetch-execute cycle, we will see how the PIC16F690 increments a value held in one of its file registers. The file registers are a set of 256 data registers, each able to hold 1 byte. The accumulator of the CPU, called the working register (W) in this controller, is also 1 byte wide.

In this controller, instructions are not sent along the data bus as stated on p. 319, but along a special instruction bus, which is 14 lines wide. This allows the instruction register to have 14 bits. Fourteen-bit instructions include all the data needed for their execution, so they are executed in a single cycle. An example of a 14-bit instruction is:

00 10 10 0100101

What actually happens is that the lines of the instruction bus are set to 0 V where there is a '0', and to the supply voltage (say, 6 V) where there is a '1'. This particular combination of lows and highs causes the CPU to perform a particular operation. Let us analyse the opcode to see what the operation is.

The opcode can be considered in three parts:

- The first 6 bits are the instruction: '001010' means 'increment a file register'.
- The next bit is the destination bit. If it is '0', place the result in the accumulator. If it is '1', place the result in the file register, replacing what was there before. In this example, the bit is '1', so the result of incrementing goes in the file register.
- The last 7 bits are the address of the file register to be operated on. '0100101' is the binary equivalent of 37 in decimal.

The outcome of this fetch-execute cycle is that whatever value is stored in file register 37 has 1 added to it, and the new value is placed in that register.

Many operations, including incrementing, affect one or more bits (often called flags) in the status register. Incrementing may affect the zero flag, Z. If the value in register 37 happens to be -1, incrementing it results in zero. In this case the Z flag is set to '1'. In all other cases it is set to '0'. Subsequent instructions may read the Z flag and take one of two actions, depending on its value.

> **Negative values**
> Representing negative values in binary is discussed on p. 369.

As another example, take the following:

01 0010 101 0 0101

This consists of three parts:

The first 4 bits, '0100', give the instruction, which is to clear one particular bit in a file register.

The next 3 bits give the number of the bit to clear. In this example '101' means bit 5.

The last 7 bits give the register number, '0100101', which is 37.

The outcome of this instruction is to clear bit 5 of register 37 to zero. For example, if the register holds 1011 1101 before the operation, it holds 1001 1101 afterward. This operation does not affect the status flags.

Bit numbers

Bits are numbered from 0 to 7, counting from the right, the least significant bit.

The two instructions outlined above are typical. They illustrate the fact that a controller operates in very small steps. A programmer has to break down what may be a very complex operation into a series of very many, very small operations.

The reason that the controller is able to perform complex tasks is that each operation takes a very small length of time to perform. If the processor is running at 20 MHz, a single cycle takes only 50 ns.

Another point illustrated by these examples is that writing a program by setting out the opcodes in binary is a slow and onerous task. There are many opportunities for making mistakes. This is why we normally use special software to help us write and develop programs (see example of programming board and software on p. 323).

Watchdog timer   Extension Box 23

In spite of all the precautions, it may still happen that a bit in memory or in one of the registers of the CPU may become altered by electromagnetic means or by a voltage spike on the supply. Substituting a 0 for a 1 or 1 for a 0 can provide the CPU with the wrong op code, or false data. Worse still, may make it jump from its present place in the program to some unpredictable place in another part of RAM, where there may be no program. The result may be errors in output or the program may crash completely.

The watchdog timer is a technique for avoiding such dangers. It keeps a watch on the CPU, making sure that it is operating correctly. The timer can be an ordinary 7555 timer circuit. More often, it is built in to the controller and runs automatically.

The CPU is programmed to trigger the timer at frequent intervals, perhaps every millisecond. The period of the timer is a little longer than the interval at which it is triggered. When the timer has been triggered, its output goes high and it should be triggered again before the period is over.

The output of the timer is connected to the reset input of the controller and, as long as the timer holds this high, the controller runs normally. If a fault develops and the CPU is no longer following the program correctly, it no longer triggers the timer.

After a millisecond or so, the timer output falls to logic low. This automatically resets the controller, which jumps back to the beginning of the program and starts again.

Activity — Controller systems

The topics in this chapter are best studied by using a microcontroller development system. Use the handbook to learn about the features of the controller that you are using. The handbook will probably contain many example programs for you to run. You may prefer to postpone this activity until after you have studied one of the program languages in the next chapter.

Questions on Processing

1 Name four kinds of microelectronic system, state who uses them and what they are used for.

2 Explain what bit, nybble, byte and word mean.

3 Describe the three types of bus found in a microelectronic system and the type of information they carry.

4 What is meant by *bus contention* and how may it be avoided?

5 List the differences in properties and uses between RAM and ROM.

6 Draw a diagram of a typical micro-computer system and explain briefly what each part does.

7 Draw a diagram of a typical micro-controller system. In what ways does it differ from a microcomputer system?

8 Draw a diagram of a typical micro-controller and explain briefly what each part does.

9 Describe what happens during a fetch-exeute cycle of a microcontroller.

10 What is an opcode? Give an example of an opcode and explain how a controller responds when the opcode is loaded into its instruction register.

Multiple choice questions

1 The unit of information is a:

    A byte
    B bit
    C nybble
    D word.

2 The number of bits in a byte is:

    A 4
    B 8
    C 16
    D 32.

3 A kilobyte is:

    A 1000 bytes
    B 1024 bits
    C 1024 bytes
    D 1000 bits.

4 The control unit of a microprocessor:

    A reads data from the data bus
    B tells the ALU what to do
    C controls the rate of the clock
    D writes addresses on the address bus.

5 If the result of an operation has the value 0, the zero flag:

    A becomes '1'.
    B does not change.
    C becomes '0'.
    D changes state.

6 The register in which the controller performs calculations is called the:

    A status register
    B accumulator
    C program counter
    D central processing unit.

7 The code used by a microprocessor when first switched on is stored:

    A in cache memory
    B in ROM
    C in RAM
    D on the hard disk drive.

8 The PIC16F690 controller has opcodes that are 14 bits wide. The first six bits of the codes for file register operations tell the controller:

    A what value is to be added to the value already in the working register.
    B the address of the register that the controller is to operate on.
    C which operation the controller is to perform.
    D the program address to which the controller is to branch.

9 If the result of certain operations is zero:

    A the program counter is reset to zero.
    B a bit in the status register is reset to 0.
    C the working register is cleared.
    D a bit in the status register is set to 1.

38 Programming

In Chapter 37 we saw that a controller is made to perform its tasks by reading and obeying a sequence of instructions held in its program memory. The instructions are stored as a binary code, a pattern of 0's and 1's. There is a different code, or **opcode**, for each operation that the controller can perform. This sequence of opcode is known as **machine code**.

The code in which a program is written is known as the **source code**. For a program written in machine code, the source code is machine code. However, most people find it difficult to use machine code as the source code. Instead, they use an easier source code.

One of the simplest source codes is **assembler**. Assembler software is available for running on a PC or other computer. Instead of using binary codes for the opcodes, the programmer keys in instructions in abbreviated form. The software turns these into machine code, which is then downloaded into the controller.

Even easier to use is a high-level language, such as **BASIC**. This name is short for Beginner's All-purpose Symbolic Instruction Code. The user types in the instructions in a form that is very similar to ordinary English.

BASIC is easy to learn, and the programs are easily understood. When used in conjunction with a programming board, the BASIC program is written and tested on a computer. Then the software compiles the BASIC program into its machine code equivalent ready for downloading into the controller. Compilers are also available for other high-level languages, such as C.

Programmable logic controllers (PLCs) are often programmed by using a system called **ladder logic**. The reason for the name becomes obvious when you look at a ladder logic program. Ladder logic is intended for use in industrial control systems. It is easy to learn as it has a small set of instructions and a graphical way of setting out the program. It is designed for use by engineers, rather than by computer programmers.

Before writing any program it is best to set out a **flowchart**. It shows the routes by which the controller runs through the program. In the next section there are flowcharts illustrating most of the actions of which a controller is capable. In Chapter 39 we turn some of these into programs written in machine code, assembler, BASIC and ladder logic.

Flowcharts

A flowchart consists of a number of boxes, representing the stages of an operation. The boxes are joined by arrows to show the sequence of the stages. The shapes of the boxes indicate the nature of the stages.

The routines described in this section are divided into seven main sections: initialisation, arithmetical and logical routines, timing routines, data handling routines, input routines, and output routines. Because they are arranged by function, a few of the more complex routines come early in the list. Skip over these the first time through.

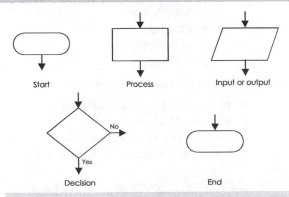

Symbols used in flowcharts. All programs begin with a 'Start' box. Many have no 'End' box.

In the flowcharts, any numerical values are given an appropriate name, written in italics. Examples are *num1*, *num2*, and *sum* used in the first flowchart. These represent two numbers and their sum.

We shall assume that all variables are integers, with values between 0 and 255. This allows them to be coded as single bytes. Working with larger values or with decimal fractions makes the programming much more complicated. In any case, integers between 0 and 255 cover most of the requirements for programming control systems.

In flowcharts (1) to (3), which involve the working of the CPU itself, we use symbol A to represent the accumulator. A left pointing arrow ← indicates copying data from one register to another. For example:

$$A \leftarrow num1$$

This means that the data held in the register which has previously been defined as *num1* is copied to the accumulator. This leaves the contents of *num1* unchanged. The left pointing arrow is also used for assignment, for example:

$$sum \leftarrow A + sum$$

This means that the content of the *sum* register is added to the content of the accumulator and the result placed in the *sum* register. The value held in *sum* is also in the accumulator.

Initialisation

It is usual for an assembler or compiler program to begin with a sequence of instructions to set up the controller for running the program. The flowchart on the right is an example.

The first thing is to declare the type of controller which is intended to run the program. This information is essential for the assembler or compiler to be able to use opcodes that the controller can understand and act on.

Usually, certain features of the contoller need to be configured for the whole of the program. We might need to declare the type of system clock (internal or external) crystal or resistor-capacitor timing, and so on).

Another feature that may need to be set up is: whether the watchdog timer is to be enabled. In the PIC controller this and other settings are coded into a configuration word, which comes next in the program.

Assemblers operate by referring to data stored at numbered addresses in memory. But names are more memorable than numbers so the next thing on the program is a list of the names and the adresses they correspond to. Some of the names may be pre-defined registers such as the status register, the program counter and the input and output ports. Others may be the names of user registers, that we allot to important program varibles.

These names, or labels are for our convenience and are used by the assembler, but the controller itself does not know anything about them. The controller works solely by numerical memory addresses.

The next items may generally be attended to in any order. Usually a port is a byte wide and can be set to to input or output bytes of data. Its bits can also be set individually. This is only the inital setting – it can be changed later. If the controller has the ability of accept analogue input instead of digital, analogue inputs can be defined next.

Usually it is safer to clear the port registers at this stage, so that external devices are not accidentally switched on. But other initial values can be sent to output ports or bit, if required.

There may be several program lines devoted to setting up on-chip devices such as timers and analogue-digital converters.

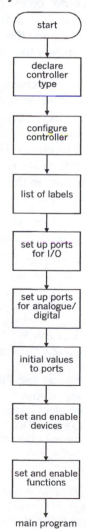

There may also be instructions to enable certain functions, such as interrupts.

Up to this point, the controller has done nothing. The assembler or other programming software has registered information (about the controller and the labels) within the memory of the PC, and set various registers in the controller. All is ready now for the program to be run. The following sections outline some of the things the controller can be instructed to do.

Arithmetical and logical routines

1) Add two bytes stored in two memory locations and place their sum in a third location in memory. This is a task to be programmed in machine code (p. 346) and in assembler (pp. 346-347). It is trivial in BASIC. It illustrates once again the small steps by which the activities of the controller proceed.

In the flowchart, the registers used for *num1*, *num2* and *sum*, have already been defined earlier in the program. Also, the two values to be added are already stored in memory at these locations. The flowchart shows how these are added. It is assumed that adding can be done only in the accumulator.

Adding together two numbers, *num1* and *num2*, to give *sum*. *Num1*, *num2* and *sum* are the labels allocated to three bytes of memory.

It is assumed that *sum* will always be a positive integer, not greater that 255, as it is held only as a single byte.

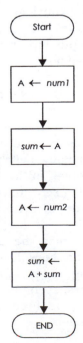

2) Repeat an action a specified number of times. The number of times, *n*, is stored in a register labelled *count*.

The structure of this program is a **loop**. It holds a routine (not specified here) which is to be repeated *n* times. The routine might be simple, such as flashing an LED on and off. Or it might be a complex one, such as the spraying paint on a car body in the assembly line.

The routine in the loop could be a simple addition, such as adding 3 to the accumulator. If A starts by holding 0, it finishes by holding $3n$.

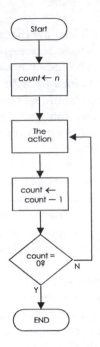

This makes a simple routine for **multiplication,** useful because controllers can add and subtract, but not directly multiply.

3) Conditional jump to another part of the program. In the flowchart below, the controller is able to read the room temperature because it receives input from a sensor through a DAC.

The diamond-shaped **condition box** is where the controller is made to take a certain action depending on whether the temperature is above the set point or not.

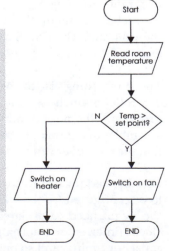

The program has the set point written into it and, at the condition box, the controller determines if the temperature is above that set point or not. If it is above (Y), the program proceeds to an output box, with an instruction to turn on a fan. If it is not above (N), the controller switches on a heater.

This flowchart is not a complete thermostat, as it takes a once-for-all decision, but it can be easily extended for complete control.

4) Logical AND is implemented by two or more decision boxes. They are connected so that the decision must be Y for both (or all) to allow the controller to get to the action.

The flowchart is used for an outdoor security floodlamp. The lamp is to be turned on only if it is dark AND the motion of a person has been detected. The lamp does not come on in daylight, even if a person is detected. It does not come on in darkenss if no motion is detected.

The system has two sensors. One is a light dependent resistor with a potential divider circuit that gives a high output (1) in darkness and a low output (0) in the light. The other sensor is a passive infrared sensor that detects changes in the amount of infrared reaching it. It is able to detect a moving person at a distance of several metres. It produces a high output when it detects a moving person.

The lamp is turned on only if both decision boxes are answered with Y.

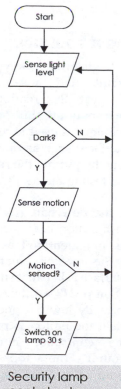

Security lamp control system.

The lamp is turned on for 30 s and then the program returns to monitor light level. The controller cycles through the program indefinitely, so this flowchart has no END box.

5) Logical OR is implemented by two decision boxes connected so that action is taken if either of them yields a Y response. The alarm is sounded if the motion sensor detects a person OR if the pressure pad is trodden on.

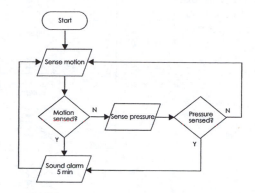

Security system with two sensors. Either one OR the other can trigger the alarm.

Timing routines

6) Delay routine. One way of creating a delay in a program is to make the controller repeat a loop operation a few hundred times. There need be nothing specific for it to do in the loop. The time taken to decrement the loop counter and jump back to the beginning of the loop is multiplied by the number of times the loop is repeated.

Overleaf is a routine for more extended delays. These are **nested loops**. The loop is repeated *m* times for every one of the *n* repetitions of the outer loop. This means that the inner loop is repeated **mn** times altogether. If m and n are both 1000, the inner loop is repeated a million times.

The times obtained with a loop delay routine are not precise. They depend on the rate of the system clock, which is not usually of high precision. Also there may be interrupts (p. 342) which increase the delay.

333

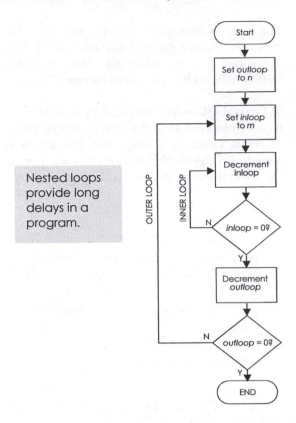

Nested loops provide long delays in a program.

Most controllers are provided with more permanent RAM in which data can be stored for many years or until it is electrically erased or overwritten with new data. Many of the newer types of controller have flash memory for this purpose.

Data is written into data RAM a byte at a time. The writing routine is given the value to be stored and the address at which to store it. The reading routine is given the address to read and the register to store it in. The routines can be placed in a loop (see (2), p. 332) for writing or reading a block of data bytes.

Data may also be transferred serially between the controller and another device. This is done by an on-chip Universal Synchronous Asynchronous Receiver Transmitter (USART). In transmitter mode, a byte placed in a register is automatically coded and transmmitted as a series of pulses from one of the controller's output pins. Conversely, when in receiver mode, a series of pulses with the correct format received from an external device is decoded and placed in its receiver register.

7) Timing an operation with a controller generally makes use of a timer register that is incremented at every fetch-execute cycle. With the high speed of the clock and the limitation of the timer register to 1 byte, the overal time period is very short. Periods can be lengthened by hardware dividers (prescalers) in the controller and by using nested program loops.

If a high-precison crystal is used for the system clock, timing can be accurate. However, the machine code or assembler programs for implementing a timing circuit are too complex to be included in this chapter. On the other hand, they are extremely simple in BASIC or ladder logic and some examples are given later for these languages.

Handling data

8) Storing data in data RAM. When the power is switched off, the values in the controller's registers are lost. This is because the data RAM is volatile – it does not store data permanently.

Input routines

9) Registering a keypress from simple input circuits like those on p. 313 is a matter of repeatedly reading the value of the corresponding bit or input pin. The same technique is used to read the input from sensor interfaces such as those on p. 314. This can be done by putting the read command into a loop and testing the result.

In the flowchart (right) the push-button is connected to the pin identified as pin 3 on the memory map. The interface is like that shown at (a) on p. 313, so the input is normally low but goes high when the button is pressed. The controller circulates around the top loop, continually reading the input, until it returns '1'.

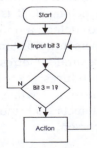

Waiting for a keypress.

The controller then drops out of the loop to perform the action. After completing this, it returns to the beginning to await the next keypress.

This routine is very simple but it has one problem. Unless the action takes several seconds to perform, the operator may still be holding the key down when the action is finished and the controller has looped back to the beginning. Present-day microelectronic systems operate so fast that they can leave the operator far behind. To prevent this happening when detecting a keypress, we add a second waiting stage to the routine.

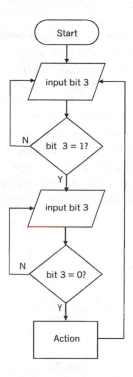

This routine waits for the key to be pressed, then waits for it to be released before proceeding.

These small program segments can be used in a number of ways. For instance, in the flowchart on the right, another version is included at several points in the program. This does not wait for the keypress, but reads it in passing. Action 5 is performed if the key is *already pressed*, but not performed if the key is not pressed.

The main program cycles through a series of four different actions that are labelled Action 1 to Action 4. At certain stages (two in this example, though there could be more) the main program calls up a **subroutine**.

The subroutine is called after Action 1 and performs Action 5 if the key is pressed. The controller is then returned to continue cycling through the other actions. It goes to the subroutine again after Action 3, and there may perform Action 5 again.

As this example shows, after a subroutine is finished, the controller always returns the processor to the point in the main program at which the subroutine was called.

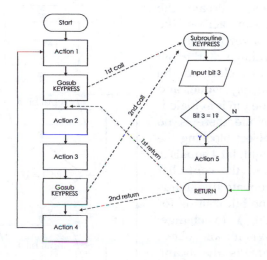

The main program calls the subroutine to find out if the button is pressed.

Action 5 could include accepting fresh data that modifies Actions 1 to 4. This is a way of allowing the user to change the operating mode of the device by pressing a button at any time.

In a longer program, the subroutine could be called many times. Repeatedly reading a register (or a bit) to determine its current state is called **polling**. An alternative to polling is to use **interrupts,** as explained later.

10) Counting events, such as the number of packages passing by on a conveyor belt, is programmed in the flow chart below. This flowchart illustrates two important principles. The first is that all stored values must be **initialised** at the beginning of a program. It is not sufficient to assume that a given bit will be '0' to begin with. It might have been set to '1' by a previously run program. Here, the first thing is to set the counter value to zero.

The sensor might be a photodiode with a beam of light shining on it across the conveyor belt. Its interface is like the circuit on p. 10, but with a connection to the 'bit 4' pin of the controller, instead of to the gate of the MOSFET.

Normally, the pin is at 0 V. When an object breaks the beam across the conveyor belt, the voltage at the pin rises close to the supply value, and the pin is at logic 1. While there is no object blocking the light, bit 4 equals 0, and the controller repeatedly reads the bit, waiting for bit 4 to change. When an object breaks the beam, bit 4 changes to 1. The controller drops through the first decision box.

It is not good enough to test the bit only once. We must check that object has passed by and the bit has returned to 0.

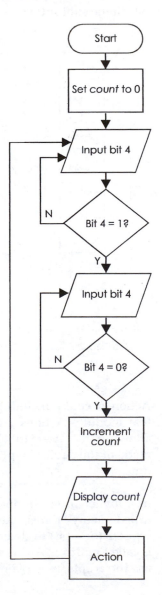

Counting events requires two checks on the input level.

The second check is needed for the reason explained on p. 335 (lefthand flowchart). Programs run very fast! If we do not wait for the bit to clear, the controller might run several times around the loop before the object eventually passes by and bit 4 goes back to 0.

Therefore, we need a second waiting loop, waiting for bit 4 to return to 0. As soon as this has happened it is safe to continue by incrementing *counter*. The next stage is to display the value in the counter byte. Finally, an action, perhaps a 'beep' completes the main program. The controller then loops back, to wait for the next object to arrive.

11) Input from an analogue sensor is converted to digital form, either by an external ADC or by an ADC on the controller chip. An external ADC is connected to a port, usually consisting of 8 input/output pins. Data from these pins, though readable individually for use with 1-bit inputs, are in this case read as a single byte. If an on-board ADC is used, there is a single pin that accepts an analogue voltage. The voltage is converted and the result appears on a special register, which can be read by the CPU.

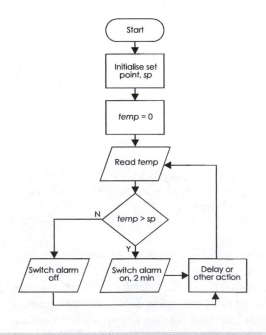

An over-heating alarm is triggered whenever the temperature exceeds the set point.

The flowchart opposite receives multi-bit input from a thermal sensor. This it compares with a set point value stored in a register. This could have been written into the program, or entered from a keyboard using a separate routine.

At the decision box, the temperature is compared with the set point. The alarm is sounded if the set point is exceeded. Comparison is repeated and the alarm switched off if the temperature falls below the set point.

Note the delay box in the flowchart. This may be a simple time delay (say 1 min) or perhaps a series of other actions that will take time to occur. This delay prevents the system from oscillating rapidly when close to the set point.

Another technique to avoid oscillations is to program with two levels, a high threshold (HT) and a low threshold (LT). HT can be a few degrees higher than the requuired temperature and LT a few degrees below it.

The temperature is read and compared with HT. If it is greater than HT the alarm is switched on. If it is lower than LT, the alarm is switched off. If it is between LT and HT there is no change: the alarm is left on if it is already on, or left off if it is already off. This incorporates **hysteresis** into the action. The action is the same as that of the non-inverting Schmitt trgger circuit described on pp. 109-110.

12) Contact bounce is a problem with some types of switch. The program responds so fast that a single key-press is read as many. The hardware solution is shown below.

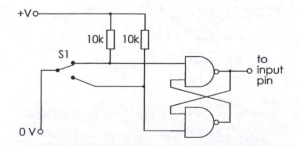

The flip-flop changes state almost instantly on first contact and remains in that state until the switch is set the other way.

The software solution is to include a delay in the input routine. This gives time for the input level to settle at its new value. The flowchart below counts how many times a button is pressed. It leaves the loop after 10 keypresses.

After detecting and registering a press there is a short delay to give the input time to settle.

The program loops back for the next press after checking that the button has been released, followed by another delay for settling.

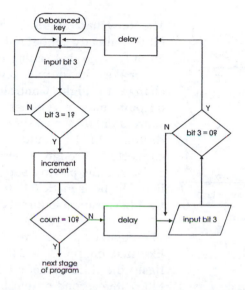

Inserting delays into a routine is a way to avoid the effect of contact bounce, though it may slow down the action of the program.

13) Comparator. Some controllers have an on-board comparator. Two input pins allow the analogue voltage from a sensor to be compared with a reference voltage from a potential divider or a standard voltage source. There may be an on-chip reference.

The output of the comparator (0 or 1) appears as a bit at a specified address in memory. This tells the controller whether the sensor output is greater or less than the set point voltage. The controller is programmed to take appropriate action.

Output routines

14) Flash an LED once, indefinitely or a given number of times. The LED is controlled by one of the single-bit outputs, or one of the lines of an output port. This can be made low or high to turn the LED off or on.

An LED can be driven directly from the output, as on p. 316. Alternatively, a high level is used to turn on a MOSFET switch or a BJT switch, as on p. 14. These routines can be used to drive LEDs, lamps, motors, relays and many other electrically powered devices. The routine below flashes the LED once.

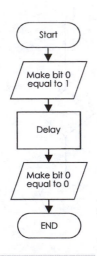

The flowchart assumes that bit 0 is already configured as an output, so that setting the bit makes its output level change to high. Controller outputs may act either as sources or sinks, so a directly driven LED could be connected through a series resistor to the positive supply line. The bit is made 0 to turn the LED on and 1 to turn it off.

This flowchart generates a single flash.

The delay could be a routine like that on pp. 333-334. To flash the LED indefinitely, follow the second output box with another delay, then loop back to the beginning.

Flashing a given number of times (right) requires a loop with a loop counter. In this flowchart the controller waits for a keypress before starting to flash. It makes the program simple and shorter if, instead of setting out the delay routine every time it is used, we make it into a subroutine which is called when needed.

15) Lookup tables are useful for holding arbitrary values, that can not be calculated from a formula. As an example, the codes for Driving a 7-segment LED (or LCD) the data for each numeral is held in memory for reference by the display routine.

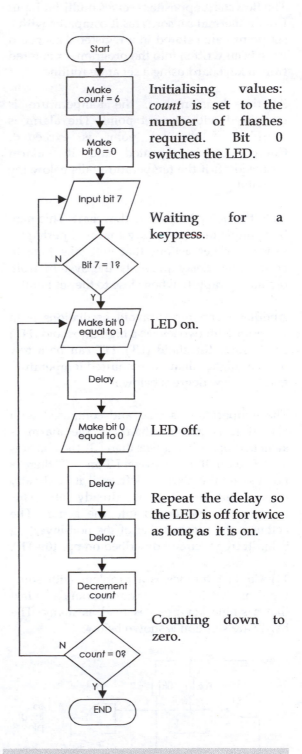

Initialising values: *count* is set to the number of flashes required. Bit 0 switches the LED.

Waiting for a keypress.

LED on.

LED off.

Repeat the delay so the LED is off for twice as long as it is on.

Counting down to zero.

A routine to flash an LED 6 times after a button is pressed.

Address (hex)	Data (binary code)
2000H	01111110
2001H	00110000
2002H	01101101
2003H	01111001
2004H	00110011
2005H	01011011
2006H	01011111
2007H	01110000
2008H	01111111
2009H	01111011

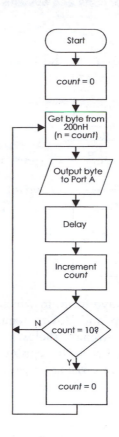

Using a look-up table to cycle a 7-segment display through numbers 0 to 9.

The base address of the table above is 2000H. In other words, the table is stored in memory at addresses 2000H to 2009H.

The codes use all except the first bit of each byte. The allocation of bits to the segments of the display is:

0 a b c d e f g

In the flowchart, we are using 7 out of the 8 bits of Port A. Seven of the pins are connected to the LED segments *a* to *g*, the eighth pin being unconnected. The segment is to be lit when the output goes high, so this routine requires a common-cathode display.

Initialisation requires two operations: setting the *count* variable to zero, and switching off all the segments of the display.

The remainder of the program is a loop that repeats indefinitely. Each time round the loop, *count* is incremented by 1. Each time round the route a byte is taken from the look-up table. The address is 200nH, where n is the current value of count. This makes the display show the digits 0 to 9, consecutively.

16) Wave forms are easily generated by a controller. A square wave output is produced by a program similar to the indefinite LED flash (14). Using a much shorter delay time produces a square wave at audio frequencies.

A sawtooth wave form is generated by the flowchart overleaf. The output is produced at an 8-bit port, that is connected to a DAC. Alternatively, we use an on-board DAC and take the analogue signal from its pin.

The program runs in a continuous loop, incrementing *count* each time. So *count* gradually increases from 0 to 255 (8 bits). At the next increment it changes to 0 again, the carry-bit being ignored. It increases from 0 to 255, repeating. The result is a rising voltage from the DAC, dropping instantly to 0, and then repeating. The frequency is set by the length of the delay.

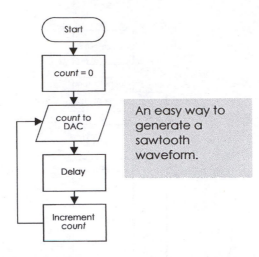

An easy way to generate a sawtooth waveform.

A triangular wave form, in which the voltage ramps both up and down, requires a more complicated flowchart (below). This uses a single bit, called *slope*, to register whether the output is currently rising (*slope* = 0) or falling (*slope* = 1).

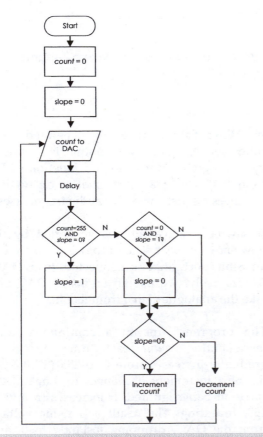

A more complicated routine is used to generate a sawtooth wave form.

This flowchart has two decision boxes to detect:

- When *count* is rising AND when it has reached its maximum value (255).
- When *count* is falling AND has reached its minimum value (0).

In practice, each of these decisions based on AND might need to be expanded to two decisions connected as in flowchart 4. The two AND decisons are connected to provide OR logic, as in flowchart 5.

The result is that if *count* has reached its maximum rising value OR its minimum falling value, the value of *slope* is changed.

The third decision box, either increments or decrements *count* according to the value of *slope*.

Wave forms that have a more complex shape can usually be generated by using a look-up table. The tabulated values are called up in sequence and sent to the DAC.

17) Stepper motor control uses a look-up table which holds the codes for energising the coils (p. 297). The flowchart (opposite) consists of a loop in which the count repeatedly runs from 0 to 3. At each stage the appropriate code is looked up in the table and sent to the coils of the motor (using four transistor switches).

The rate of rotation of the motor depends on the length of the delay. This is a subroutine (not shown here) similar to flowchart 6. The instructions to increase or decrease the delay change the values of *n* or *m*.

The circuit has two pushbuttons, 'Faster' and 'Slower'. Each time round the loop, the states of these buttons are read. If either one is pressed, the delay if shortened or lengthened.

18) Automatic door control, provides an example of program development and illustrates some general points about programming control systems. The doors are driven by a reversible motor that can open or close them.

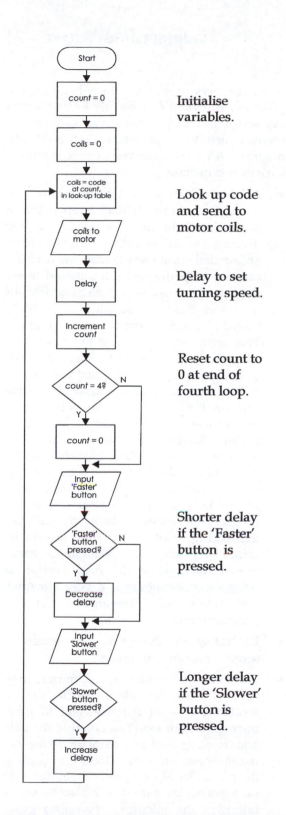

Initialise variables.

Look up code and send to motor coils.

Delay to set turning speed.

Reset count to 0 at end of fourth loop.

Shorter delay if the 'Faster' button is pressed.

Longer delay if the 'Slower' button is pressed.

A routine to control the speed of a stepper motor by using two buttons.

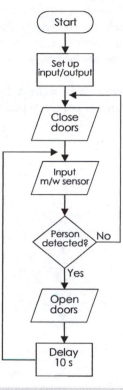

Controlling the automatic doors of a supermarket.

The sensor in this system is a microwave device that detects persons approaching the door. There are two outputs to control the motor. One is a relay switched by a transistor switch. This turns the motor off or on. The second output goes to a second relay wired to operate as a reversing switch. In this flowchart we do not show these outputs separately, but simply state 'Close doors' and 'Open doors'.

After initialising the pins as inputs and outputs and closing the doors for security, the controller waits for a person to approach the doors.

The controller cycles round the waiting loop until a person approaches. Then the doors are opened. There is a 10 s delay to allow the person to enter, then the doors are closed. If the person is walking slowly and has not passed through the doorway, the system detects this and keeps the doors open. After the doors are closed, the system waits for the next approaching person.

This flowchart is a good beginning for a program as it covers the essentials, but there are details that need to be thought about. A practical point is that when several people approach the doors in quick succession, the doors do not need opening if they are already open. Another point is that just switching on a motor for a given time does not guarantee that the doors are fully open. We need positive input about whether the doors are fully open or not . This kind of data, about what has (or has not) *actually* happened is a kind of feedback.

341

To obtain this feedback, the system needs two more sensors. A set of microswitches are placed so that one switch closes when the doors are fully open and another switch closes when the doors are fully closed.

We write these into the previous flowchart by including the program segment below in place of the 'Open doors' stage.

The first step is to check if the doors are already open. At the 'Doors shut' box the controller reads the microswitch that indicates this condition. If the doors are not already shut, they are taken to be open and the program skips directly to the end of the open doors routine. If they are already shut, the motor is turned on to open the doors. Then the 'doors open' microswitch is read to find out if the doors have opened as wide as possible. If not, the routine loops around, with the doors motor still running, to check again. When the 'doors-open' sensor confirms that the doors are open, the motor is switched off.

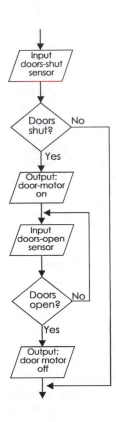

This expansion of the 'Open doors' routine gives better control of the doors.

A similar routine is needed to expand the 'Close doors' stage.

There are many other ways in which the system can be expanded and improved. For instance, as in many control systems, it is useful to have a manual override button. In this case it would allow the doors to be locked shut when the store is closed. It would also keep the doors closed while the window-cleaner is at work.

Interrupts

A miicroprocessor is always busy handling the program, proceeding from one instruction to the next unless told to stop. Events sometimes occur which need to interrupt this steady progress. An immediate response is essential. Such events include:

- **Arithmetic errors**: a typical example is 'division by zero'. It may happen that a particular term in a calculation is the divisor in an equation. If this term occasionally evaluates to zero, the division can not be performed. This should never happen in a correctly written program, but may sometimes occur during the early stages of development of a new program. This error would be detected inside the controller and causes it to interrupt itself. Instead of going on to the next instruction, it jumps to a special **interrupt service routine** (ISR). If the program is in BASIC, this causes a message to be displayed saying 'Division by zero error'. The microprocessor is then halted and the programmer checks to find why such an error occurred. Acting on reported errors of this kind, the programmer revises the program to eliminate them. We call this **debugging**. Failure to find arithmetic errors may cause the program to 'hang' or to 'crash' when the final version is being run. Debugging is a very essential, and often time consuming, part of programming.

- **Logical errors**: these may have effects as serious as arithmetic errors.

- **Clocked interrupts**: A computer may contain a real-time clock, which can be written to (to set the time and to issue instructions), or read from (to find the date and time). As well as general time-keeping, it can act as an alarm clock, interrupting the controller at any preset time, daily, or on a particular date. It can also be set to interrupt the microprocessor at regular intervals, such as once a second, or once a minute or once an hour.

The microprocessor is programmed to perform a particular action when it is interrupted. In a data logger, for example, it could be programmed to measure and record a temperature at 1-minute intervals, but to continue with other tasks in between. After completing the task that is programmed as an interrupt service routine, the controller returns to the point at which it left its main program. In the data logger, the measurements could be taken so quickly that the user would be unaware that the program had been interrupted.

- **Peripherals**: In a computer system there are several devices that frequently need to interrupt the microprocessor. These devices include the keyboard, the hard disk drive and the printer.

It is possible that a microprocessor receives interrupt signals simultaneously from more than one source. Each device that may interrupt is allocated a priority. This is done by programming or by connecting the devices to different pins with different priorities.

Microcontrollers too may have interrupts, including:

- **Change of input**: The voltage at a particular input has changed from high to low or low to high. This could be the result of opening or closing a switch.
- **Timer overflow.** A timer has reached the end of its set period. Also there may be a built-in watchdog timer (p. 328).
- **USART** has received a new byte.
- **Writing data** to EEROM is completed.

As an alternative to interrupts, a system may rely on **polling**. In some computer systems each input device (which may be something complex such as a keyboard or input port IC) has an **interrupt flag** in one of its registers. In between its main activities, the controller interrogates each device to find out which one or more of them has its interrupt flag set. It then takes action, servicing those with the highest priority first.

Similarly, certain registers in a microcontroller may hold **flag bits**. Each bit is normally 0 but goes to 1 when a given event occurs. The controller is programmed to poll these bits and to act accordingly.

Polling is generally slower than using interrupts because a device or flag bit wait to be interrogated and, by then, it may be too late to avoid a program crash or a loss of data. It is possible to program some controllers to ignore interrupts during certain critical stages in their operations.

Many programmers use interrupts but there are some who prefer polling. They say that interrupts introduce uncertainty into the way the program runs.

Direct and indirect addressing

Normally a program addresses stored data directly. For example, the controller might be instructed to increment a given value that is stored at a given address. The opcode includes all the essential information – it might say 'Increment the value stored at 037h and place the result in the working register'.

As explained on p. 327, the address is included in the 14-bit opcode. It is the actual address at which the data is stored. This is **direct addressing**.

Because the address is part of the opcode, it can not be altered while the program is running. The instruction will always result in incrementing the value stored at that same address. It can not be changed so as to increment the value stored at any other address. What if we want to increment the values at several different addresses? We use **indirect addressing**.

With indirect addressing the opcode says something slightly different, It might say 'Look in the file select register (FSR) and there you will find an address. Go to that address, increment the value stored there, and place the result in the working register'.

Note that this instruction does not include the address to be incremented, but tells the controller to look in the FSR to find it. The content of the FSR can be changed while the program is running. After each change FSR sends the controller to a different register to perform is action.

Indirect addressing is particularly useful for operating on a block of stored data. The flowchart below shows how.

Given the first and last addresses in the block, the routine clears them all to 00h, one at a time. A loop is by far the shortest way to program a repeated operation.

Other block operations can be programmed like this, such as adding, say, 20 to each stored value. Or a block of data can be searched to find a register or registers holding a given value and to perform a given operation on it. For example it could find all registers that hold more than 100 and subtract 30 from them.

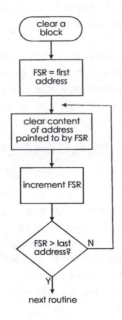

Indirect addressing simplifies operations on blocks of data.

The Stack

In most controllers a small area of RAM is set aside for the stack. This is used for the temporary storage of data. It works in a 'last-in-first-out' manner, like the plate dispenser sometimes seen in canteens. The stack pointer holds the address of the top of the stack. If a byte of data held in the working register is 'pushed' on to the top of the stack, its value is transferred to the RAM location immediately above the current 'top of stack'.

The stack pointer is incremented to point to the new top-of-stack.

The data is held there until it is removed by the 'pop' operation. It is then transferred back to the working register and the stack pointer is decremented to indicate the new location that is top-of-stack. Several data bytes can be pushed or popped between the working register and the stack.

The stack is often used for storing an intermediate result in a calculation. It is also used for registering the state of operations when the CPU jumps to a subroutine, or when the controller is responding to an interrupt.

Immediately before it leaves its current place in the main program, the CPU pushes on to the stack such information as the current contents of the working register, the present value of the program counter, and the states of the flags in the status register. It then jumps to the subroutine or to an ISR. When it returns from the subroutine it pops the data back (in the reverse order) so that the CPU recovers the state that it was in before it jumped. It is then ready to continue with the main program.

Summary

The flowcharts in this chapter demonstrate the following program features. The numbers in brackets list flowcharts that show each feature:

- Initialise values of variables, look-up tables, define pins and ports as inputs or outputs. (p. 331)

- Program a sequence of actions. (p.331 and 1, 17)

- Perform elementary arithmetic and logical operations, including counting. (1, 2, 4, 5, 10, 14B)

- Perform timing operations, including creating time delays. (4, 5, 6, 7, 14, 16)

- Accept input from switches and sensors. (almost all)

- Take decisions as a result of input, arithmetic or logical data. (3, 4, 5, 9A, 9B, 10, 11, 12, 14B, 17, 18)

- Create a sequence of outputs to LEDs, lamps, motors and other actuators, using look-up tables. (15-18)

- Use polling to detect external events. (4, 5, 9, 10, 14B, 17, 18)

- Call on subroutines. (9C)

Most programs incorporate loops, of which there are several kinds:

- Indefinitely repeating loops. (almost all)

- Loops that run for a predetermined number of times. (2, 6, 14B)

- Loops that run until a given condition is true. (almost all)

- Nested loops. (6, 10, 18B)

Where a number is qualified by 'A', 'B' or 'C' it indicates that the feature is present in either the first, second or third flowchart.

The chapter also discussed the importance of interrupts, direct and indirect addressing, and stack operations.

Activities — Programming

Look for ways of extending and improving some of the flowcharts in this chapter. Examples of changes you might make are:

Flowchart (3): add a routine to allow a set point to be keyed in.

Flowchart (15): modify this to count *down* from 6 to 1, repeating.

Flowchart (17): adapt the routine to allow the direction of rotation to be reversed.

Flowchart (18): extend this to provide for manual override.

Plan programs for the following tasks and draw the flowchart. Outline what input and output hardware would be required, but do not describe circuit details:

- To control a heater in a thermostat program (see the analogue thermostat on p. 295).

- To operate a security system for a lock-up garage that has a broken-beam intruder detector outside the main door, a pressure mat inside the side door, and a vibration detector on the main door. The alarm is given by a siren and a flashing beacon.

- A traffic light controller, that produces the standard sequence of red, amber and green lights. The lights are at a crossroads (below), but your program need show only the program for the light indicated by the arrow. Then modify the program so that traffic waiting at the indicated light is not kept waiting if there is no approaching traffic on roads A and B.

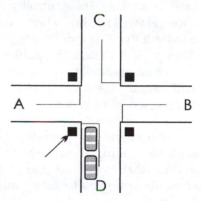

- Program the following sequence. When a car is waiting at a car park exit barrier, an illuminated sign requests the driver to put 50p in the slot. If this is done the barrier is raised and the car may drive away. The barrier is lowered as soon as the car has gone.

- Design a program to automatically control the points of a model train layout, so that the train circulates in a regular way around three rail loops.

- Design a program for controlling the motion of the conveyor belt at a supermarket checkout.

- Outline a program for timing the laps and displaying the times on a model racing car layout.

39 Programming languages

The purpose of this chapter is to outline the distinctive features of four different ways of programming a controller. The examples are based on some of the programming flowcharts in Chapter 38. This chapter does not set out to teach the reader how to program. The only way to learn this is by hands-on experience.

Practice programming on a programming board, using software running on a personal computer. Start with a simple programming task, such as making the controller flash an LED once (Flowchart 14). Then extend the routine to flash the LED indefinitely. Then add input from a push-button. By starting with a simple task and gradually adding refinements you will soon learn the essentials of programming a control system.

As an alternative to working with a programming board, there are a number of programs that simulate the action of controllers, allowing you to set up inputs and to read outputs.

Programming in machine code

Machine code for the PIC16F690 controller has already been described on p. 327.

In the example below, we list a machine code program for flowchart 1, p. 332, which adds together two bytes of data. We assume that the two bytes are already stored in Registers 20h(*num1*) and 21h (*num2*). Their sum is to be placed in Register 22h (*sum*). This is only a segment from a longer program, so it does not start at address 00h. Assume that the sequence of binary instructions is stored in program memory at addresses 70h to 74h. The program is listed in the table on the right.

When programming the controller, it might be necessary to key in the binary code, digit by digit. Or perhaps the programmer would have a hex keypad for entering the code in hex.

Program memory address	Binary code	Equivalent in hex
70H	00100000100000	0820h
71H	00000010100010	00A2h
72H	00100000100001	0821h
73H	00011110100010	07A2h
74H	00000001100011	0063h

In either case, the machine code is extremely difficult for a human to read and to check for correctness.

The effect of these instructions is as follows:

At 70H: move the contents of Register 20h to W (the working register, or accumulator).

At 71H: move the contents of W to Register 22h.

At 72H: move the contents of register 21h to W.

At 73H: add Register 22h to W, and place the result in Register 22h.

At 74H: go into sleep mode.

This program demonstrates the 'many simple steps' by which the controller operates.

Programming in assembler

Writing programs in machine code is possible but difficult and is very subject to human error. Although using hex makes it easier to enter the instructions in compact form, a microprocessor may have a hundred or more different opcodes in its instruction set. With such a large number it becomes difficult to remember them.

The solution to this problem is to write the program in **assembler** language. Assembler 1replaces opcodes with **mnemonics**.

Mnemonics are a set of abbreviations usually of three letters, sometimes more, one for each opcode. They are intended to help the human programmer to distinguish the different operations and remember what each operation does.

Example

The instruction 00100000100000 on the instruction bus of a PIC16F690, equivalent to 0820h, tells the controller to move the contents of register 20h to W (the working register). Instead of 0820hH or its long binary equivalent, we use the mnemonic MOVF, which is short for 'MOVe F', where F is a register file. This is represented by the first 6 bits in the instruction.

The mnemonic is followed by one or two **operands**, that in this case tell the controller which file register (F) to use and where to place the result. As explained on p. 297, the seventh bit is the destination bit and the last 7 bits are the register number. The assembler enables the program to use labels to identify registers, which makes it easier to remember what is stored in the register. It also makes it unnecessary to remember the hex address at which the data is stored. In this example the label for register 20h has already been initialised as '*num1*'. The full instruction (opcode mnemonic plus operands) is:

MOVF num1, 0

The assembler listing for the machine code program opposite is shown below, as it would appear on the computer screen.

In this example, we are using the labels *num1*, *num2* and *sum* for registers 20h, 21h, and 22h respectively.

The example illustrates another helpful feature of assembler. It is possible to add **comments** to the program. These are typed in on the right, after a semicolon. Comments help other people to understand how the program works. They are also very useful to the original programmer too. On coming back to a program a few weeks after it was written, perhaps to correct or extend it, it is surprisingly difficult to recollect how it works. Comments are helpful reminders.

An assembler allows us to run the whole or sections of a program, or to go through the program one line at a time. At every stage it displays the contents of the accumulator or working register, the flags and all the registers, so that we can see exactly what is happening. We can also put a temporary break in a program so that it stops at a given **breakpoint** to allow us to examine the contents of all the registers at that point. When a program is finished and debugged, we can download it into the program ROM of the controller, or save it on to the hard disk.

Programming in assembler is a slow matter because we have to issue the controller with instructions one at a time. As this example shows, it usually takes several instructions to perform an elementary action, such as adding two numbers together.

```
MOVF     num1, 0       ; move contents of 20h to W

MOVWF    sum           ; move contents of W to 22h

MOVF     num2, 0       ; move contents of 21h to W

ADDWF    sum, 1        ; add contents of W to 22h and
                       ; place result in 22h

SLEEP                  ; go into sleep mode
```

Programming a PIC in assembler

An assembler program may be a part of a collection, or suite, of programs for use in developing software for a controller. All the programs are linked so that the programmer can pass from one to another and possibly back again as the software is developed. A suite such as this is often called an **integrated development environment**, or IDE.

A view of an IDE in action is seen on p. 323 and an IDE for the PIC is displayed below

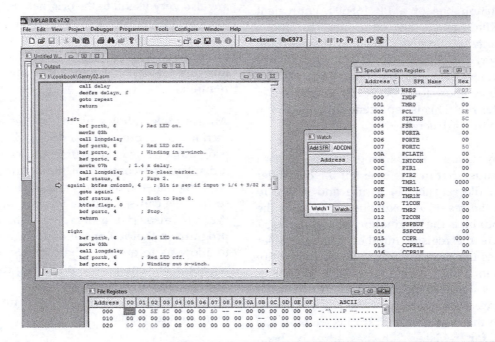

MPLAB™ is an IDE from Microchip, for programming the PIC series of microcontrollers. Here we are displaying a few of its many windows. The largest window is a text editor where the programmer keys in the program that is being developed. The window on the right shows the contents of the Special Function Registers, including WREG (working register), the STATUS register, and the input/output ports.

MPLAB comprises:

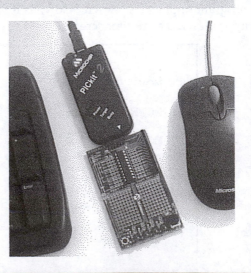

- A text editor.

- A debugger, seen in operation above.

- A simulator, which is software running on the PC but behaving like a PIC.

- Programmer software to download the program into the program memory of the PIC. This uses (among others) the PICkit™ 2 programming device (right). It has a socket for the PIC and an area of board where simple input and output circuits are wired up for testing. It also has LEDs, a push-button and a potentiometer for analogue input.

If we want to add numbers that total more than 255, or have many more significant figures, or that are negative, or to add three or more numbers together, the routines become even more complicated when written as single steps. In addition to this, there are dozens of routines that may be used several times in a program. It may be possible to save programming time by including these as subroutines that can be called whenever they are required. Even so, writing a moderately complex program in assembler is a major task.

Programming in BASIC

Assembler is often used for writing programs for control systems. Such programs are usually short and simple, and the need for single-instruction steps is acceptable. Assembler is less often used for writing programs for computers because their requirememts are generally more complex. Most programmers use a **high-level language** for this purpose. A high-level language is one or two stages removed from assembler because a single command in the high-level language is often the equivalent of many dozens of assembler instructions.

Example

In the popular high-level language known as BASIC, we can perform the addition of two numbers (flowchart 2) by typing:

30 LET sum = num1 + num2

This is a typical line from a BASIC source program. It has the same result as the machine code sequence and the assembler program from pp. 346-347. The '30' is the line number, because the instructions are numbered and executed in numerical order in most versions of BASIC.

A BASIC statement or command looks very much like an ordinary sentence written in English. The meaning of the instruction is clear, even to a person who does not know BASIC. It is much shorter than the machine code into which it has to be converted before the controller can get to work on it.

To be able to create a BASIC *source* program, you need a BASIC *language* program in the computer to accept the commands you type in, to display them on the screen, and to allow you to edit it, to test it, to check for programming errors, and to save it on to disk.

A BASIC source program is stored in RAM or saved on to disk as a sequence of bytes, which are coded to represent the words and numbers in the program lines. In this form, it is meaningless to the controller that is intended to run it. It must be **compiled**. A BASIC **compiler** is a program that takes a source program written in BASIC and converts it into machine code, ready for downloadimg into the ROM of the controller. Each type of controller or family of controllers needs a special compiler that is geared to using its particular set of instructions.

As an example of a BASIC program, we will look at a program written for one of the more complex flowcharts, the program for cycling a 7-segment display from 0 to 9, repeating (pp. 338-339). Here we show the listing written in BASIC for the Stamp 1. This consists of a PIC microcontroller on a very small circuit board, with a special ROM to allow it to run BASIC programs. Stamp BASIC does not have line numbers. Any line to which a jump may occur is given a label. The label name is followed by a colon to define it as a label the first time it is used. The program is called 'Counting':

```
'Counting
DIRS = %01111111
LOOP:
LOOKUP
B0,(63,6,91,79,102,109,125,7,127,1
11),B1
PINS=B1
PAUSE 1000
B0=B0+1
IF B0=10 THEN ZERO
GOTO LOOP
ZERO:
B0=0
GOTO LOOP
```

Programming a PIC in BASIC

There are several IDEs available for programming PICs. One of these is Proton™ IDE, from Mechanique, which has all the expected features of an IDE. A typical working screen is seen below. Another similar IDE is Microcode Studio©, also from Mechanique. There are also compilers such as the one used for programming the BASIC Stamp® controller board from Parallax Inc.

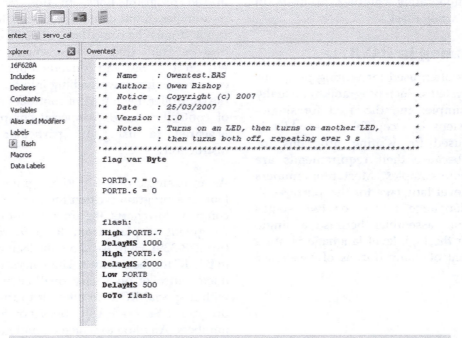

```
'***************************************************************
'*   Name    : Owentest.BAS                                   *
'*   Author  : Owen Bishop                                     *
'*   Notice  : Copyright (c) 2007                              *
'*   Date    : 25/03/2007                                     *
'*   Version : 1.0                                             *
'*   Notes   : Turns on an LED, then turns on another LED,     *
'*           : then turns both off, repeating ever 3 s        *
'***************************************************************
flag var Byte

PORTB.7 = 0
PORTB.6 = 0


flash:
High PORTB.7
DelayMS 1000
High PORTB.6
DelayMS 2000
Low PORTB
DelayMS 500
GoTo flash
```

A short test program listed in the editor window of Proton IDE is ready to be compiled and run.

Similar software is available for programming in other high-level languages such as C and its variants. C is widely used by professional programmers but is more difficult to learn than BASIC.

This program is intended to be run with seven of the Stamp's input/output pins (P0 to P6) connected to the anode pins of a 7-segment common cathode LED display. There is a 1 kΩ resistor in series with each segment to limit the current. The program begins by initialising the input/output pins as inputs or outputs. The bit values in the variable DIR (direction) are 0 for input and 1 for output. Here, pins 0 to 6 are all defined as outputs, while the unused pin 7 is defined as an input.

There is no need to initialise the contents of storage memory bytes B0 and B1, as these are automatically set to zero when the controller starts to run.

Next comes a line label, 'LOOP:' This is followed by a specification of a look-up table. The numbers in brackets are the bit patterns for the 7-segment digits. B0 detemines which value is selected from the look-up table. B1 is the byte to which the value is copied.

Example:

On count 3, B0 = 3 and B1 takes the value 91. In binary, this is 0101 1011, the code for the numeral '2'.

The value of B1 is placed on the pins. Where there is '1' the pin goes to logic high, turning on the corresponding segment of the display. After a pause to allow the number to be displayed, B0 is incremented. It is tested to see if it has reached 10 or not. If it has not reached 10, the program jumps back to LOOP to display the next numeral. If it has reached 10 there is a jump to the ZERO label, where B0 is made equal to zero. Then the program jumps back to LOOP to begin again.

Programming in VPL

Visual Programming Language® is one of the newest languages for programming controllers. It is part of the *Microsoft Robotics Studio*.

It operates in an entirely different way from assembler and BASIC. For one thing the language is expressed visually as flowcharts. But these flowcharts are not the same as the flowcharts used in Chapter 37.

The flowcharts of Chapter 37 illustrate a flow, or sequence, of *commands*. The program is a series of commands – *load* this, *increment* that, if this is greater than that *goto* there. The flowcharts simply put this command structure into a different form. The arrows in the flowchart show the way from one command to the next. The programs and the flowcharts are based on a *flow of command*.

A VPL program is based on a *flow of data*. In the sample program shown below, data is typed into the Data activity box. In other programs the data may originate from, say, a sensor belonging to a robot. When the program is run, the data follows the route described in the caption, and is processed on the way. Eventually, one of two messages is displayed, depending on the value of the data.

VPL is intended for programming robots. A wide range of activity and service boxes is provided to interface the same programs to particular robots, such as those controlled by a BASIC Stamp, or by the Lego Mindstorms bricks. It also can produce programs for controlling a number of commercial robots such as the iRoomba *Create* robot and the *Traxter*. The underlying aim is to unify the approach to robotic programming.

In this screen shot, The data, '20', has been typed into the Data bpx. When the program is run, the data is passed to the Variable box, where it is given a name, *count*.

Count is passed on to the 'If' decision box, where it is tested. If count is greater than 5 it is passed to the upper Calculate box. There a string value is calculated, 'Input too high'. This is passed to the dialogue box, and an Alert message is displayed.

If count is 5 or less, it is passed along the other ('else') route to the lower Calculate box where it is added to the string 'Count = '. The resulting message is sent to the lower dialogue box.

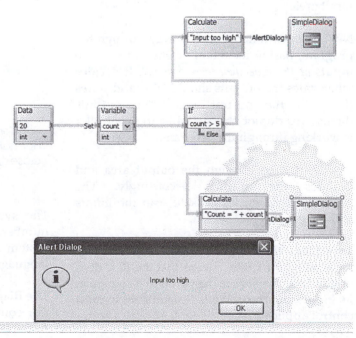

Programming in ladder logic

Ladder logic is a special high-level language, used with programmable logic controllers (PLCs). It is particularly suitable for programming control systems. There are several different forms of ladder logic. Here we illustrate ladder logic using the symbols for the Mitsubishi PLCs.

It is important to be clear from the start that processing a ladder logic program is very different from processing a program written in assembler, BASIC or any other high-level language. In these, the controller starts at the beginning of the program, fetching and executing instructions as it moves from address to address in program memory. Occasionally it may meet a branch instruction and skip to some other part of the program. It may jump to a subroutine and then return to the main program. Once it has begun, the path it takes is determined by the instructions it meets on its way.

In a PLC, the memory is divided into three main areas: the input area, the program area, and the output area. The PLC receives inputs from (usually) a large number of sensors. The first stage in its operation is to scan all these inputs and register their current states in the input area.

Next, the controller works its way through the program, basing its action on the states of the inputs *at the time they were scanned*. It decides what states the outputs should take, and stores these in the output area. The outputs themselves do not change while the controller is working through the program.

Finally, it runs through the output area and updates the outputs accordingly. The controller then cycles back to scan the inputs and the process is repeated.

Typically, the 'input-process-output' cycle takes about 0.1 s. This means that the PLC does not make an *immediate* response to a change in its inputs, but the delay is insignificant in most control applications.

Like other microcontroller systems, a PLC is programmed by software running on a special programmer device or on a personal computer. Ladder logic is represented by a distinctive set of symbols so it is more conveneient to use a special PLC programmer, which has keys marked with these symbols. When a program has been entered and debugged it is downloaded into the memory of the PLC.

Ladder logic is derived from **relay logic**. Relays can be wired together to build logic circuits that can control industrial machinery. With the coming of microelectronics, relay systems have been mainly (though not totally) replaced by PLCs. However, the symbols used for writing ladder logic programs are derived from the symbols used in drawing relay systems. The basic symbol for inputs resembles the symbol for a pair of relay contacts. These may be 'normally open' or 'normally closed'. The symbol for an output is a circle, similar to the symbol for a lamp or a motor.

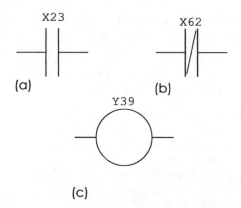

Ladder logic symbols for (a) input contacts normally open, (b) input contacts normally closed, and (c) an output device.

The symbols are labelled with a reference number, X for input, Y for output. The labelling system is different in some versions of the language.

The diagrams opposite show how the symbols are connected together to represent one or more logical operations.

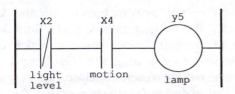

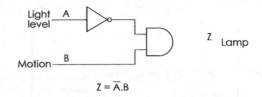

To understand the meaning of the diagram, think of the vertical line on the left as a positive supply line, and the vertical line on the right as the 0 V line. Then, if the two sets of contacts are closed, current flows through them and through the output device, which is a lamp in this example. The lamp comes on.

The set of normally closed contacts represents a sensor that produces a logic low level in darkness. The logic high does not energise the imaginary relay, which has normally closed contacts. In the light, the output of the sensor is logic high, the imaginary relay is energised and the contacts are open.

The set of normally open contacts in the diagram represent a sensor that detects a moving person. When a person is detected, the sensor produces a high output and the 'relay' contacts close.

Thus, if it is dark AND motion is detected, both sets of contacts are closed. Imaginary current flows through both sets of contacts and the lamp comes on.

Remember, there are no actual sets of contacts, and no currents flow through them. Instead the logic of the software turns on the lamp if the input from the light sensor is 0 AND the input from the motion sensor is 1. This instruction produces AND logic.

Some PLCs are programmed as connected logic gates. In such a PLC we represent the logic of the previous diagram with the standard logic circuit diagram shown at the top of the next column.

The action of this logic is not a satisfactory substitute for programming a security lamp referred to in flowchart 4, p. 333. The input from a motion detector is momentary and the lamp would go off after a second or two. Here is an improvement to the system.

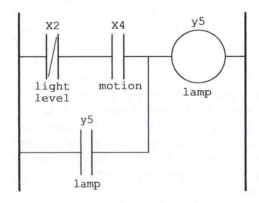

The additional 'set of contacts' have the same reference number as the lamp. They are normally open but close when the lamp is on. Once the lamp has been turned on by motion detected in darkness, these contacts close. The lamp is latched on. It stays on, even if motion is no longer detected, and even in the light.

This is not a satisfactory circuit, but it does illustrate how ladder logic can obtain the OR operation by connecting sets of contacts in parallel.

To prevent the lamp staying on indefinitely, we can include a manual reset button:

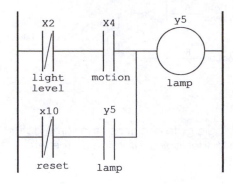

A reset button unlatches the 'circuit'.

The reset button input is programmed as a set of normally closed contacts. When the button is pressed, the contact opens. Assuming it is dark and no motion is detected, there is no way 'current' can flow to the lamp, which goes out.

To produce the same action as in flowchart 4, we can use an additional imaginary device, a timer. Several different types of timer are available in ladder logic software and the one we use here is a pulse timer. Its contacts are normally open, but close for a period when triggered. The length of the pulse can be preset. In the diagram, the legend 'k30' sets it for a 30-second pulse.

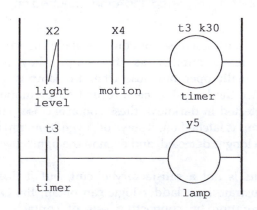

This is the ladder logic equivalent of the security lamp flowchart on p. 333.

Instead of the sensors turning on the lamp directly, they trigger the timer. The timer has a pair of normally open contacts, which close when the timer is triggered. The lamp comes on when these contracts close. The contacts stay closed and the lamp stays on for 30 s, even thought the light level and motion inputs may change.

This program has the same action as the flowchart on p. 354. We could add more instructions to this program so that the PLC could control an extensive factory security system. The logic diagram will eventually consist of two vertical lines with numerous 'rungs' joining them. This is why this form of programming is called ladder logic.

A counter is another useful item in some ladder logic programs. On the screen it is similar to the drawing below.

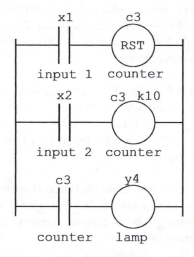

A counter program. Different symbols may be used in different PLCs.

A pulse applied to input 1 resets the counter; its contacts are open and the lamp is off. The counter registers each pulse arriving at input 2. On the tenth pulse, the counter contacts close and the lamp comes on.

After the count has reached 10 the counter contacts stay closed until it is reset.

Multiple choice questions

Questions on assembler

In the assembler routine below, the meanings of the instructions are:

movlw k   Load numeric value k into the working register w (value in hex).

movwf f   Copy value in w to file register f.

incf f, d   Increment register f and place result in destination register (w or f).

decf f, d   Decrement register f and place result in destination register(w or f).

bcf f, b   Clear bit b of file register f.

portc is an 8-bit port set as outputs with an LED driven by bit 3.

Here is the routine:

```
movlw 16h
movwf num  ; f label is 'num'
incf num, w
bcf num, 4
decf num, f
movwf portc
```

1   The value in num after the 2nd line is executed is:
   A 0.
   B 16.
   C undecided.
   D 16h.

2   The value in w after the 2nd line is executed is:
   A 16h.
   B 0.
   C 17h.
   D undecided.

3   The value in w after the 3rd line is executed is:
   A 16h.
   B 0.
   C 17h.
   D undecided.

4   The value in num at the end of the routine is:
   A 05h.
   B 16h.
   C 0.
   D 06h.

5   At the end of the routine the LED is:
   A On.
   B Flashing.
   C Not known.
   D Off.

6   After the end of the routine the instruction to turn on an LED connected to portc, bit 4 without affecting the first LED is:
   A bcf portc, 4.
   B incf num, f.
   C bsf portc 17h.
   D bsf portc, 4.

Questions on BASIC

In the following PicBASIC routine, the meanings of the instructions are:

Poke address, value   Write value into address. Value is in hex ($) and address is a label.

Pause period   Delay, milliseconds.

Goto   Go to labelled line.

Two LEDs are driven by the outputs at Port C, bits 4 and 5.

Here is the routine:

```
lediode;
    Poke Portc, $20
    Pause 1000
    Poke Portc, $0
    Pause 500
    Goto lediode
```

7   When the routine is running, the LED at bit 4 is:
   A always off.
   B glowing dimmly.
   C always on.
   D flashing on and off.

8 When the program is running, the LED at bit 5 is:
    A flashing alternately with the other LED.
    B always off.
    C glowing dimly.
    D flashing.

9 The program stops running:
    A after 1500 ms.
    B until the powerr is switched off.
    C after the LED has flashed on and off.
    D after 500 ms.

10 To program the LEDs to flash on and off together, change $20 in the second line to:
    A $30.
    B $33.
    C $40.
    D $60.

11 The number of times the LED flashes in a minute is:
    A 500.
    B 1000.
    C 60.
    D 40.

12 The mark-space ratio of the signal switching the LED is:
    A 1.
    B 1.5.
    C 2.
    D 0.67.

Questions on ladder logic

13 The logical operation performed by this instruction is:

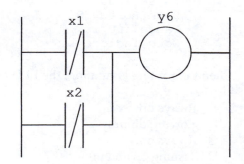

A NOT.
B OR.
C NAND.
D Exclusive-NOR.

14 The logical operation performed by the instruction below is:

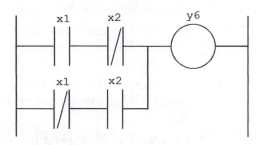

A Exclusive-OR.
B NOR.
C AND.
D Latch.

15 In the instruction below, the lamp comes on when:

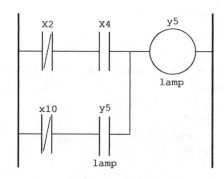

A X2 and X4 are closed.
B X10 is opened.
C X10 is closed and the lamp is off.
D X2 is opened and X4 is closed.

40 Neural networks

Neural networks may be used for controlling or analysing highly complex systems. A neural network might be used for controlling a power station, for example, for landing a lunar module, or for driving a car. Other neural networks are concerned with recognising patterns, as described later.

Animals too have neural networks. They reach their highest state of development in the human brain. A neural network, whether animal or electronic, consists of an assembly of **neurons**. As a very simple example of an electronic neuron, consider the op amp circuit below.

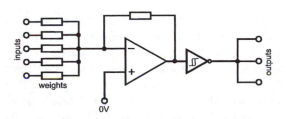

An op amp inverting amplifier with weighted inputs is an electronic model of a neuron. The Schmitt trigger inverter ensures that the output is either high or low.

The essence of this circuit is that it:

- accepts **inputs** from a number of sources,

- **processes** them by weighting, summing, and amplifying them, and

- sends **output** to several other circuits.

The weighting is the means by which the behaviour of the neuron is modified. To make the neuron respond more strongly to a given input, the resistor of that input is reduced in value. Then larger current flows to the op amp from that input, increasing its influence on the output of the op amp. To allow this to be done, the input resistors are variable.

In the brain, behaviour in response to various inputs depends on previous experience. In other words, the brain can *learn*. In particular, it usually learns to respond in a way that is beneficial to the animal. To enable the op amp neuron to learn, it must be able to adjust its own weightings. Each neuron needs a small memory in which it stores the results of previous settings of its input resistors. It needs a way of adjusting its input resistors to modify its behaviour in the right directions.

The idea of weighting inputs is fundamental to neural networks. This is because it is not the neurons *themselves* that store information: it is the *connections* between them. As the brain learns a new behaviour, some connections are strengthened, while others are weakened. Using the analogy of the op amp neuron, some resistors are decreased in value while others are increased.

In the previous paragraph we used the word 'analogy'. It is important to remember that the electronic neuron is in many ways *analogous* to an animal neuron, but works in an entirely different manner. In the electronic circuit, the measure of the signal is its *voltage*. In an animal neuron the signal is a series of electrical pulses and the measure of its strength is the *rate of pulsing*. So we must not think of an electronic neural network as a model of a brain.

> **Analogous**
>
> Having a similar action, but operating in a different way or using different structures.

Early **artificial neural networks**, or ANNs, as we call arrays of electronic neurons, were based on op amps. With the development of digital electronics, ANNs are built as VLSI integrated circuits. They have at least the complexity as a Pentium or similar microprocessor and are known as **neurocomputers.** A neurocomputer differs markedly in architecture and operation system from a digital computer.

A digital computer is *programmed* to process input data *serially*. A neurocomputer *accepts data* from the outside world and sends in to be processed by its neural network. It processes the data in *parallel* (all input signals are processed simultaneously). Each time it processes data it may modify the outcome as a result of what it has been taught, or experienced.

The program of a digital computer must necessarily be based on a known set of rules and equations. However, in a neurocomputer, it may be that the values and rules are not completely or precisely known.

For instance, the conditions in the fireball in a furnace are highly complex, and there are many interactions between the combustion of fuel, the flow of air and the generation of heat that are difficult or impossible to predict. In such a case, a neurocomputer would produce more effective control than a digital one. The neurocomputer does not need to know the rules; sometimes there may not be any. All it has to do is to link a required set of outputs to a given set of inputs. It varies its weightings until it achieves as close a match as possible between what the outputs are supposed to be and what they actually are.

Mathematical models

The action of an op amp and its weighted inputs is essentially mathematical. Mathematical models can be programmed and run on a digital computer. This means that ANNs can be simulated effectively on an ordinary personal computer.

There are several different forms of ANN, many of these still being at the research stage. Their capabilities are being investigated. A form that is often used is the **feedforward network**, shown in the diagram at top right. Data is input on the left and fed through three or more levels of neuron to the outputs. Other forms of ANN have different patterns, some including feedback from the output level to an earlier level.

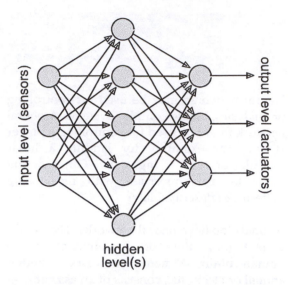

input level (sensors) — output level (actuators) — hidden level(s)

In a feedforward ANN, each neuron sends signals to the next level in the input-output direction.

The input level of an ANN consists of neurons that receive input from **sensors**, or perhaps data of other kinds. These send signals to each neuron in the **hidden level**. The ANN can decide how many neurons are needed in this level, and also how many hidden levels there need to be. The last hidden level sends signals to each of the neurons of the output level. These is a neuron for each of the possible output signals and these connect to **actuators**.

As a simple example of an ANN, consider a industrial reaction vessel in which a mixture of substances is heated and stirred until a reaction has occcured and the mixture has been converted to a viscous compound.

There are four sensors in the vessel, each of which gives a binary output:

- **Cold**: its output = 1 when the temperature is below 40°C.

- **Hot**: its output = 1 when the temperature is above 45°C.

- **Viscosity**: its output = 1 when the coefficient of viscosity exceeds 100 Pa s.

- **Full**: a level sensor with output = 1 when the tank is full to capacity.

The system has four actuators:

- **Heater**: a valve that opens (=1) to allow hot water to circulate in a jacket surrounding the vessel.
- **Cooler**: a valve that opens (=1) to allow cold water to circulate in the jacket.
- **Stirrer**: a motor that drives (=1) a stirrer.
- **Done**: an indicator lamp that lights (=1) when the process is finished.

This system is to be **trained** by **supervision**. We tell the ANN what output is required for every combination of inputs. The required behaviours can be set out as a truth table (bottom of page). Because the hot and cold sensors can not both produce a '1' output at the same time, there are only 12 combinations of input conditions.

The truth table determines what happens at each stage in processing. For example, the stirrer is switched on only when the vessel is full.

The 'done' lamp comes on only when the vessel is full of a viscous mixture between 40°C and 45°C.

The truth table is typed into the computer, and the instruction is given to generate the ANN. The ANN is then put through a series of training runs. To start with, the inputs of each neuron are usually weighted at random. They are varied at each run and, if a given change causes the actual outputs to match more closely to the expected outputs, the weightings are changed further in the same direction. Gradually the system discovers the most effective weightings. The mathematical techniques used for changing the weightings is beyond the scope of this account.

In another learning procedure, **reinforcement training**, the network is not told the expected outputs. Instead, its performance is **scored**, depending on how successful it is. It uses this score to modify its weightings and gradually acquires the desired behaviour.

Row	Inputs				Outputs			
	Cold	**Hot**	**Viscous**	**Full**	**Heater**	**Cooler**	**Stirrer**	**Done**
0	0	0	0	0	0	0	0	0
1	0	0	0	1	0	0	1	0
2	0	0	1	0	0	0	0	0
3	0	0	1	1	0	0	1	1
4	0	1	0	0	0	0	0	0
5	0	1	0	1	0	1	1	0
6	0	1	1	0	0	0	0	0
7	0	1	1	1	0	0	1	0
8	1	0	0	0	0	0	0	0
9	1	0	0	1	1	0	1	0
10	1	0	1	0	1	0	0	0
11	1	0	1	1	1	0	1	0

The result is a displayed map of the ANN showing the signal strengths for a given combination of inputs. The map in this example has four input neurons, six neurons in the hidden layer and four in the output layer. After a number of training runs, the four output neurons generate signals as specified in the truth table for each combination of input signals.

This is a simple example, intended to demonstrate what an ANN does. It is obvious that the same result could have been obtained by connecting a few logic gates together. But to connect logic gates correctly we would need to know the logic for every step in the processing. With an ANN we need to know only the initial input and the final output.

Pattern recognition

Some of the most successful applications of ANNs have been in the field of pattern recognition. The example already quoted can be taken as a kind of pattern recognition, for the ANN is taught to recognise the input pattern of temperature not over 45°C, temperature not under 40°C, viscosity over 100 Pa s, and vessel full. When it recognises this pattern it turns on the 'done' sign.

There are other kinds of patterns, such as patterns in financial data. ANNs have been used in various kinds of financial forecasting. Another application is in looking for patterns of spending that indicate fraudulent use of a credit card.

Visual patterns are readily recognised by the human brain. The patterns below are all recognisable as the capital letter 'A'.

We are also good at recognising partial patterns:

The message is not difficult to read, even though we can see less than half of each letter.

Visual patterns include alphabetic characters, faces, signatures, finger prints and palm prints. There is much use for ANNs in crime prevention and discovery, and in security systems generally.

A digital camera is used for inputting visual patterns. The image is enlarged so as to fill the whole frame. The image is broken down into pixels. In a letter identification system, the input is binary with no shades of grey. A pixel is either 0 (white) or 1 (black). Each pixel provides input to the ANN. In a fairly low-resolution system, the image may be 10 pixels by 10 pixels. This provides 100 inputs to the ANN. There are 26 outputs, each corresponding to a letter of the alphabet.

The ANN is 'shown' specimens of the letters in various styles, including handwritten versions. As it is shown each specimen, the ANN is told which letter it represents. After a series of learning trials, the ANN gradually learns to recognise the letters with a reasonable degree of reliability.

Quantitative input and output

Input to an ANN is not limited to binary values. Inputs can be taken from analogue sensors and converted to binary. They can also be values expressed on arbitrary scales. For example, an ANN to control a washing machine may have dirtiness ranked on a scale 'clean', 'slightly soiled', 'soiled' and 'heavily soiled'. We do not need to specify exact borderlines between the degrees of soiling.

ANNs have been used in vapour detection systems. A typical input to the system is an array of gas sensors.

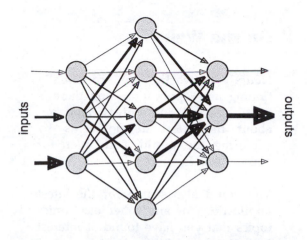

inputs

outputs

Software may display an ANN in this form, with the width of the arrows proportional to the strengths of the signals received by and generated by each neuron. Numerical values are also displayed, though not shown in this diagram. Some weightings may be negative (that is, inhibitory).

This net has three neurons in the input layer, five hidden neurons, and three neurons in the output layer. The combination of inputs shown in the diagram (slight, medium, strong) produces a strong output from the middle neuron, and a weak output from the other two.

Depending of the mixture of vapours present in the air, each sensor produces an output ranging from 0 V to 10 V. Usually, different types of sensor are most sensitive to one particular vapour, such as ammonia, acetone or methanol. However they respond as well to several other vapours, though to a lesser degree. As a result, a given vapour or mixture of vapours results in a unique combination of voltage signals from the array. The voltages are converted to digital form and fed to the ANN.

After an array has been trained, it can reliably distinguish which vapours are present, either singly or in a mixture in a range of proportions. For example, an array of 9 sensors has been used to distinguish clearly between acetone, ammonia, isopropanol, lighter fluid and vinegar and mixtures of these.

ANNs sensitive to odours have applications in the food industry. These include quality control, an ANN being able to detect rancid odours against a background of other acceptable odours. ANNs can check that packaging is effective in containing strong flavours such as onion.

In medicine, ANNs have application in detecting odours from the body and its fluids. This aids diagnosis of diseases and infections.

Questions on neural networks

1 Describe a simple electronic neuron based on an op amp. How does its action different from that of a neurone in the brain of an animal?

2 Outline the features of a neurocomputer and how it differs from a typical digital computer.

3 Draw a diagram to show the structure of an artificial neural network. Explain how it is trained to produce a particular response to a given combination of inputs. Illustrate your answer by describing a practical example of an ANN.

Multiple choice questions

1 A neurocomputer is distinguished by having:

    A memory.
    B a microprocessor.
    C parallel processing.
    D VLSI.

2 The input level of an ANN often consists of:

    A neurons.
    B data.
    C actuators.
    D sensors.

3 Training an ANN alters its:

    A weightings.
    B input.
    C connections.
    D hidden level.

4 If an ANN is trained by 'rewarding' it for producing the expected outputs, the training is described as :

    A feedforward.
    B feedback.
    C reinforcement.
    D an expert system.

5 In contrast to a digital computer, a neurocomputer is:

    A programmed.
    B trained.
    C faster.
    D without a memory.

On the Web

Neural networks are continually finding new applications and there is much research in progress. Find out about the latest developments by exploring the World Wide Web for news of ANNs.

You could also search for the latest applications of any other electronics topics that you have found of interest whilst studying this book.

A Useful information

Electrical quantities and units

Current: This is the basic electrical quantity. Its unit, the **ampere**, is defined by reference to the mechanical force a current produces between two parallel conductors.

Symbol for the quantity, current: I.

Symbol for the unit, ampere: A.

> **Symbols**
>
> Symbols for quantities are printed in italics (sloping) type. Symbols for units are printed in roman (upright) type.

Charge: The unit of charge is the **coulomb**, defined as the amount of charge carried past a point in a circuit when a current of 1 A flows for 1 s.

Symbol for the quantity, charge: Q.

Symbol for the unit, coulomb: C.

Equation: $Q = It$.

Potential: The unit of potential is the **volt**. The potential at any point is defined as the work done in bringing a unit charge to that point from an infinite distance. This is rather an abstract definition.

In electronics, we are more concerned with **potential difference,** the difference in potential between two points relatively close together, such as the two terminals of an electric cell. Potential difference is usually measured from a convenient reference point, such as the Earth (or 'ground') or from the so-called negative terminal of a battery. The reference point is taken to be at 0 V. Potential difference is often called **voltage**.

> **Quantity or unit?**
>
> It is easy to confuse V and V. Some books use U for the quantity instead of V to avoid this.

Symbol for the quantity, potential: V.

Symbol for the unit, volt: V.

Electromotive force: The unit of emf is the **volt**. It is a measure of the ability of a source of potential to produce a current *before* it is connected into a circuit The emf of an electric cell depends on the chemicals it is made of. The emf of a generator depends on such factors as the number of turns in its coils, the magnetic field and the rate of rotation. When the battery or generator is unconnected, the potential difference between its terminals is equal to its emf. When the battery or generator is actually connected to a circuit and current flows into the circuit, the potential difference between its terminals falls because of internal resistance, but its emf is unaltered.

Symbol for the quantity, electromotive force: E.

Symbol for the unit, volt: V.

Electric field strength: see p. 52.

Resistance: The unit of resistance is the **ohm**. When a current I flows through a conductor there is a fall in potential along it. The potential difference between its ends (or, in other words, the voltage across it) is V, and its resistance R is defined as:

$$R = V/I$$

This equation expresses Ohm's Law (but this is not the form in which Ohm originally stated it). Other useful versions of the equation are:

$$V = IR \qquad I = V/R$$

Symbol for the quantity, resistance: R.

Symbol for the unit, ohm: Ω.

The ohm is also the unit of reactance (symbol, X) and impedance (Z), both of which vary with frequency:

Reactance and impedance of a capacitance C at frequency f are given by:

Reactance and impedance of an inductance L at frequency f are given by:

$$X_L = Z_L = 2\pi f L$$

Conductance: The unit of conductance is the **siemens**. It is defined as the reciprocal of the resistance. If a conductor has a voltage V across it and a current I flowing through it, its conductance G is:

$$G = I/V$$

Symbol for the quantity, conductance: G.

Symbol for the unit, siemens: S.

> **S or s?**
>
> This is another source of confusion. The symbol for siemens is a capital S, but the symbol for second is a small s.

Power: The unit of power is the **watt**. In electrical terms, the power P dissipated in an electrical device is proportional to the current flowing through it and the voltage across it:

$$P = IV = I^2R = V^2/R$$

Symbol for the quantity, power: P.

Symbol for the unit, watt: W.

Capacitance: The unit of capacitance is the **farad**. If the charge on the plates of a capacitor is Q coulombs and the potential difference between them is V volts, the capacitance C in farads is given by:

$$C = Q/V$$

Symbol for the quantity, capacitance: C.

Symbol for the unit, farad: F.

> **The farad**
>
> This unit is too large for most practical purposes. In electronics, we more often use the submultiples: microfarad (μF), nanofarad (nF) and picofarad (pF).

The **relative permittivity** of the dielectric is:

$$\varepsilon_r = C_d/C_0$$

where C_d is the capacitance with a given dielectric, and C_0 is the capacitance with a vacuum as the dielectric.

Inductance: The unit is the **henry**. The inductance of a coil in which the *rate of change* of current is 1 A per second is numerically equal to the emf produced by the change of current. This applies to a single coil and is more correctly known as **self-inductance**.

If two coils are magnetically linked, their **mutual inductance** in henries is numerically equal to the emf produced in one coil when the current through the other coil changes at the rate of 1 A per second.

Symbol for the quantity, self-inductance: L.

Symbol for the quantity, mutual inductance: M.

Symbol for the unit of either, henry: H.

Frequency: The unit of frequency is the **hertz**. It is defined as 1 repetition of an event per second.

Symbol for the quantity, frequency: f.

Symbol for the unit: Hz.

Decibel: This is a quantity with no units, expressing the ratio between two *identical* quantities. It is fully discussed on p. 47.

Symbol for the quantity: dB.

Laws for circuit analysis

Ohm's Law: The most frequently used of all laws in electronics is expressed by any of the three equations listed on p. 363.

Kirchoff's Current Law (KCL): At any instant the sum of currents arriving at and leaving a node is zero. This can be expressed in another form as: the total current entering a node equals the total current leaving it. In these definitions, the term 'node' means a point in a circuit where three or more conductors join.

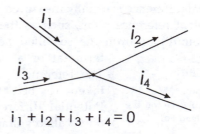

$$i_1 + i_2 + i_3 + i_4 = 0$$

Illustrating Kirchoff's Current Law. Given all but one of the currents, we can calculate the unknown one.

Kirchoff's Voltage Law (KVL): At any instant the sum of the potential differences across the branches of a loop of a network is zero. A **loop** is any closed path through a network.

There are three loops in the figure below, ABCD, BEFC, and ABEFCD. Going round each loop in a clockwise direction, the equations for each loop are:

For ABCD: $v - v_1 - v_2 = 0$

For BEFG: $v_2 - v_3 - v_4 = 0$

For ABEFGD: $v - v_1 - v_3 - v_4 = 0$

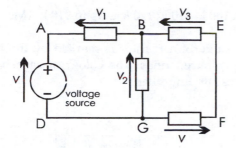

The arrows indicate the positive direction for the pds across the source and each resistance. Calculated values may be positive (with the arrow) or negative (against the arrow.

Sinusoids: A sinusoidal signal is one that, when plotted to show how the voltage or current varies in time, has the shape of a sine curve. Its amplitude and frequency define a sinusoid. Its period is equal to 1/frequency.

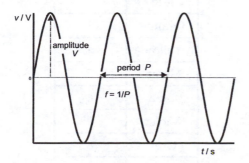

When plotted as a graph, a voltage sinusoid has the same shape as a sine curve.

The amplitude of the sinusoid is the **peak** voltage or current. The **root mean square** (rms) is a measure of its average value. The rms value of a sinusoid is 0.707 times its amplitude. Or the amplitude of a sinusoid is 1.4 times its rms value.

Networks

Resistance networks: When two or more resistances are connected in series, their total resistance is equal to the sum of their individual resistances. When two or more resistances are connected in parallel, their total resistance R_T is given by the equation:.

$$\frac{1}{R_T} = \frac{1}{R_1} + \frac{1}{R_2} + \frac{1}{R_3} + K + \frac{1}{R_n}$$

(a)

(b)

Two or more resistances may be connected (a) in series, or (b) in parallel.

For two resistances in parallel, this equation simplifies to:

$$R_T = \frac{R_1 \times R_2}{R_1 + R_2}$$

Capacitance networks: When two or more capacitances are connected in series, their total capacitance C_T is given by the equation:

$$C_T = \frac{1}{C_1} + \frac{1}{C_2} + \frac{1}{C_3} + K + \frac{1}{C_n}$$

When two or more capacitances are connected in parallel their total capacitance is equal to the sum of their individual capacitances.

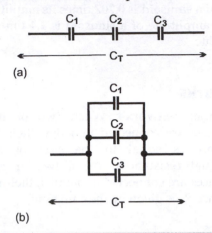

(a)

(b)

> An uncommon capacitance value can often be made up by wiring two or more capacitances in parallel.

Resonant network: If a capacitor and inductor are connected in series or in parallel the resonant frequency occurs when $X_C = X_L$. Then:

$$f = \frac{1}{2\pi\sqrt{LC}}$$

If the capacitor and inductor are in parallel, the impedance of the network is a maximum at the resonant frequency. If the capacitor and inductor are in series, the impedance is a minimum.

Transformers: In an ideal (no-loss) transformer that has n_p turns in the primary coil and n_s turns in the secondary coil, the ratio between the rms voltage v_p across its primary coil and the rms voltage v_s across its secondary coil is:

$$\frac{v_p}{v_s} = \frac{n_p}{n_s}$$

The inverse relationship applies to rms currents:

$$\frac{i_p}{i_s} = \frac{n_s}{n_p}$$

Components

Resistors: The preferred values of the E24 series, in ohms, are:

1.0 1.1 1.2 1.3 1.5 1.6 1.8 2.0 2.2 2.4 2.7 3.0

3.3 3.6 3.9 4.3 4.7 5.1 5.6 6.2 6.8 7.5 8.2 9.1

After these 24 values, the sequence repeats in multiples of ten:

10 11 12 13 ... up to ... 82 91,

then 100 110 120 ... up to 820 910,

then 1 k 1.1 k 1.2 k ... up to 8.2 k 9.1 k, ('k' means kilohms)

then 10 k 11 k 12 k ... up to 82 k 91 k,

then 100 k 110 k 120 k ... up to 910 k 1M.

The value of a resistor is marked by three or four coloured rings. The colour of each band represents a number.

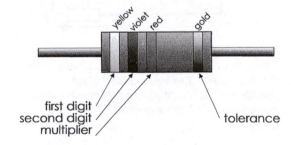

first digit
second digit
multiplier
tolerance

The table shows the meanings of the colours:

Colour	Number
Black	0
Brown	1
Red	2
Orange	3
Yellow	4
Green	5
Blue	6
Violet	7
Grey	8
White	9

This is how to read the bands, starting from the end:

First band	First digit of resistance
Second band	Second digit of resistance
Third band	Multiplier — a power of 10, or the number of zeros to follow the two digits.

In the four-band code, the first three bands represent the three digits and the fourth band is the multiplier.

Tolerance is coded by an additional band:

Colour	Tolerance	Printed code
Brown	±1%	F
Red	±2%	G
Gold	±5%	J
Silver	±10%	K
No band	±20%	L

The **resistor printed code** uses one of three letters to indicate the unit and the position of the decimal point:

R = ohms,   K = kilohms,   M = megohms.

Example: 5K6 means 5.6 kΩ.

A letter at the end of the code indicates tolerance, as in the table above.

Example: 4R7K means 4.7 Ω ±10%.

Temperature coefficient (or **tempco**) is the change of resistance in parts per million per kelvin or degree Celsius. The coefficient is generally positive. Thermistors have negative tempco (ntc). The tempco is indicated by a 5th (or 6th) coloured band, as listed on the right.

Colour	Tempco
Black	200
Brown	100
Red	50
Orange	25
Yellow	15
Blue	10
Violet	5
Grey	1

Capacitors: High tolerance capacitors have values (in μF, nF and pF) in the E12 series:

10 12 15 18 22 27 33 39 47 56 68 82

and multiples of this range.

Low tolerance capacitors (such as electrolytics) are usually made only in:

10   15   22   33   47   68 and multiples.

Small value capacitors may be marked with a three-digit code. The first two digits are the first two digits of the value in picofarads. The third digit represents the number of zeros following.

Example: '473' means 47 000 pF = 22 nF.

Symbols used in circuit diagrams are shown on the right and overleaf. The logic gate symbols are those defined by the American National Standards Institute (ANSI).

Conductors

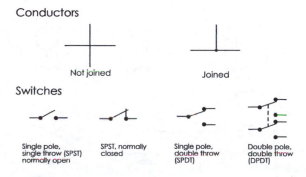

Not joined   Joined

Switches

Single pole, single throw (SPST) normally open   SPST, normally closed   Single pole, double throw (SPDT)   Double pole, double throw (DPDT)

Push-buttons

Push to make (PTM)   Push to break (PTB)

Power supply

Cell (positive on left)   Battery of cells (positive on left)   Power supply (voltage levels usually marked)

Resistors

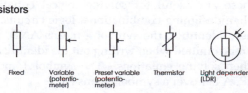

Fixed   Variable (potentiometer)   Preset variable (potentiometer)   Thermistor   Light dependent (LDR)

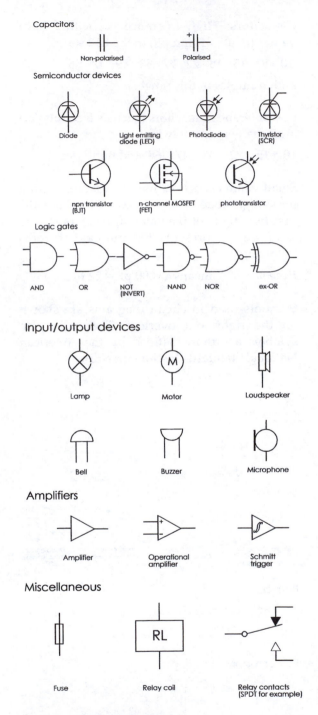

Capacitors

Non-polarised   Polarised

Semiconductor devices

Diode   Light emitting diode (LED)   Photodiode   Thyristor (SCR)

npn transistor (BJT)   n-channel MOSFET (FET)   phototransistor

Logic gates

AND   OR   NOT (INVERT)   NAND   NOR   ex-OR

Input/output devices

Lamp   Motor   Loudspeaker

Bell   Buzzer   Microphone

Amplifiers

Amplifier   Operational amplifier   Schmitt trigger

Miscellaneous

Fuse   Relay coil   Relay contacts (SPDT for example)

Logical identities

These are useful for solving logical equations when designing combinational logic circuits. In these identities, the symbol • means AND, and can be omitted when writing out the identity and when solving equations. The symbol + means OR. A bar over a symbol means NOT.

The order of ANDed and ORed terms makes no difference:

$$A \bullet B = B \bullet A$$
$$A + B = B + A$$

Brackets can be inserted and removed just as in ordinary arithmetic:

$$A \bullet B \bullet C = (A \bullet B) \bullet C = A \bullet (B \bullet C)$$
$$A + B + C = (A + B) + C = A + (B + C)$$
$$A \bullet (B + C) = A \bullet B + A \bullet C$$

ANDing A with itself and with true and false:

$$A \bullet A = A$$
$$A \bullet 1 = A$$
$$A \bullet 0 = 0$$

ORing A with itself and with true and false:

$$A + A = A$$
$$A + 1 = 1$$
$$A + 0 = A$$

Negation:

$$A + \overline{A} = 1$$
$$A \bullet \overline{A} = 0$$
$$\overline{\overline{A}} = A$$

De Morgan's Laws:

$$\overline{A + B} = \overline{A} \bullet \overline{B}$$
$$\overline{A \bullet B} = \overline{A} + \overline{B}$$

Miscellaneous:

$$A + A \bullet B = A$$
$$A + B \bullet C = (A + B) \bullet (A + C)$$
$$A + \overline{A} \bullet B = A + B$$

Number systems

The table opposite lists the equivalents of numbers written in the decimal, binary, hexadecimal and binary coded decimal systems.

Decimal	Binary	Hex	BCD
0	0000	0	0000 0000
1	0001	1	0000 0001
2	0010	2	0000 0010
3	0011	3	0000 0011
4	0100	4	0000 0100
5	0101	5	0000 0101
6	0110	6	0000 0110
7	0111	7	0000 0111
8	1000	8	0000 1000
9	1001	9	0000 1001
10	1010	A	0001 0000
11	1011	B	0001 0001
12	1100	C	0001 0010
13	1101	D	0001 0011
14	1110	E	0001 0100
15	1111	F	0001 0101
255	1111 1111	FF	0010 0101 0101

Negative binary numbers

The most common way of representing negative values in binary is the **twos complement**.

We must first decide the number of digits in which we are working. Microcontrollers often use 7 digits plus a **sign digit** on the left. In this example we have four digits plus the sign digit. To form the twos complement of a number, write the positive number using 4 digits. Then write a 0 on the left to represent a positive sign. Next write the ones complement by writing 0 for every 1 and 1 for every 0. Form the twos complement by adding 1 to this number. Ignore any carry digits. This is the binary equivalent of the negative of the original number.

Example

Find the 4-digit equivalent of −3.

Write +3 in binary	0011
Write the sign digit for +	00011
Find the ones complement	11100
Add 1	1
Result is twos complement	11101

11101 is the equivalent of −3, the 1 on the left meaning that it is negative.

Check by adding −3 to +3

+3	00011
−3	11101
Add	[1] 00000

Ignore the leading digit. The result is zero.

Twos complements may be used for adding a positive number to a negative number (as above) and for adding two negative numbers.

Acknowledgements

The author thanks Eastern Generation Ltd., Ironbridge, Shropshire, for permission to take photographs on their site (p. 293).

Information about the use of a neural network to detect smells (pp. 360-361) is based on 'Electronic noses and their applications' by Paul E. Keller et al., Pacific Northwest National Laboratory.

All photographs and all drawings (except those on pp. 30 and 55) are by the author.

B Answers to self-test questions

Some answers show the actual calculated value (to 3 significant figures) and, in brackets, the most appropriate standard value.

Page 1: Conductors are copper wire, brass switch and breadboard contacts, metal film of resistor, electrodes and electrolyte of cells. Insulators are PVC insulation of wire, plastic parts of switch and breadboard, plastic diode body and silicon substrate of diode.

Page 11: $0.886 \, \Omega$.

Page 14: 4.35 A (5 A).

Page 15: 1(a) $350 \, \Omega$ ($360 \, \Omega$), **1(b)** $1.47 \, k\Omega$ ($1.5 \, k\Omega$), **2** 6.67 mA.

Page 19: 13.4 mA.

Page 27: a $240 \, \Omega$ and $300 \, \Omega$, 4.69 V, **b** $2.7 \, \Omega$ and $3.3 \, \Omega$, 1.33 V.

Page 30: 1 4.17 mF, **2** 20 V.

Page 42: 1 3.63 ms, **2** $103 \, k\Omega$ ($100 \, k\Omega$).

Page 44: 407 Hz, 2.47.

Page 45: a $48.2 \, \Omega$, **b** $816 \, \Omega$, **c** $2.2 \, k\Omega$, **d** 329 Hz.

Page 46: a $106 \, k\Omega$, $10.6 \, k\Omega$, $1.06 \, k\Omega$, **b** 9646 Hz, **c** At 10 kHz, $R = 470 \, \Omega$, $X_C = 7.23 \, \Omega$; at 500 kHz, $R = 470 \, \Omega$, $X_C = 0.145 \, \Omega$.

Page 55: 5.04 V.

Page 64: 1 125 mS, **2** 91.5 mV, $200 \, k\Omega$, $750 \, k\Omega$.

Page 65: $60 \, \Omega$ ($62 \, \Omega$).

Page 73: 412.

Page 74: $3 \, k\Omega$.

Page 75: 5.89 pF (5 pF).

Page 78: 10 000.

Page 82: a $X_C = 2022 \, \Omega$; $X_L = 1991 \, \Omega$, **b** $X_C = X_L = 2006 \, \Omega$, **c** $X_C = 1981 \, \Omega$; $X_L = 2032 \, \Omega$.

Page 93: 1 Positive, **2** Negative.

Page 95: 373, $2.2 \, k\Omega$.

Page 96: $2.19 \, k\Omega$ ($2.2 \, k\Omega$).

Page 97: 1 9.18, **2** $1 \, T\Omega$, **3** $29.4 \, k\Omega$ ($30 \, k\Omega$).

Page 105: -0.106 V.

Page 109: Possible values: $R_1 = 3 \, k\Omega$, $R_2 = 2k\Omega$, $R_3 = 1 \, M\Omega$, $R_4 = 200 \, k\Omega$.

Page 112: Possible values: $R_1 = 240 \, k\Omega$, $R_2 = 270 \, k\Omega$, $R_3 = 240 \, k\Omega$, $R_4 = 100 \, k\Omega$.

Page 118: 15.4 kHz.

Page 127: source follower, emitter follower.

Page 151: 6.39 V, 5.95 V.

Page 152: a 400 mA, **b** 11.1 V, **c** 7.5 V, 5 W, **d** $9.1 \, \Omega$, 2 W.

Page 162: (a) Practically unlimited, **(b)** 1.

Page 166: (a) 0 for A = 1, B = 0, C = 0, 1 for all other inputs, **(b)** As in this table:

Inputs			Outputs			
C	B	A	1	2	3	4
0	0	0	1	1	0	1
0	0	1	0	0	0	1
0	1	0	1	1	1	0
0	1	1	0	1	1	0
1	0	0	1	1	1	0
1	0	1	0	0	1	1
1	1	0	1	1	1	0
1	1	1	0	1	1	0

Page 179: (a) 1, **(b)** 00

Page 241: 14.7 dB.

Page 264: The maximum rate at which the output voltage can change.

Page 270: None.

Page 2276: 67 cm.

Page 283: (a) 1205 kHz. **(b)** 455 kHz.

Answers to other questions with numerical answers, to questions with very short answers, and to multiple choice questions are on the Companion website.

The site also has interactive questions.

Index